CAMBRIDGE LIBRARY COLLECTION

Books of enduring scholarly value

Physical Sciences

From ancient times, humans have tried to understand the workings of the world around them. The roots of modern physical science go back to the very earliest mechanical devices such as levers and rollers, the mixing of paints and dyes, and the importance of the heavenly bodies in early religious observance and navigation. The physical sciences as we know them today began to emerge as independent academic subjects during the early modern period, in the work of Newton and other 'natural philosophers', and numerous sub-disciplines developed during the centuries that followed. This part of the Cambridge Library Collection is devoted to landmark publications in this area which will be of interest to historians of science concerned with individual scientists, particular discoveries, and advances in scientific method, or with the establishment and development of scientific institutions around the world.

Life and Letters of James David Forbes

First published in 1873, this co-authored biography of the Scottish physicist, Alpine explorer, and university leader James David Forbes (1809–1868) includes extracts from Forbes's letters. John Campbell Shairp, Forbes's successor as principal of the United College of the University of St Andrews, writes of Forbes's personal, family, and professional life, including his years at St Andrews. Forbes's student and his successor in the Natural Philosophy chair at the University of Edinburgh, Peter Guthrie Tait, himself an accomplished mathematical physicist who co-wrote, with Lord Kelvin, *Treatise on Natural Philosophy* (1867), discusses Forbes's scientific achievements and contributions. A. Adams-Reilly, a celebrated Irish mountaineer, cartographer, and friend of Forbes, writes of the latter's Alpine travels and his work and interest in glaciers. In Shairp's words, in addition to all of his academic accomplishments, Forbes was also Britain's 'father of Alpine adventure'.

Life and Letters of James David Forbes

John Campbell Shairp
Peter Guthrie Tait
Anthony Adams-Reilly

CAMBRIDGE UNIVERSITY PRESS

Cambridge, New York, Melbourne, Madrid, Cape Town, Singapore,
São Paolo, Delhi, Dubai, Tokyo

Published in the United States of America by Cambridge University Press, New York

www.cambridge.org
Information on this title: www.cambridge.org/9781108014069

This edition first published 1873
This digitally printed version 2010

ISBN 978-1-108-01406-9 Paperback

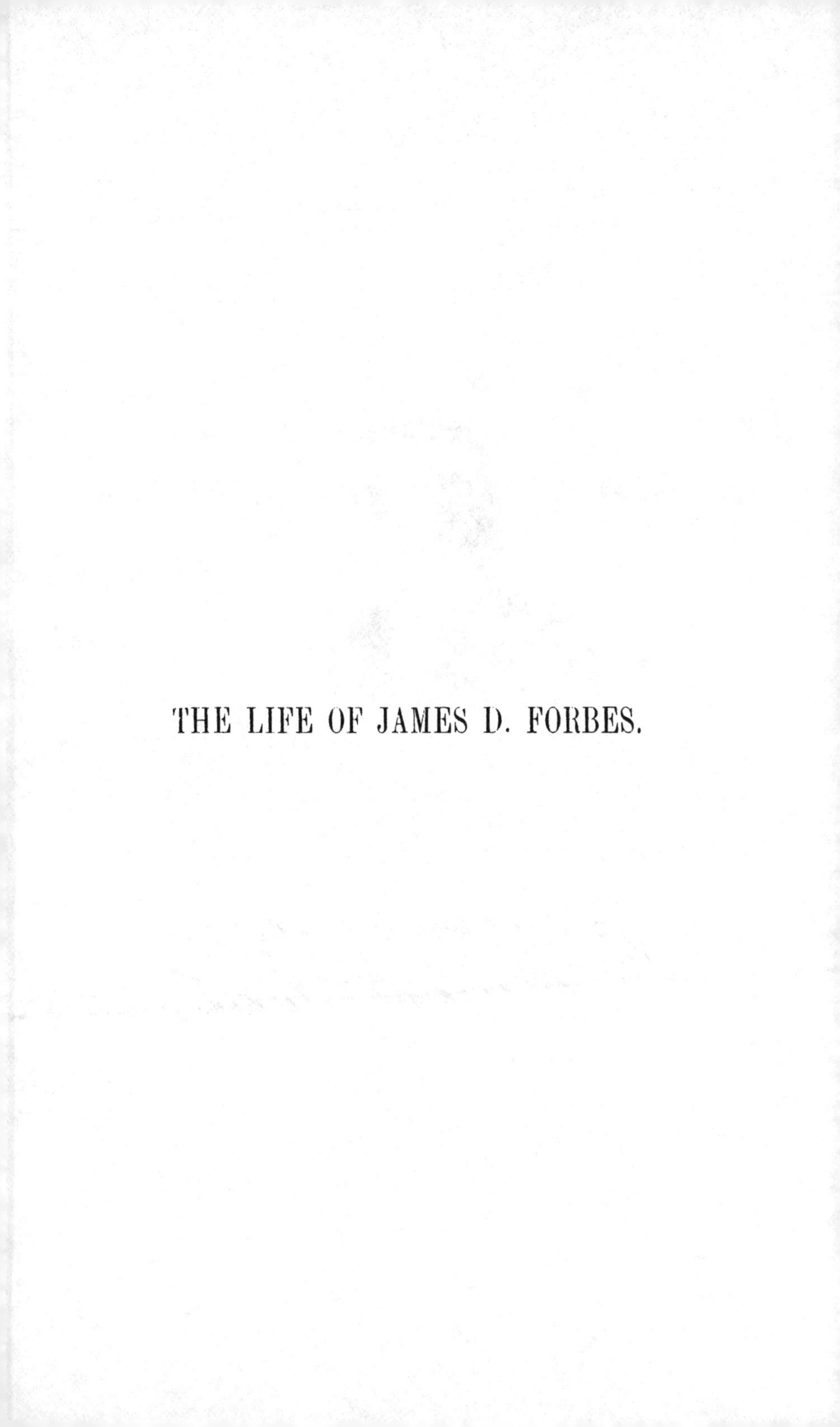

THE LIFE OF JAMES D. FORBES.

Yours sincerely
James D. Forbes

LIFE AND LETTERS OF

JAMES DAVID FORBES, F.R.S.,

D.C.L., LL.D.,
LATE PRINCIPAL OF THE UNITED COLLEGE IN THE UNIVERSITY OF ST. ANDREWS,
SOMETIME PROFESSOR OF NATURAL PHILOSOPHY IN THE UNIVERSITY OF EDINBURGH,
FORMERLY SECRETARY R.S.E.,
CORRESPONDING MEMBER OF THE INSTITUTE OF FRANCE,
ETC. ETC. ETC.

BY JOHN CAMPBELL SHAIRP, LL.D.,
Principal of the United College of the University of St. Andrews,

PETER GUTHRIE TAIT, M.A.,
Professor of Natural Philosophy in the University of Edinburgh,

AND

A. ADAMS-REILLY, F.R.G.S.

WITH PORTRAITS, MAP, AND ILLUSTRATIONS.

London:
MACMILLAN AND CO.
1873.

PREFACE.

This biography is the work of three writers, each working separately and, in a great measure, independently of the other two. This subdivision of labour has, I am aware, serious drawbacks, and would not have been resorted to, had any one person been found who could have undertaken adequately to describe the various aspects of Forbes' life and work. But, in default of a writer who could do this, it was thought better to portion out the work to three writers, than to have it inadequately done by one. No doubt the book has in this way lost something in symmetry, but it is hoped that it may have gained more in thoroughness and completeness.

In thinking over Forbes' life and work, these naturally appeared in several aspects, all harmoniously combined, yet each distinct in itself. There was his work as a scientific investigator and discoverer; his work as an Alpine explorer, and, as far as Britain is concerned, the father of Alpine adventure; his work as a professor and a university reformer; lastly, his character as a man. In this last aspect he was no less worthy of regard than in his other and more public capacities.

1. The description and estimate of Forbes in the first of these aspects has been undertaken by one who was his student, and is his successor in the Natural Philosophy chair of Edinburgh University, Professor P. G. Tait. The chapters headed 'Forbes' Scientific Work' are from his pen. Before writing them, however, he had the advantage of discussing the whole subject fully and frequently with his friend and Forbes' friend, Sir William Thomson of Glasgow University, who has also, after these chapters were in proof, carefully gone over them and weighed their contents. Every statement which they contain may therefore be regarded as not merely proceeding from Professor Tait, but as endorsed by Sir William Thomson—a double guarantee for accuracy, which in delicate matters of discovery is of high value.

2. The description of Forbes' travels and labours among the Alps is written by Mr. A. Adams-Reilly, himself a well-known Alpine traveller, whose conversation and letters on his favourite subjects were to Forbes, in his later years, like a renewal of his own

youth, and to whose achievements, as an explorer and surveyor of the Alps, Forbes in his letters bears so strong a testimony. Mr. Reilly has, with much toil and ingenuity, made Forbes' journal, letters, and jottings in a great measure tell their own story. The map of the Mer de Glace here given has been done under Mr. Reilly's eye and guidance.

3. It has fallen to my share to give some account of Forbes' early life, his professoriate, and his later years at St. Andrews. It would have been well, if this could have been done by some one whose friendship with Forbes dated from at least his vigorous prime, for my acquaintance with him began only with his arrival in St. Andrews. Once begun, however, it soon became intimate and friendly. Though this late beginning of intercourse has been, no doubt, a disadvantage, yet I do not feel as if I had been a stranger to him even before our acquaintance commenced: so familiar to me were the scenes and some of the persons that surrounded Forbes' earlier years.

Of this tripartite work Chapters XIV. and XV. have been written by Professor Tait; Chapters VIII., IX., and X. by Mr. Adams-Reilly; and Chapters I., II., III., IV., V., VI., VII., XI., XII., and XIII. by me. Each writer is responsible for that part which he himself has written or put together, and for that alone.

As far as possible, I have tried to carry on the narrative by means of the copious letters and journals which Forbes left. And here let me state how large a share of the necessary labour Mrs. Forbes has taken on herself. She has selected, extracted, copied, and in many cases re-copied, all those portions which are here given from her husband's vast and methodically-preserved correspondence. Indeed, but for her untiring exertions in this way, I could never have overtaken my part of the task.

Our best thanks are due to those pupils and friends of Forbes who have furnished letters conveying their recollections of him, or have forwarded letters of his which they had preserved. Among these I would offer special thanks to the Rev. Professor Kelland for the full account which he kindly supplied of Forbes' work in Edinburgh University.

With these remarks this Preface might have ended. It had been our hope that we might have been allowed to tell our story, without reverting to controversies which, we had thought, had been long since extinguished. But after most of these sheets were in the press, a book appeared, in which many of the old charges against Principal Forbes in the matter of the glaciers, were, if not openly repeated, at least not obscurely indicated. Neither the interests of truth, nor justice to the dead, could suffer such remarks to pass unchallenged. How it has been thought best for the present to meet them I must leave my friend and fellow-labourer Professor Tait to tell. Without pretending to judge of the details of controverted questions which belong to a field of study with which I am unfamiliar, I desire here to express the absolute faith which I, in common with all who really knew Forbes, have in his scrupulous and chivalrous honour, as well in every other matter, as in doing justice to the claims of previous discoverers. Since the subject has been re-opened, it is well that it should be sifted to the bottom. The claims of Forbes we will gladly leave to the verdict of those who are competent and impartial judges, in full confidence that—

> 'Whatever record leap to light,
> He never shall be shamed.'

J. C. SHAIRP.

ST. SALVATOR'S COLLEGE, ST. ANDREWS,
February 8th, 1873.

CONTENTS.

CHAPTER VIII.

CHAPTER IX.

CHAPTER X.

CHAPTER XI.

CHAPTER XII.

CHAPTER XIII.

CHAPTER XIV.

CHAPTER XV.

APPENDIX A.

APPENDIX B.

APPENDIX C.

APPENDIX D.

LIST OF ILLUSTRATIONS.

ERRATA.

Page 143, line 5, *for* 'Albyne' *read* 'Altyre.'
„ 224, „ 2, „ 'Collinton' *read* 'Colinton.'
„ 243, „ 35, „ 'less' *read* 'more.'
„ 258, „ 19, „ 'Switzerland elsewhere,' *read* 'Switzerland or elsewhere.'
„ „ „ 29 (and throughout), *for* 'Voght' *read* 'Vogt.'
„ 259, last line (and throughout), 'Abswung' *read* 'Abschwung.'
„ 376, line 29, *for* 'Pelroux' *read* 'Pelvoux.'

THE LIFE OF

JAMES DAVID FORBES.

CHAPTER I.

PARENTAGE AND BOYHOOD.

JAMES DAVID FORBES, the youngest child of Sir William Forbes of Pitsligo and Williamina Belches or Stuart, was born in his father's town-house, 86, George Street, Edinburgh, on the 20th April, 1809.

Each of his parents belonged to an ancient Scottish race. The Forbeses of Monymusk and Pitsligo, in Aberdeenshire, of whom his father was the lineal representative, are an offshoot from the Lords Forbes, the chiefs of the House of Forbes. On turning to Douglas's 'Baronage,' and to Burke, I find that the second Baron Forbes, who died about 1460, had a second son, Duncan Forbes of Corsindie, who was the ancestor of the Forbeses of Monymusk and Pitsligo. A grandson of this Duncan, also named Duncan, was the first of his name who got a charter of the lands of Monymusk, from which his descendants for generations took their designation.

William, the son and heir of Duncan, the first laird of Monymusk, married Lady Margaret Douglas, a daughter of the powerful House of Angus; and their son and heir, William, was in 1626 created a Nova Scotia Baronet.

The great-great-grandson of the first Baronet of Monymusk, John Forbes, married Mary Forbes, daughter of Alexander Lord Pitsligo, through whom, on the failure of the male line of Pitsligo in 1781, her descendants became nearest heirs and representatives of that noble and attainted House.

John Forbes, however, died before his father, and therefore never succeeded to the Baronetcy; but the grandson of John was the Sir William Forbes, who by his energy and character raised the family to an eminence which it had never before attained. Sir William was the grandfather of James David Forbes, whose life the following pages will attempt to describe.

The father of Sir William had married a lady of his own kindred and name, and died early, leaving his family young and poor. Left thus, while still a boy, the young Sir William migrated with his widowed mother from Aberdeenshire to Edinburgh, and there took to banking; in which business he showed such energy and enterprise, that he in time founded the well-known banking-house in Parliament Square, which long bore his name. His birth and circumstances combined the conditions of the two extremes of society—the high bearing of an old race with that discipline of poverty and thrifty training which belong to the humblest. And perhaps no circumstances are more fitted to form an energetic and noble character. Sir William not only succeeded in restoring the decayed fortunes of his family, but was known as one of the most influential and beneficent men of Scotland in his day. His high character and well-used influence are not even yet forgotten in Edinburgh and in the South of Scotland.

His paternal grandmother was, as has been said, a daughter of one of the attainted Lords of Pitsligo, famous for the part they took and the sufferings they endured in the '15 and the '45. On the failure of the male line of Pitsligo in 1781, Sir William, as we have seen, in right of his grandmother became representative of that ancient Jacobite House, and was henceforth known as Sir William Forbes of Pitsligo. The Jacobite blood in their veins was visible in Sir William and every one of his descendants, and coloured all their views and feelings on matters of Church and State. Old Tory and strong Episcopalian principles were part of the family inheritance. These old-world views, however, did not interfere either with Sir William's business talent, the fascination of his manners, or the warmth of his heart.

Sir Walter Scott makes feeling allusion to Sir William, immediately after his death, in the Introductory Epistle to the Fourth Canto of 'Marmion,' which was addressed to one of his sons-in-law. The passage beginning—

> 'Far may we search, before we find
> A heart so manly and so kind'—

is by no means in Scott's most finished style, but it contains a fine tribute of gratitude and affection. In the Notes he speaks of Sir William Forbes as a man 'unequalled, perhaps, in the degree of individual affection entertained for him by his friends, as well as in the general respect and esteem of Scotland at large. His "Life of Beattie," whom he befriended and patronized in life, as well as celebrated after his decease, was not long published before the benevolent and affectionate biographer was called to follow the subject of his narrative.'

The second Sir William was like the first in all but this: that the father had made by his own exertions the fortune and position which the son worthily upheld and used. Of this gentleman, much as he was beloved and looked up to by the more intimate circle of relatives and friends, that which the world will now most care to remember is his lifelong friendship with Sir Walter Scott, and the strange way in which their fortunes were intertwined. Walter Scott and William Forbes had known each other in boyhood; in opening manhood they had, with a number of other comrades, helped to raise a corps of volunteer cavalry, and both served together in what, from Scott's record of it, must have been a very jovial company. But just before this the paths of these two had crossed in a more delicate and tender region. Most readers of Lockhart's 'Life of Scott' will remember the allusions it contains to a 'first love,' which ended unfortunately for the poet. It is there told how the acquaintance began in the Greyfriars Churchyard, where rain happening to fall one Sunday after church-time, Scott offered his umbrella to a young lady, and the tender having been accepted, escorted her to her home, which proved to be at no great distance from his own. To return from church together had, it seems, grown into something like a custom, before they met in society. It then appeared that the mothers of the two young people, Lady Jane Stuart and Mrs. Scott, had been companions in their youth, though, both living secludedly, they had scarcely seen each other for many years. The two matrons now renewed their former intercourse. For long years Scott nourished this dream, but it was doomed to end in disappointment. 'The lady,' we are told, 'preferred a friend of Scott's, who was in this also a rival,—a gentleman of the highest character, to whom some affectionate allusions occur in one of the greatest of the poet's works, and who lived to act the part of a most generous friend to Scott throughout the anxieties and distresses of 1826 and 1827.' That lady was Williamina Belches, sole child and heir of a gentleman, who was a cadet of the ancient family of Invermay, and who afterwards became Sir John Stuart of Fettercairn. The more fortunate rival was Sir William Forbes, who married the lady whom Scott so loved. The youngest child of this marriage was James David, whose life is here to be recorded. Lockhart adds that he has no doubt that this deep disappointment had a powerful influence in nerving Scott's mind to face steadily and perseveringly those legal studies which were to fit him for being called to the bar. Perhaps it may have had this effect. More subtle observers have traced to it another result deeper and more lasting. Keble in a beautiful essay on Scott more than hints a belief that it was this imaginative regret haunting Scott

C H Jeens

Lady Forbes

all his life long which became the true well-spring of his inspiration in all his minstrelsy and romance. And there is evidence to prove that Keble divined aright. Certainly there is a purity and elevation in Scott's conceptions of female character which would well accord with such an experience idealized. One instance more of the old truth that poets

'Learn in suffering what they teach in song.'

However this may have been, it is certain that the success of his rival made no break in Scott's friendship with Forbes. Immediately after Sir William's marriage with Miss Belches Stuart, we find him serving along with Scott in that band of mounted volunteers which counted on its roll most of Scott's most intimate friends. This was in 1797, when Scott was yet an unknown man. About thirty years afterwards, when Scott's fame was at the full and his dark days had begun, the following entries occur in the poet's diary:—1826, January 20, 'Sir William Forbes called,—the same kind, honest friend as ever,—with all offers of assistance.' Again, January 26, 'Sir William Forbes took the chair, and behaved as he has ever done, with the generosity of ancient faith and early friendship. In what scenes have Sir William and I not borne a share together,—desperate and almost bloody affrays, rivalries, deep drinking matches,—and finally, with the kindest feelings on both sides, somewhat separated by his retiring much within the bosom of his family and I moving little beyond mine! It is fated our planets should cross, and that at the periods most interesting for me. Down—down—a hundred thoughts!'

Farther on in Lockhart's Life we read: 'Sir William Forbes, whose banking-house was one of Ballantyne's chief creditors, crowned his generous efforts for Scott's relief by privately paying the whole of Abud's demand (nearly £2,000) out of his own pocket, ranking as an ordinary creditor for the amount, and taking care at the same time that his old friend should be allowed to believe that the affair had merged quietly in the general affairs of the trustees. It was not until some time after Sir William's death that Sir Walter learned what he had done on this occasion.'

In a letter dated Abbotsford, October 28, 1828, Scott on hearing of the death of his early friend thus writes: 'Your letter brought me the afflicting intelligence of the death of our early and beloved friend Sir William. I had little else to expect from the state of health in which he was when I last saw him; but that circumstance does not diminish the pain with which I now reflect that I shall never see him more. He was a man who from his habits could not be intimately known to many, although everything which he did partook of that high feeling and generosity which belongs perhaps to a better age than that we live in. In him I feel I have sustained a loss which no after years of my life can fill up to me; and if I look back to the gay and happy hours of youth, they must be filled with recollections of our departed friend. In the whole course of life our friendship has been uninterrupted, as his kindness has been unwearied. Even the last time I saw him (so changed from what I knew him), he came to town when he was fitter to have kept his room, merely because he could be of use to some affairs of mine. It is most melancholy to reflect that the life of a man whose principles were so excellent, and his heart so affectionate, should have in the midst of external prosperity been darkened, and I fear I may say shortened, by domestic affliction. But those whom He loveth He chasteneth.' And then follow some reflections on the thought of meeting departed friends hereafter, more serious than Scott generally indulged in.

There is but one allusion, as far as I know, made by Sir Walter Scott to the lady who became Lady Forbes. It occurs in his diary about a year before the above letter regarding Sir William was written. In July 1827, Sir Walter, while on a visit in the neighbourhood, drove over to St. Andrews, not having seen it for many years. And thus he notices it: 'I did not go up to St. Rule's

Tower, as on former occasions: this is a falling off, for when before did I remain sitting below when there was a steeple to be ascended? . . . I sat down on a gravestone and recollected the first visit I made to St. Andrews, now thirty years ago. What changes in my feelings and my fortunes have since then taken place!—some for the better, many for the worse. I remembered the name I then carved in Runic characters beside the castle gate, and asked why it should still agitate my heart. But my friends came down from the tower, and the foolish idea was chased away.' What name that was may easily be divined. An old sexton still lives who remembers that day, and points to the spot within the roofless Tower of St. Rule where Sir Walter sat on a stone, 'with a rough hairy cap on his head.'

The above extracts contain incidents and allusions of so private a nature that scruples might have been felt about noticing them here, had they not already been made public, in so famous a biography.

The marriage of Sir William Forbes with Williamina Belches or Stuart took place on 19th January, 1797. Their family consisted of four sons and two daughters, and of these James David was the youngest child. The following pleasing record of his childhood and early years was written by Miss Forbes, the last survivor of the family.

Miss Forbes's Sketch of her Brother's early Life.

He was born 20th April, 1809, at 86, George Street, Edinburgh. Though Colinton House, within four miles of Edinburgh, was the usual home of the family, his mother had been recommended to remove into town for the winter, on account of her health; and before the end of the year she was ordered by her physician to Lympstone near Exeter, and was accompanied by Sir William and their infant son, the rest of the family being left at home. They took with them from Scotland two faithful attendants, for whom their young charge ever retained the most tender regard,—his mother's favourite maid and his own nurse, Lizzie Jervis, a remarkable woman in her way; of the old Scotch type, stately and reserved, strict and conscientious, one who never went out to walk without her small, well-worn Bible in her pocket.

Writing to her daughter, his mother says: 'Dear James is a sweet, thriving, merry pet;' and again: 'He is really the dearest and the best baby you ever saw.'

But he was never to be conscious of the blessing of such a mother's love and care. She died 5th December, 1810, when he was little more than a year and a half old. But though too young to know his irreparable loss, or to retain the slightest recollection of that lovely spot where his mother died, he never ceased to regard it with a sacred and almost romantic veneration. Witness the pilgrimage made to Lympstone in his twenty-sixth year, so touchingly described in a letter to his sister, which few could read unmoved. An intense sensibility to the associations of the past, a clinging fondness for the scenes of his youth and the memories of his early days, were among his most amiable and abiding qualities. Little relics, which many men would have despised as childish, were preserved by him with a loving appreciation of the value once attached to them. And this taste for the simplest and most innocent pleasures he retained unimpaired to the last.

The return to Colinton House was melancholy in the extreme to the bereaved husband, and such of his family as could sympathise, young as they were, in the overwhelming sorrow that clouded his life even to its last day.

Under these circumstances, that fair young face and joyous spirit seemed sent to brighten the nursery like a sunbeam. His father idolized him as the last precious legacy of a beloved wife; while his two sisters and three elder brothers welcomed him home as the cherished Benjamin of the family: no shade of jealousy was ever awakened by the peculiar place he held in his father's affections; it was his acknowledged right, and was secured to him by the gentle sweetness of disposition that remained unspoiled by it all.

After this sad year Colinton, charming residence as it was, was rather endured than enjoyed by its owner; endured for the sake of his children, who found health and freedom in its airy situation, and endless amusement among the beautiful wooded banks of the water of Leith. Never were these days forgotten.

Sir William's health suffered at this time so severely from the painful associations connected with his home, that his friends anxiously pressed the necessity of a change of scene, and his sister Mrs. Skene at length succeeded in persuading him to spend a few weeks of the summer with his children at Invercanny, a quiet spot in a lovely country near Inverye on Deeside, where his sister and her husband then resided with their family. This led to a visit to his own paternal property, Pitsligo, in Aberdeenshire, where the natural tone of feeling was in some degree restored, by the interest awakened in the wants of his tenants, and the improvements on the estate.

This summer (1811) was rendered memorable by the appearance of the far-famed comet, which no one who ever saw can forget. Its sojourn was long, and night after night was the infant philosopher danced at the window, in full view of the glorious phenomenon, its splendid tail seeming to sweep across a quarter of the heavens. His brothers and sisters pleased themselves in after years with the fanciful imagination, that some secret inspiration was thus communicated which never lost its influence, but continued ever after to lend an irresistible attraction to the study of the heavens.

These visits to Pitsligo were repeated for many successive summers with the best effects on the whole family, and among its moors and mosses and simple rural population the younger branches enjoyed a liberty admirably calculated to promote health and independence. Nor did young James fall behind his elder brothers, in the feats of those happy holidays; creating miniature waterfalls, damming burns, and leaping drains, little less formidable in proportion to his size than the glacier crevasses in after years. Meanwhile his early education was carried on along with that of his brother Charles (nearly three years older) in the school-room at Colinton, under the care of their sisters' governess, who had acted a mother's part by them all, ever since the loss of their own; and when arrived at the age of manhood he never forgot what he owed her. A keen sense of gratitude and a tenacious memory of past kindnesses were marked features in his character. And there was little in the recollection of his childhood to mar those pleasing impressions. Gentle and docile, he was seldom in fault; and that inborn love of truth, perhaps his most distinguishing moral characteristic through life, was conspicuous from earliest infancy. No childish excuse or prevarication ever passed his lips; in the minutest detail he was accuracy itself; James's word was never questioned.

His brother Charles and he shared the same lessons and the same sports. They were not brothers only, but bosom friends, scarcely ever parted for a day, till the marriage of the elder of the two; and even then the tie remained unbroken till severed by death. They both acquired the rudiments of their classical education from the village schoolmaster, Mr. Hunter, the simple warm-hearted tutor, who afterwards gloried in having had the smallest share in the instruction of such a pupil.

Many were the early indications of the bent of his future genius, such as keeping, while quite a child, a constant watch over the variations of the thermometer and barometer, with such a singular delight in the study of the almanack, that it went in the family by the appellation of 'James's red brother.'

In order to beguile the long winter evenings, and to afford his children a rational amusement at home, Sir William had fitted up a room with chemical apparatus, an air-pump, and an electrical machine, which no doubt must have had considerable effect in stimulating the latent taste for natural science; but with James it could not long remain a latent taste, it soon became a passion, though no one suspected how strong. The village schoolmaster, although amply endowed for his then position, was no genius; so, when the brothers had gone through their English education, and were sufficiently advanced in Latin, they were transferred to the College of Edinburgh; and it may be imagined with what zest the lectures of Professors Leslie, Jamieson, and Hope would be listened to. But other classes besides these had to be attended. James was designed for the Bar as a profession, and in deference to his father's wishes he applied himself to it accordingly; went through the usual routine of study, and in due time passed as advocate. But his heart was not there, and he contrived all the while to promote his favourite studies at the same time with indefatigable diligence, though always 'under the rose.' For in those days law was considered so dry a study, and so often prescribed merely as the most eligible mode of securing an independence, that parents looked with jealousy on any tastes that might prove too attractive, either in art or science, and so interfere with the chance of success in a profession generally acknowledged to be a hard struggle at best. A simple incident may be mentioned in illustration, related many years after by an attached old servant, who happened to be in the nursery when Sir William came in, and asked all his children what he should bring them from London. The youngest instantly replied: 'Bring me a telescope, Papa.' 'Ah, Jemmy,' was the reply, 'you'll never make salt to your porridge.' 'But he has made salt to his porridge' was the triumphant comment, for he was the pride and darling of the whole household.

He had therefore to labour against every discouragement and disadvantage in his own field of study, and must have worked double tides to secure time for both objects, since the proficiency in mathematics, which he considered so important, must have been entirely the result of self-education.

All this accounts sufficiently for the recluse life he led at this time, hardly ever mixing with his family,—unlike as it seemed to his natural disposition and earlier habits, which were cheerful and sociable to a degree. He had a keen sense of humour, and as a boy, when anything tickled his fancy, it was joyous to hear his ringing laugh. You might have heard him, too, whistling, as he rushed upstairs, two steps at a time,—a favourite accomplishment of his, even in after-life, and resorted to occasionally as a pleasant method of throwing off the impression of any little *contretemps*. During the hard battle between law and philosophy, however, all this light-heartedness seemed to forsake him, and frequent fits of abstraction surprised his sisters and disturbed his father. He had been indulged with a separate study of his own, apart from his brothers, and no one but Charles knew or guessed one-half of what was going on there. On account of his early classes he breakfasted there alone, before the early ride into town. After dinner he returned to his study, and declined appearing at supper, in those days the most cheerful and sociable of all the family meals. He required a good deal of sleep, and on principle did not indulge like many other hard students in sitting up at night, but generally went to bed early.

Such was the ordinary tenor of his winter life, till the memorable visit to Italy, 1826–7, which brought things to a climax.

The results are before the world, and the steps of his progress from that time, clearly recorded by himself.

His father died in 1828.

To this interesting record little need be added. It will be supplemented by some early recollections which James Forbes wrote in his later years.

The high quality of his race was conspicuous in him from the first. Strong and sterling fibre, and a lofty bearing, were combined with tender and delicate feelings not often found in so vigorous a character. But whatever may have been the hereditary basis of his energetic nature, it was no doubt intensified by the peculiar surroundings of his childhood and youth. Seldom is a boy of his birth brought up in more complete domestic seclusion. The widowed and heart-broken father, after his wife's early death, withdrew to almost absolute retirement, which, while it was congenial to his own feelings, enabled him to devote himself more entirely to the rearing of his children. This devotion was especially concentrated on the youngest, who became his father's idol as the last legacy bequeathed to him by her whom he had lost. His home at Colinton House enabled him to seclude himself more completely than most places so near a great capital. Lying not far from the base of the Pentlands, approached by tree-shaded lanes and by-roads, itself shut within its own park by old trees, with a garden screened by still older holly hedges wonderful for their size, with an ivy-mantled ruin in the grounds said to have been rent in pieces by Cromwell's cannon, and with a retired background which sloped to the water of Leith, as yet undisturbed by the railway whistle, it would be hard to find in so peopled a neighbourhood so solitary a retirement. The boy's not too strong health and the father's anxious watchfulness secluded him still farther, and made school a thing not to be thought of. The family at Colinton were their own society, the visits even of relations were rare, and there was little or no contact with the world beyond. Such a mode of life, acting on a nature itself strong and intense, could not but increase and concentrate still more that intensity which was ever afterwards its most marked characteristic. The stream, confined from the very first within high and strong embankments, channelled still deeper the groove in which it naturally ran.

James Forbes never went to any school, but was entirely a home-trained boy. Up to the time he entered College at the age of 16, his only teachers had been his sisters' governess and the village schoolmaster, Mr. Hunter, of Colinton Parish School. Fear for his boy's health made his father nervously anxious lest he should overwork himself. In this easy and desultory home teaching he picked up the necessary reading, writing, arithmetic, and the first elements of Latin. What more was needed he was able to teach himself. Equipped with these necessary elements, his mind was free to follow its own bent, and to turn his small acquirements to the uses which his own nature prompted. By his twelfth year that bent had clearly declared itself in favour of physical science, although from deference to the known wishes of his father he was obliged long to conceal it.

The first dawnings of his scientific bias are well described in a retrospect of his early years, which he drew up at St. Andrews, in 1861.

Portion of Journal written in St. Andrews in 1861.

[1821—1824.]

It is a very curious occupation to reflect how the tastes and destinies of after-life are foreshadowed in the tendencies of childhood and youth. Some of my most childish and immature occupations and fancies have a close relation to the serious occupations to which nearly the whole of my middle and after-life has been devoted. And any success which has attended these efforts is in a measure probably due to my having been suffered to follow the strong bent of a natural predisposition for certain employments.

A love of collecting curiosities is too commonplace to deserve mention. A love of mechanical contrivance is also common, and was not earlier developed in me than perhaps in most boys with equal advantages. The wright's shop at Colinton was from (I suppose) my ninth or tenth year a source of great pleasure; and by the aid of lead castings I succeeded in making trains of toothed wheels, in which I took particular delight. An apparatus of this kind intended to be applied to measure the distance moved over by a velocipede was an occupation for a long period. At a later time, perhaps when I was thirteen or fourteen, a pasteboard semicircle, with sights, clamp, and tripod support, for measuring vertical angles, was a source of great delight to me; and at a later period I made a more elaborate metal quadrant for astronomical purposes. I keep the semicircle still as one of the relics of my youth.

About the same and at a later time I took great pleasure in organizing and conducting societies, partly composed of imaginary members, a certain number of names being apportioned to myself and to my kind brother Charles, for debates and reading papers on literary and scientific subjects, with occasional experiments, and all the formalities of meetings and constitutions, &c. This taste found its full development when I became member and secretary of the Royal Society.

Another early habit was that of imagining and delivering lectures. These took their rise in the circumstance that for a winter or two I rode alone regularly from George Street to Colinton and back, taking lessons of Mr. Hunter. To amuse the tedium of the way, I delivered aloud on horseback, to an entirely imaginary auditory, a regular, pre-arranged course of lectures on certain topics, in short, extemporized essays. This must have been about 1824, or earlier. As far as I recollect, these were conducted very methodically, and no doubt contributed to give me some facility of composition and expression. Within less than ten years after (little did I then anticipate it!) I was lecturing, not in sport, in the University.

Another amusement of about the same time was writing periodicals: at first a sort of newspaper, called the *Telescope*,—a miserable thing. I was then about twelve (in 1821-2). Later a weekly essay-paper, called the *Hebdomadist*,—a good deal composed of attempts at Science. Also a magazine (with Charles), called the *Adelphi Magazine*, and others which I have still preserved.

In February 1823, I commenced keeping a register

of barometer and thermometer, and took an interest in meteorology, which has been continued till now.

All these employments occupied no small share of my time and thoughts between my twelfth and sixteenth years, and served to give habits of industry, facility of writing, and of recollecting what I read. Of sermons I wrote plenty : but that is too ordinary to deserve notice. I may here mention that from 1821 (but probably a good deal earlier) I was enthusiastically attached to the English Church as a profession, and I was only withdrawn from it by my father's expressed wish in 1824, that I should follow the profession of the Bar. I may be allowed to say that my father was wrong. Had I been brought up at Cambridge, I should have done well : as it turned out, a kind Providence provided for me well. But it is another instance that it is usually well to allow a boy to follow his bent, when it takes any steady direction.

The following reminiscences occur in one of his journals : '1817—1818, living at Colinton House.—Began Latin early in 1817. In September was reading Phædrus with my father. April 1818, began lessons with Mr. Hunter, schoolmaster at Colinton, reading Cæsar. My brother Charles was also instructed by him. I remained sincerely attached to this worthy man till his death, which took place somewhere about 1840, I think.' Ideas of Inventions.—'2nd June, 1825 : My reason in making this little book is, that so many more ideas (chiefly mechanical) occur to me than I can properly execute, that, in order to keep them accurately in memory, I shall write them down here, taking the most special care to distinguish which are truly original, and what parts are borrowed, and from what sources.' Here follow suggestions for—'A Pluviometer or Rain Gauge. A Universal Balance. A Velocipedometer. A Machine for finding a heavenly Body by its Arc of Right Ascension and Declination. Barometer by compression. Anemometer.'

The above extracts clearly exhibit the two dominant

tendencies of his nature. The one is seen in the wish, thus early formed, to take orders in the Church of England, in which he was brought up, and to which he remained ever strongly attached. From this aim he was not without reluctance turned by the express desire of his father that he should study for the Scottish Bar. This accordingly from his fifteenth to his twenty-first year was his ostensible purpose in life. But under this professional surface was working all the while the other and deeper tendency of his nature. His head was busy with mechanical contrivances,—a new velocipedometer, an anemometer, and a metal quadrant made by himself for astronomical purposes. At the same time he was devouring every scientific book he could lay hands on, from the Nautical Almanac to Woodhouse's Astronomy. But all this devotion to Science was kept strictly secret; he laboured at it in private and said nothing. His father would have objected to these pursuits as too great a strain on his young brain, too severe a tax on his health, and as likely to divert him from the dry studies of the Bar for which he was destined. Amid his contrivances and his reading, his pen was not idle. He had begun to keep for himself a meteorological journal, another journal of observations in astronomy and geodesy, and a record of ideas and inventions.

One more extract from his autobiographic memoir of 1861, illustrative of these things, may here find place :—

'With 1825 commences a personal record of a peculiar kind, a journal of observations in astronomy and geodesy, continued with scarcely any interruption for seven years to the end of 1831, and extending to 651 pages, 8vo. On looking carefully over it now (December 1861), after a lapse of exactly thirty years from its close, during which time I have scarcely if ever opened it, I find in it a curious record of the development of my mind and tastes, by a process of the most purely individual and unaided study, which has a sort of interest of its own, independently of the charm of personal recollection with which of course I regard it. I first commenced a rude meteorological register, which I believe is preserved, I think in February 1823; and from this time scientific pursuits occupied much of my private thoughts, though they were carefully concealed from the members of my family. Astronomy presently became a passion with me, and, having exhausted some trifling children's and school books to which I had access, I read Ferguson's Astronomy, of which my father gratified me by presenting me with a handsome copy, and also a small orrery, in 1825. The astronomical journal to which I have referred opens with the mention of a visit to the Edinburgh Observatory. The first year's record (1825), containing a great deal of puerility and showing much inexperience, is marked at the same time by a devotion to the subject and an ardour for learning which is sometimes touching. My calculations and observations (with a good achromatic of my father's, two and three-quarters inch aperture) were pursued at late hours, and also on journeys. Many of the pages are filled with elaborate (though often very inaccurate) arithmetical calculations of the positions of planets and satellites for a planisphere I was constructing. The books I consulted were Ferguson's Astronomy, "Wonders of the Telescope," Ewing's Astronomy, with tables for calculation of planetary places (this book was lent me by Mr. Hunter, schoolmaster, Colinton), Keith on the Globes (Mr. Hunter's), Chambers's (folio) Dictionary; and I seem to have ventured on an occasional consultation of Woodhouse's Astronomy, to which I had access at home. In July 1825 I commenced reading through the astronomical papers in Hutton's abridgment of the Philosophical Transactions. In September 1825 I became possessed (by Miss Ballingall's kindness) of a Nautical Almanac, which gave me a thrill of pleasure which I yet vividly remember.

'In October 1825 I began making rude angular instruments. In December 1825 I made an observation on a peculiarity of oblique vision in apparently increasing the number of the stars, which in 1826 became my first published communication to Brewster's Journal.

'In January 1826 I began to make a more elaborate quadrant of copper. In May I got hold of the tables in Vince's Astronomy. For long previously I had been in the habit of picking up scientific information in Jameson and Brewster's Philosophical Journals, which lay regularly on my father's table.'

CHAPTER II.

YOUTHFUL TRAVELS.

TILL he had completed his sixteenth year Forbes was entirely home-trained and self-educated. The home training was such as belonged to a Scottish patrician house, whose head was not only highly honourable and refined, but more tender-hearted and serious than most men of his own order at that day. The early death of his mother, and the secluded life which he led from childhood till manhood, were influences which entered deeply into James Forbes's character, and were apparent to the last. Pure, truthful, tender-hearted to all within his own immediate circle; not diffusive of sympathy, nor given to fraternize with those beyond it; serious, scrupulous in all duty, intense and concentrated on whatever he had on hand,—these qualities, which in after life distinguished him, grew naturally in such a home as his. In boyhood he had no jostling with his fellows and no schoolboy friends; and when he went to College, the daily ride home to Colinton left small time for College acquaintanceship.

For his intellectual training he was indebted mainly to the processes he has himself described in his journal. His real and early educators, more than the home governess and the village schoolmaster, were his meteorological and astronomical records and his 'Ideas of Inventions.' By these, continued some of them for years, he trained himself to be the patient and accurate observer of nature and the cautious and correct reasoner which he afterwards became.

In November 1825 he for the first time entered College in the University of Edinburgh, and joined the classes of Latin and of Chemistry. The latter class was then taught by Dr. Hope; and Sir William Forbes attended it that session along with his son. It is noteworthy that only three months before this entrance at College, viz. in August 1825, the entry in his journal 'Began Euclid' occurs.

So the young physicist had been for years groping his way into all the wonders of nature and astronomy by self-prompted experiments and self-taught calculations without any knowledge of regular geometry. Nothing can show more how strong was the native bias, how deep the craving to pry into the secrets of the physical world. Of his first session at College no record remains. Two things only are noteworthy: first, that in attending during this winter a private Greek class he made his first friendship beyond his own family circle; the other, that he must have made good use of his winter in the Latin class, as is proved by the familiarity with which he turned to the pages of Virgil and Horace when the next winter found him in Italy.

About the close of his first year at College—that is, in the spring of 1826—he entered on another stage of his self-education, which was destined to have important results. He commenced an anonymous correspondence, unknown to all his friends, on scientific matters with the late Sir David Brewster. That celebrated man of science, then in the full swing of his long and untiring energy, was at this time among other works carrying on his well-known *Philosophical Journal*. The young enthusiast of seventeen ventured to offer him a paper for insertion in the Journal, and was warmly welcomed by the Editor.

'On the 29th May, 1826, is recorded my first communication under the signature of Δ to Sir D. Brewster. It was a calculation of a conjunction of Jupiter and

Venus on July 31st, but owing to an accident it was not printed. All my subsequent communications were printed, and the titles and copies of most of my correspondence with Sir D. Brewster are duly recorded in the Journal. My letters are mostly very puerile in expression, and altogether unworthy of preservation. Dr. Brewster's on the contrary show great kindness, thoughtfulness, and a warm estimation of my enthusiasm for science. Of course I have rigorously preserved the originals of these.'

This interesting correspondence, once begun, went on for successive years. Two months, however, after it began—that is, in July 1826—James Forbes started with his father and family on a journey to the Continent. Foreign travel did nothing to interrupt the correspondence. Rather it increased in young Forbes the desire to record in so well-known a journal some of the marvels of nature and art which opened before him. A journey abroad was not then the common or easy affair it has nowadays become. Very few Scottish youths at that time had such an opportunity of self-improvement. James Forbes was not one to let the opportunity pass unused. The journey lay by Paris, Strasburg, Innsprück, on through Venice, Padua, Bologna, Florence, to Rome.

On Rome he looked with young and unsophisticated eyes. Learned people may smile at the entry in his journal as trite and puerile, yet it breathes a fine boyish ardour, which those who make that visit later in life might well envy. When men see those famous places for the first time in ripe manhood, learned and fastidious criticism often will not let their feelings go freely forth. They are so repelled by the uncertainty of the old sites, by the present social degradation, and by the effects of long priestly domination, that they have no heart for the ancient glories. Not so the ardent and intelligent youth of seventeen, alive through all his pores, to see, think, and feel.

'We are now in Rome, which it has been my

anxiety to see ever since I set out from home: all that surrounds me is interesting, from 2,000 years of antiquity. We may walk with Horace, and frequent imperial residences, or stray with Virgil into his pastoral shades. They are for centuries dead, but seem to live again in their works when we at the same time behold the objects they describe. Could I drink the Sabine wine with indifference which grew where Horace drew his favourite beverage? or see the "flavus Tiberis" still run yellow in its ancient bed? or could Virgil in vain sing the "Clitumnique greges," while we have under our eyes the verdant pastures?'

In November he twice visited Vesuvius. Here is the account of his second ascent:—

'At half-past two we reached the bottom of the cone. At the top the scene was quite different from last day, the whole of the large crater being full of smoke and fumes, while the strong sulphurous smells prevailed. We rested and refreshed ourselves on a hot crevice for some time, where we left several eggs to roast. Advancing now by the south edge of the crater, we had a tolerably easy walk about half-way round, during which we heard occasionally noises like thunder proceeding from rocks every now and then giving way from the side in great masses, whose fall is reverberated and renewed by the echoes of the vast cavern. At length the edge of the crater grew much lower; and here we not only went down that descent, but in towards the crater all at once, to the extent of 1,000 feet, exceedingly steep, but passable in walking. In this abyss Nature presented herself in a new form; all was unlike the common state of things; we were in truth in the bowels of the earth, where her internal riches were displayed in their wildest form: the steep we descended was composed of minerals of the most singular yet beautiful description.

* * * * * * *

'After retracing our steps in this descent, which proved rather laborious, we visited a crevasse some way down on the outside of the hill, opened within the last forty days, whence very hot air rushes with great rapidity and in such a degree as to make ferns blaze, though they were quite damp. Resuming the edge on the summit, we returned as we had come, to the top of the descending path; and having eaten our roasted eggs, which were excellent, we went down to the hermitage, where we took the cold dinner we had brought with us. Vesuvius is now about 3,400 feet high; formerly it was 4,250, but in the prodigious eruption of 1822 more than 800 feet of the cone were thrown off. At eruptions in general, and that of 1822 in particular, the most tremendous thunder and lightning ever remembered by Salvatore, or his father before him, attended to give more force to the scene; the sea retires from its troubled bed, while shells in masses are thrown from the volcano. The weight of the air diminishes by whole inches of mercury. There is a vast unison and sympathy of all the powers residing in the earth, and of those most especially affecting the constitution of our globe derived from the heavens. Electricity, and consequently magnetism, the equilibrium of the atmosphere, the level of the ocean, the effects of weather, the reservoirs of internal fire in all parts of the globe, the influence of the moon itself, are all employed, affected by, or brought to the aid of this vast production of the mechanism of nature. Surely it is not too much to say that some indissoluble bond unites these various agencies, which perhaps it is the lot of this age to discover. If connection exists, and should the latent principle be discovered, it is impossible to foresee how great may be the extension of human intellect, how deep our insight into the physical economy of all that surrounds us.'

On November 29th he watched and recorded the appearance of an eclipse of the sun at Naples, and December 2nd has the following record:—

Temple of Jupiter Serapis.

'*Dec. 2nd*, 1826.—Being very fine, we set out to Solfaterra. . . . We then went to the Temple of Jupiter Serapis, a most interesting specimen of antiquity. Three pillars only stand, but the groundwork of the temple is beautifully distinct. It is a very singular fact that the sea-water stands in the court, and actually raises tides in the building now about half-way up the standing pillars: the sea has at one time worn circular holes, where even now shells seem to lie; the lower part exhibits no such friction, but a pillar which lies on its sides in the bottom has its whole surface bored in the same fashion, and full of sand and fragments of shells; the present tide does not nearly go to the level of this fallen pillar. There are three totally different levels of the sea at different times, not a little singular in the physical history of our globe; and to these we may add the fact of the sea having retired from the lighthouses now in the heart of Naples, and of its having encroached (apparently very considerably) on some ancient buildings near Sorrento. By a careful combination of the different levels with the several antiquities of the edifices, we may form some idea whether the sea has regularly encroached and retired on the parts of the shore elevated and depressed by earthquakes.'

'ASTRONI, *December 7th*, 1826.

'We returned to the road along the side of the lake, and, continuing it, went up a steep hill to a gateway entering the Royal Hunting Park of Astroni, the crater of an extinct volcano. We immediately and suddenly descended a steep riding road, and one of the most interesting and romantic scenes I ever beheld was presented to our view. It was indeed a volcanic crater, as the precipitous sides of overhanging rocks sufficiently testified; but, instead of the horrid bare chasm which the summit of Vesuvius presents, the cliffs are wooded in the richest manner with the evergreen olive, which at this season shines to the utmost advantage; and only here and there were disclosed the towering and romantic crags which had originally been the barrier of this convulsed spot.

'We were just entering the more gloomy and sunless portion of the path, when a scene shot across us so rural, so enchanted, that it seemed one of the fond images of fancy while contemplating the bliss of Paradise. A lake placid and lovely was before us, the one of whose banks, less interesting than the other, was dimmed by the mountain shadow; while on the opposite side, where the overhanging branches kissed the water, with small verdant banks between, the sun shot down his full enlivening beams. All around is rich, lonely, romantic. While we gazed on the scene with rapture, each feature was rendered more appropriate by the majestic eagle, which, flapping his extended pinions across the lake, alighted with that grand and solemn motion peculiar to the royal bird. The scene was one which cannot be described, scarcely imagined; which Milton might have infused with effect into his Eden or Dante into his Paradise. It reminded me of the description in the minstrel's lay in "The Lady of the Lake"—

> '"Where shall he find . . .
> So lone a lake, so sweet a strand?
> There is no breeze upon the fern,
> No ripple on the lake,
> Upon her eyrie nods the erne,
> The deer has sought the brake,
> The little birds scarce sing aloud,
> The springing trout lies still."'

Tomb of Virgil, 1826.

'*Dec.* 8*th.*—Visited to-day the small humble edifice which the general voice of the most learned men has pronounced to be the tomb of Virgil. Hallowed spot, let reverence be paid to thy humble remains! 'Twere pity that moderns should cast off every interesting association by doubting its authenticity. The building is a square chamber, at the side of the entrance to the grotto of Pausilippo; it is surrounded with niches for urns rising upwards straight to a flat roof, though most of the exterior stones have been removed. In the interior are ten niches for urns, and in the centre was one containing the ashes of the poet, supported by nine small pillars, as some believe; on it was the inscription dictated by Virgil immediately before his death, and so well known.

* * * * * * *

'*Dec. 10th.*—I shall add one or two remarks as to the authenticity of this famous spot. In the first place, tradition is unanimous in marking this place, which on such a point is of no small importance; but the most distinct written account of antiquity is that of Silius Italicus, to whom the spot actually belonged;—that passage is in his obscure style, and has been well discussed by Eustace. . . . On the whole, it is my clear opinion that we may not only frequent the place now pointed out with the fond credulity of imagination, but that there are sufficient grounds for recognizing it on an unprejudiced consideration.'

'POMPEII, *March 22nd*, 1827.

'*March 22nd.*—In Pompeii we see what could not have been presented to us without some very peculiar deviation from ordinary events, such as an eruption of Vesuvius overwhelming a city not with lava but with ashes. In Rome, as in other cities of Italy and those of Greece, we see the remains of antiquity as Time has left them, destroying all the mean or individual houses, preserving only the finest fragments of the noblest edifices. What would Pompeii have been under such a fate? Nothing probably might have remained, as it contains nothing of singular notice in an architectural view, and all might now have been levelled with the ground. But in Rome what light have we into the manners of the ancients? What we know of their private customs, inventions, and civilization is only from their works, which often must mislead from a change of terms. And is it not most extraordinary to find all that ancient writers mention as the custom of their time realized in an actual Roman city laid open to our senses? Who could have been so sanguine as to expect to see the chalked advertisements, the titles of the shops, the tickets of the theatre, the gentlemen's names on their houses, the loaves, honeycomb, soap, grain, raisins, plums, nay, the very fresh olives which the ancients had. Pompeii differs from all other towns, inasmuch as we see the smallest houses, the meanest shops, and the most trifling manners and customs of a nation of two thousand years since, of which we are ignorant and whose name carries no charm in its sound, but of a people whose history has fortunately been handed down to us, and which we are taught from our infancy as that of one of the most extraordinary and powerful nations that ever existed. These ideas struck me particularly on a second visit. The views of the streets of Pompeii are such as to our almost certain knowledge are unique, with the exception of one other city, which from its peculiar situation may probably never be excavated in the same manner.'

'ROME, *Good Friday, April* 15*th,* 1826.

'In the afternoon heard the service of Vespers in the Sistine. The whole, as well as last night, was in mourning, viz. purple; the canopy and covering removed from the Pope's throne, which was plain wood with purple cushions; the tapers on the altar and above the great rail which divides the chapel were dull red, and the Cardinals' seats entirely deprived of their rich carpets; the Pope himself in red and the Cardinals in purple robes. When the Tenebræ begins (I allude to all three days), all the candles were lighted, including fifteen small ones on a branched candlestick, which are extinguished one by one as the chanting proceeds, which is varied every day with suitable Psalms and Lessons chiefly from the

Lamentations of Jeremiah. It is very interesting to accompany the service with a book. At the same time that artificial darkness is spreading by the extinction of the candles, the twilight increases into the shades of night, and the effect of the sun's parting rays over the frescoes of the chapel is very fine; and by the time that the Benedicitus begins, and the fifteen candles are all out but one, the sun is probably set. At this moment two persons with extinguishers commence putting out the corresponding tapers on the altar and rail of the chapel, one at each verse of this beautiful Psalm; and when all are gone, the fifteenth candle, which still remains, is removed from its place—probably as a symbol that our Saviour, though apparently dead, after all the desertion of His friends and the darkness which overspread His soul, still lived, though unseen. At this juncture the Pope descends from his throne, and kneels before the altar, the Cardinals and the whole court doing the same at their stations. The chapel is in perfect darkness, except some lights from the music gallery, the last shades of twilight gleaming most softly through the northern windows, and the single taper throws a faint glow from behind the deserted altar on the priests who kneel before it. At such a crisis how does the first thrilling swell of Allegri's Miserere burst upon you! how silent the moment, when the finest piece of vocal music ever composed, and performed by the best vocal choir in the world, seems to draw to earth the notes of the angelic chorus, and exalt your thoughts to higher spheres! When the last words, "Tunc imponent super altare tuum vitulas," have died on the ear, the Pope in a low tone reads a short prayer, still kneeling; a rumbling noise is then heard, which represents the confusion attendant on the Redeemer's death; after which the flambeaux are introduced, and all depart in silence.'

These descriptions of scenes so often before and since described will show what opportunities James Forbes had presented to him, and how he profited by them. For a boy of seventeen his mind leapt out with very remarkable force to meet whatever he saw that was wonderful in nature or great in history. That one so keenly appreciated day at Pompeii as a lesson in Roman manners and customs was to an eye observant as his worth several winters poring over Roman antiquities. And on one more or less familiar with the Roman poets the days he wandered over the Sabine hills and Vale of Horace, or visited first the tomb, then the birthplace of Virgil, impressed the spirit of their poetry more powerfully than several sessions in the best taught Latin class in any University would have done. Nor was he, as his journals show, less open to impressions from the mediæval and modern aspect of Italy. On the homeward route there are some very characteristic entries.

'MANTUA, *June 9th.*

'The town has a fine appearance on entrance, yet nothing which can exactly satisfy the poet or gratify the curious traveller full of recollections of the master mind who immortalized his birthplace,—"Mantua me genuit." I regretted that I overlooked visiting the ancient head of Virgil, fondly cherished as original, which has been restored by the French, and is here preserved. With regard to the place of his birth, universal tradition has assigned it to the modern Pictole, and Dante, who from his early date ought to have been well informed, countenances this opinion. But Eustace is disposed from a consideration of localities to give the honour to Vallagio. In such a state of uncertainty as to what concerns the great man, we find little which can be called tangible to rouse our recollections. The most pleasing way of supplying them is by the perusal of his works, that monument which will endure as long as any work of human fancy, or while a spark of civilization remains to be cultivated. I, who have lately perused his works and recited his last verses over the moss-grown ruins of his tomb, now over his favourite plains and among his hallowed groves, hail his shade where it may still be supposed to delight to roam.'

It could not be that one so young and enthusiastic could then feel the full force of Virgil's 'single words and phrases, his pathetic half-lines, giving utterance, as the voice of Nature herself, to that pain and weariness, yet hope of better things, which is the experience of her children in every time.' It is only one who has seen much of life, and felt it deeply, who finds such meaning in his words. But in the purity and gentleness of the Roman poet there was that which came home to the tender nature of this youth of seventeen.

When they had crossed the Alps, we find he made this entry at Geneva :—

'Went to Ferney, a beautiful drive of about a league ; two rooms, as it is well known, are preserved as the philosopher Voltaire left them. I took little pleasure in seeing the recollections of this detestable man, of whom perhaps one may say, the sooner he is forgot it will be the better for his character and the peace of the world in general.'

Here is an anticipatory glance of what was to be :—

'*July* 1*st.*—Reached Chamouni, this most interesting spot, which was unknown even to the natives of Geneva, though distant only eighteen leagues, till discovered in 1741 by two English travellers, Pococke and Wyndham. . . .

'*July* 2*nd.*—We set off from the inn (the Hôtel de l'Union, a very comfortable house) at half-past nine to visit the Mer de Glace. We were furnished with excellent mules, spiked poles, and three guides—viz. Michel Cachet, surnamed "Le Géant" from having spent seventeen days with Saussure on the tremendous ridge of the Col de Géant, where more than 10,000 feet above the sea he made a long series of experiments. He ascended Mont Blanc with Saussure three times since. The second guide was Simond, who is called "Des Dames;" and he has once gone to the top of Mont Blanc. The third was Jean Marie Coutet, son of the famous Coutet, lately dead. We had a delightful ride up to the Montanvert. . . . In one instant we were presented with the singular view of the Mer de Glace. Its surface has much the appearance of tempestuous waves frozen; but the wonders of its form have been too much spoken of. . . . We walked some distance on the Mer de Glace, which appeared much rougher on a near approach. . . . In the course of our walk I conversed with the memorable guide "Le Géant." I considered his information as peculiarly valuable, for when he and one or two others are dead (and he is nearly 70), all living record of some of the most daring and interesting adventures that have ever been made will be gone. He was of the second party that ever reached the summit of the highest European Alp, and accompanied Saussure in thirty or forty other excursions. He took four days to the expedition of Mont Blanc, while most others take three. He said he did not suffer much from the rarefaction of the air, but that Saussure felt it considerably, especially in using his instruments. He said that on the Col de Géant they really saw red snow, in patches about two miles in length but very narrow; but the philosopher could not find the cause of the phenomenon. In that expedition he described the scenery to have been the very horror of nature, a wild fortnight abode for human beings. Saussure (he said) when he visited Monte Rosa believed the highest summit to be quite practicable, but did not care to perform it, as he had a view of the peaks from one of the lower ones; and as he had already succeeded in attaining the highest point of Europe, and proved it to be so, he did not care to accomplish one less lofty. I requested "Le Géant" to say what he thought of the practicability of the ascent of Mont Blanc, and whether a man of ordinary strength might achieve it. He said, "Men vary so much, it is not easy to say;" but he added, "Avec du beau temps, des bonnes jambes, un bon tête pour passer les fentes, et surtout avec le bon plaisir du Dieu, on peut le faire." . . . . We did not leave Chamouni till half-past six, and as we drove slowly down the valley I felt that it was with more regret that I left this interesting spot than any town or any scene that we had visited in our travels.'

The following is the record of arrival at home :—

'We left Dunbar at twelve; we proceeded at a rapid pace with happy hearts towards home. Before reaching Musselburgh our eyes were greeted with the sight of Edinburgh, Arthur Seat, the Calton Hill, and Inch Keith, noble objects which never can be mistaken. Without prejudice I consider that Edinburgh combines more beauty and majesty as a town, than any five cities we have visited put together. I cannot conclude this journal without observing how grateful I felt for this happy return to home and all its peaceful joys, and how much more grateful for that protection and guidance from hour to hour bestowed upon us in every change of time and circumstance, to which during our long absence we have been subjected.'

A year of finer, more varied, and impressive education than the one thus passed, could hardly be conceived. Forbes was at the very age when impressions come most quickly, and stamp themselves most deeply into the imagination and character. The above extracts show how open his mind lay to influences from every side, how it enriched itself by imagery, gathered from every quarter, classical, mediæval, and romantic. But while he was thus alive through all his being, the one master bias was never forgotten; the undercurrent of his thoughts kept flowing strongly and steadily in that one direction in which it had long since set. The first sight of Chamouni, as we have seen, called forth that scientific interest combined with love of adventure which was so deeply rooted in him. But a more definite record of the fixed purpose of his life is found in the following entry in his astronomical journal :—

'ROME, 31*st December*, 1826.

'These two months have passed over my head without abating one jot those charms, or rather endearments, which the noble study of astronomy holds out to me. My utmost earthly wish for my own happiness, as far as I am personally concerned in it, is to devote my days to the study of my beloved pursuit. My imagination is not heated, nor are my fancies distended or distorted, while I write this. It is my earnest wish and prayer in my soberest moments. I shall be more sensible of my advantage when I return home, than I was when I left it. I have opened a source of great satisfaction, and what one day may be most advantageous to my advancement, in my anonymous communications to Brewster's Journal, which, to my almost inconceivable pleasure and surprise, appear to have been well received. This opens a vista before me, which I would scarcely have looked for in my time of life and situation. I have kept this a secret to the present time from every one. I am now preparing a packet of papers, chiefly on meteorological subjects about Naples, which I shall soon send.

'My labours recorded in the past pages have chiefly been directed to the calculation of the conjunction of Jupiter and Venus, the eclipse of the moon in November, the eclipses of the sun, and the formation of my quadrant. I have also written an essay on the apparent number of the stars, ascribing their immensity to the effect of oblique vision.'

The scientific correspondence with Dr. Brewster alluded to above, which had commenced before James Forbes left Scotland, and which formed so important a stage in his scientific education, was never lost sight of during his Italian sojourn. The ascent of Vesuvius, the solar eclipse at Naples, and other sights or events presented during his travels, were recorded almost on the spot, and speedily despatched to Dr. Brewster, to appear in his Journal. On the return of the family to Scotland, among the first things that attracted the eyes of James

Forbes were successive copies of the Journal lying on his father's table, with his own papers inserted in them.

How he was affected by seeing these first products of his pen in print appears by the following extracts from his Meteorological and Astronomical Journals in 1827 :—

'COLINTON, *August 22nd*, 1827.

'I had yesterday the pleasure of seeing Brewster's Journal and reading the two articles of mine he inserted in October last, on the apparent number of the stars, the heats and colds of last year, and elements of the lunar eclipse among the celestial phenomena, together with all the other papers inserted or favourably noticed which I have since sent. Annexed to the first paper I sent are these words: "We should be glad to hear again from the author of this article, and if possible learn his address."

'*August 26th*, 1827.—In Brewster's April Journal the following is among the notices to correspondents: "We have received Δ's very excellent set of observations made at Rome on the 15th January, and also his interesting *Observations on the Climate of Naples* and his *Remarks on Mount Vesuvius*, which will appear in the next number. We shall be glad to hear from him as often as convenient." Accordingly my paper on Vesuvius appeared in the July number. It would be difficult to express the satisfaction I feel at the success attending my first published essays. I have communicated the secret to Charles, but to no one else. I am now busily preparing a paper on the horary oscillations of the barometer at Rome, and reducing about 500 observations made there, which is rather laborious, but in about ten days I hope it will be ready.

'*September 8th*.—Finished this morning a complete memoir on the horary oscillations of the barometer at Rome, which I sent to Brewster's Journal. It amounted to twenty-three large quarto pages, and contained the three preceding tables at length.

'*October 4th*.—To-day saw the new number of Brewster's Journal: it contains my remarks on the climate of Naples. The notices to correspondents stated that the paper on the horary oscillations of the barometer came too late, but will appear in the next number.

'*December* 18*th.*—Brewster's Journal has been published to-day: only half of my paper on the horary oscillations has been published, owing to a small error in the table of final results, and the remainder postponed till next month, together with my paper on Monte Testaccio, which is approved and will be inserted.'

CHAPTER III.

COLLEGE COURSE.

WHEN the Edinburgh University session of 1827-8 opened, James Forbes, after an absence of one whole year from College, re-entered it as a student in the Moral Philosophy and Natural History classes. The former class was then taught by Professor Wilson, the latter by Dr. Jamieson. These two teachers were both eminent in their way, but very different from each other. Professor Wilson discoursed at large on his subject, if with no vast learning or keen edge of subtlety, yet with a flow of fervid and poetic eloquence, which fired youthful imaginations, and helped to let loose over Scotland those floods of turbid rhetoric, which from pulpit and platform have deluged it for a generation, and are only now abating—if indeed they are abating. Dr. Jamieson, plain, practical, not to say prosaic, but accurate, painstaking, and diligent as an observer, so rarely ventured on figures of speech that the one or two metaphors in which during the whole session he indulged were well known and waited for, and when produced were welcomed with annual rounds of applause. To both of these so different men James Forbes had a side of affinity.

To the work of the Moral Philosophy he gave himself in earnest, but he contrived to weave into the subjects proper to it some of his own favourite knowledge. We find him recording in his astronomical journal, December 31st, 1827:—

'One considerable astronomical work I have been engaged in which I have not yet recorded, an essay for

Professor Wilson on the influence and advantage of the study of astronomy on the mind, with scientific illustrations and notes at the end. It is almost finished, and will amount to sixty large quarto pages.'

The sixty quarto pages must, if the poetic Professor took time to read them, have conveyed to him a mass of useful information, which was probably new to him. At the close of the session James Forbes stood first and obtained the medal in Professor Wilson's class, and the acquaintance thus begun ripened, I believe, into a lifelong friendship. At the close of the session he thus records what was probably a reward for his winter's labours and success:—

'*May* 14*th*.—To-day Papa made me a present of a splendid theodolite, just come down from Troughton, with five inches' arcs divided on silver and single minutes. It is a splendid instrument every part, and I hope to put it to some astronomical as well as trigonometrical purposes. Dr. Troughton himself said this theodolite was packed as the best of three, and it was made by the best workman he ever had, who is now dead. It is therefore probably as fine an instrument as ever came from the hands of a maker.'

The father whose kindness is here recorded was not long to be spared to his family. For some time the health of Sir William Forbes had been declining, so that even friends and acquaintances observed the change. We have seen how Sir Walter Scott amid his own calamity had been struck by seeing it. The end is thus noted in the journal of his son:—

'24*th October*, 1828, 1-2 P.M.—By the grace and power of God Almighty I resolve that the firstfruits of this tremendous blow shall be a determination to keep steadily in view, as a tone of mind, the existence of a future state, where my beloved father is now gone—a consideration which bears in its consequence on every iota of our actions, and of which I presently acknowledge myself wickedly forgetful. I make the resolution under

Sir William Forbes

7th Bart of Pitsligo

a thrilling impression of mind, yet am not insensible to its real difficulty in general life, though now so appallingly thrust upon me. Yet with the aid of the Omnipotent nothing is impossible; and I crave His power to enable me to perform this matter, so that when my time comes I may die the death of the righteous and my last end be like his, who but an hour ago since left me his example as a legacy.

'JAMES D. FORBES.

'*Sunday, November* 22*nd*, 1828.—This day I had the satisfaction of receiving the Holy Communion with our whole remaining family, being the first time of our being at chapel since our father's death. Such a judgment, if anything, was fit to prepare me for the due reception of that sacrament, and I took it with contrition and subsequent satisfaction. I proposed to myself the following resolutions, the first of which took its origin in the hour of my father's death:—

'1st. To keep stedfastly in view, as a tone of mind, that I am created for a future and eternal life.

'2nd. What naturally flows from the former: to curb pride and over-anxiety in the pursuit of worldly objects, especially fame.

'3rd. To be diligent in the pursuit of my winter studies.'

'COLINTON, *Sunday, November* 16*th*, 1828.

'Feeling myself now also on my own footing, I have prepared a letter to my excellent brother Charles, my best friend in our present state, as to the disposal of my property in event of my death. In looking to future arrangements, I mean not here to express the pangs which parting from the members of our family and leaving the spot so long our paradise fix on my mind.

'I am anxious to carry on my literary career in such a way as may possibly one day make me independent of a profession, which is undoubtedly a matter rather of necessity than satisfaction. I have commenced my law studies, and am not without hopes that that Power which sees my smallest griefs will give me steadiness to pursue them with advantage and with honour. Be this as it may, in losing my father I have lost my principal aim in the concealment of my name from Dr. Brewster. But an ambition has come over me, I hope not a criminal one, to be early a member of the Royal Society, which would be of advantage to my future fame, which I look to principally as the means of acquiring a name in the literary world, of which every one knows the advantage. This is one point. But another use I think of making of my incognito before I part with it is in a lucrative view, as I propose writing letters to the conductors of each of the foreign Reviews, offering to write.'

The winter duties of this session (1828-9) consisted in attendance on the classes of Natural Philosophy and Civil Law in Edinburgh University. Sir John Leslie still taught the Natural Philosophy class, though now somewhat past his prime. This was the class in which were handled all those subjects on which almost from his tenth year Forbes' thoughts had mostly dwelt. All these years he had clung to them with wonderful pertinacity.

His meteorological and astronomical journals, as well as other observations, had been carried on almost from childhood with unflagging zeal.

That a student, entering the class of Natural Philosophy with such a natural genius for the subject, so well prepared, and so persevering, should succeed, was a matter of course. Here is his own record of the results:—

'Regarding my past winter's studies at Civil Law, I have got on better than I expected, nor can it be said to be very repulsive. I expect to pass my trial at the beginning of the session in May. At Leslie's class I got on well, and easily got the first prize, for which I believe I shall get a gold medal. I have also attended Jamieson's very pleasantly.'

But this session contained one event more important and interesting to the young votary of science than the attendance on any class in College.

During the middle of the session Forbes resolved to throw off the disguise he had so long worn, and to reveal to Dr. Brewster whose was the real countenance concealed under the mask of Δ.

In the following letter he introduces himself to Dr. Brewster in his own name:—

'*December* 1*st*, 1828.

'SIR,—Many causes may have combined to inform you of my name and circumstances; and if it has been so, I need hardly apologize for the long anonymous character which I have adopted. When I commenced my communications to the Journal, and they experienced a reception so much warmer than I expected or deserved, I had just completed my seventeenth year, and at present am in my twentieth; so that it can scarcely surprise you that sooner than this I did not choose my name should be brought before the public. So strong was my feeling, that at this moment no one is acquainted with my correspondence but yourself and one of my brothers. My youth you must have well known by innumerable marks of internal evidence in my compositions, and I gladly seize this opportunity of thanking you in the most heartfelt manner for your continued attention, and the many marks of kindness you have shown me,—not the least so unqualified a recommendation as your last contained.

'With regard to the Royal Society, I should not wish to be balloted for till the first February meeting, and till my name is proposed I should prefer that my anonymous character should not be brought to light. I have thought it fair, however, after such candid and gentlemanly behaviour as I have experienced, that from you my real name should no longer be concealed.

'Very sincerely yours,

'JAMES D. FORBES.'

The elder of the two correspondents welcomed the revelation in the following terms:—

'*December* 19*th*, 1828.

'DEAR SIR,—I did not know who Δ was till your letter informed me. Various circumstances rendered it probable that you were under the mask, and I once asked your uncle, Mr. George Forbes, if he knew any person who used a seal such as yours. But he did not. The transmission of your packet from Rome through Sir A. Wood and some other circumstances were strongly against you, but these were always overpowered by the belief that Δ must be older than you could be.

'Many circumstances conspired to make your letter more than agreeable to me, though, like all our pleasures, it had its alloy of sorrow. It was a matter of real satisfaction to me that Scotland possessed a young man capable of pursuing science with the ardour and talent of Δ; and that he should have belonged to a family from whom I received much kindness, and for whom I feel the most sincere affection, was a ground of the most unfeigned satisfaction.

'I trust you will not allow any professional pursuits to interrupt your studies and researches. The cultivation of science is a luxury of no common kind amid the bustle and vexation of life, and is quite compatible with the most active professional duties. Your education and the example you have had to copy will, I am sure, guard you against those presumptuous and sceptical opinions which scientific knowledge too often engenders. In the ardour of pursuit and under the intoxication of success scientific men are apt to forget that they are the instrument by which Providence is gradually revealing the wonders of creation, and that they ought to exercise their functions with the same humility as those who are engaged in unfolding the mysteries of His revealed will.

'I hope I shall soon have the pleasure of being personally acquainted with you, and it will at all times give me the truest pleasure to be able to be of any service to you in your scientific studies.—I am, my dear Sir,

'Ever most faithfully yours,

'D. BREWSTER.'

It is thus that James Forbes notes in his astronomical journal the conclusion of his anonymous correspondence:—

'My anonymous correspondence with Dr. Brewster is now ended, and a happy termination as well as continuance it has had. In the beginning of March I hope to find myself a member of the Royal Society, a sort of consummation I have long looked to with faint hope. And here it may be proper to give a list of everything I have transmitted to Dr. Brewster for publication; any particular account of my transactions being unnecessary.

'*October* 1826.—On the Apparent Number of the Stars.

'1826.—Heat and Cold.

'*July* 1827.—On Mount Vesuvius.

'*October* 1827.—On the Climate of Naples.

'*January* 1828.—On the Horary Oscillations of the Barometer at Rome, Part I. Eclipse of the Sun.

'*April* 1828.—Caves Mount Testaccio. Horary Oscillations, Part II.

'*July* 1828.—Materials and Styles of Buildings in Italy. Meteorology, 1826-7. Notices in Astronomy and Meteorology. Solar Spots. Radiation.

'*October* 1828.—Physical Notices of the Bay of Naples. —No. I. On Mount Vesuvius. Self-registering Thermometers.

'*January* 1829.—Physical Notices of the Bay of Naples, II.—Buried Cities. Self-registering Thermometers. Notice of Auroral Arch.'

Shortly after this recognition Dr. Brewster proposed his newly discovered correspondent for admission to the Royal Society of Edinburgh. That society is the most ancient and distinguished scientific society in Scotland, and has long numbered among its members all that the country contains of scientific eminence, and not a little of the literary genius of each age. To become a member of it is regarded as an honour by men of ripe age and reputation. But for a youth to be elected before he had completed his twentieth year was very unusual, if not unprecedented.

It is thus that Dr. Brewster introduced James Forbes:—

'Mr. Forbes has a family claim of an irresistible kind; but, even if he were the son of a peasant, his learning, his intuitive genius, and his absolute devotion to science entitle him to the warmest reception which the Royal Society can give him. He will be one of its greatest ornaments. I must also assure you that for years he has been an anonymous correspondent of mine, and that I promised solely from his talent to recommend him, when I did not know whether he was the son of Sir W. Forbes or of a peasant. He is also a travelled person, and has surveyed with a scientific eye the classical and volcanic wonders of Italy. If you can find men of his stamp, you will do well to secure them.'

'*April* 19*th*, 1829.—This day is the last of another year. It carries me from my teens, from the important season of youth between twelve and twenty. I trust I am indeed entering on a new way of life. May God apply these serious recollections to my heart!'

In the same month he notes that he shook hands with Dr. Brewster for the first time at the Royal Society. In August of the same year (1829), Forbes visited Dr. Brewster at his villa of Allerly, a little above Melrose, and on the opposite bank of the Tweed. Journeys in those days were not made so easily as now, and therefore young persons valued them more highly, and were all the more deeply impressed by them.

It was his first intercourse with a philosopher of kindred talents, and his first sight of the Vale of Tweed and its ruined abbeys. The conversations he had with Dr. Brewster and the new sights he saw all round equally delighted him.

'We drove to Abbotsford; unfortunately did not find Sir W. Scott at home, but saw the lobby, a perfect picture of the wonderful owner's mind. I was extraordinarily sorry at not seeing Sir Walter, as I knew from his own authority he would have treated any of the family with more than usual kindness. And Dr. Brewster wrote me that, having met Sir Walter at dinner, he expressed his regret at not being able to see me, as during my stay at Allerly he was continually occupied with company.'

On turning to the Life of Sir Walter, we find that it was during this same month that he received a visit from the historian Hallam with his son Arthur, who not long afterwards was cut off in the bloom of opening genius, and whose memory is embalmed for ever in Tennyson's 'In Memoriam.' Sir Walter had, during the preceding June, suffered from the first symptoms of that disease which at last proved fatal. Yet in August he was able to accompany the Hallams to see the sights of his neighbourhood. Of their visit together to Melrose an affecting memorial remains in the fine lines by young Hallam, beginning,—

'I lived an hour in fair Melrose.'

Probably this was the company with which Sir Walter was engaged that day when young Forbes called at Abbotsford and did not find the poet.

The following letter contains some notice of this visit to Brewster:—

'ALLERLY, MELROSE, *September* 1*st*, 1830.

'DEAR SISTER,—The day I arrived, Dr. Brewster took me to Chiefswood, Mr. Lockhart's cottage, which is really a little paradise, and would, I am sure, hit your fancy. Mrs. Lockhart was not at home, for we had met her driving her own drosky. As we were going up the approach Dr. Brewster said, "We shall find him smoking;" and to be sure the first symptoms we saw of the great man were certain puffs rising above the bushes. He was walking backward and forward on his little lawn, with a little table standing upon it, with parliamentary papers, proof sheets, &c. lying upon it, and a glass of something or other. I was very much surprised and unexpectedly pleased with him. He was a much younger and finer-looking man than I expected, strikingly like F. Grant. He has a tremendous pair of eyes, but his manners were highly affable and pleasing. He began abusing the plague of writing on the Greek question, and envying Dr. Brewster's Life of Newton, whilst the Doctor declared he thought nothing could be more delightful than to write what everybody would appreciate and admire, and on such interesting subjects. You may suppose I took Lockhart's side. . . .

'I went yesterday to leave my card at Abbotsford. I saw the great man hobbling up a plantation, apparently frightened at a visitor, a class which indeed he has reason to fly. . . . So I did not see him, but I dare say that if he has time he will let me hear of him. All his family are with him just now. I met with his factotum, William Laidlaw, who wrote many of the novels to his dictation, and who seems a remarkable man. . . .'

The following are some additional notes written by young Forbes during his visit:—

'After leaving Abbotsford, drove by the south side of the Eildon Hills, where the Doctor took me to see a very interesting quarry, exposing most beautiful prismatic colours of compact felspar, an uncommonly fine display. Again set off with Dr. Brewster in his gig for Jedburgh at half-past nine o'clock. . . . Left Dr. Brewster at Dr. Sommerville's, the old minister of the place, rather a remarkable man in the literary world, and now in his ninetieth year. I persuaded the Doctor to let me see the house in which he was born, which is near the Roman bridge. I afterwards returned and sketched the house on the back of a letter of his to me, dated August 1st, 1829, marking the window of the room in the roof at which he tried his telescope.'

Again, 'Dr. Brewster talked very kindly and freely of my prospects, and strenuously advised me to direct my efforts to some specific end. He assured me that, if I kept one object in view, I should soon know more about it than anyone else. I said that I had lately thought the relation of bodies to heat a very fertile subject; in which he perfectly agreed, and commended the idea. I said that experiments took more time than writing. He agreed with me, but said that nothing could be more delightful than the prosecution of successful ones.'

This conversation contains the first hint that young Forbes's thoughts were already travelling towards that subject to which he afterwards gave so much attention and with such good results.

Thus it was that the intercourse begun by anonymous correspondence, and maintained so honourably to both sides for several years, at last passed into personal acquaintance. And the acquaintance was by this visit to Allerly cemented into a friendship which stood the strain of severe trials, and lasted as long as life.

But this friendship, though interesting in itself, and helpful to his scientific pursuits, could not make him feel less acutely the change which the last year had brought to him and his family. Though the brothers and sisters clung to each other and to their domestic life with the tenacity that belonged to their race, every one of them felt that when the head of the house had been removed, an irreparable change had come. The thought was now forced on them that the family home must soon be broken up, that they must soon quit the old house at Colinton, endeared to them by all the recollections of childhood and boyhood. He says that 'the idea of leaving this endeared spot was like a dagger near his heart.' And though for a while some of them might live together and make a family home, yet it never could be with them as it was before. For each one had become his own master, and must regulate his movements more or less on his own responsibility, without that advice and guidance on which they had hitherto in common leaned.

No one felt these things more than the youngest and favourite son. There are many allusions to it in his diaries during this period. Thus he writes in Fettercairn greenhouse:—

'Every succeeding week tells me how changed is my situation since this time last year. I believe I have scarce made any notes this year on my professional views, but they have much and deeply occupied my attention, till, having done everything which information and advice could accomplish, with long and anxious consideration, I have taken my general principles of viewing it, and leave it in humble confidence to the wise Disposer of events. My present thoughts are as follows: to devote a moderate but steady share of attention to law, which, as far as I understand, may never occupy nearly all my time, and at first will leave me very great leisure. It is my idea not to pass my Scotch law trials till spring 1831. Then, if things go on well, to go abroad for a considerable time. I confess my private views in this tour to be to examine with a much more scientific view than last time the south of Europe, to form scientific connections in France and Italy, and—I have never before had courage to record it either by word of mouth or on paper—to write a personal narrative like Humboldt, and Travels in Italy.'

When the year brought round the anniversary of his father's death, his grief came back as bitterly as at the first: a whole week seems to have been kept as a solemn season of remembrance and self-examination.

'*Saturday, October* 24, 1829. ¾ *past* 12.

'With a bleeding heart I have just come out of my late father's bedroom, where I resolved to spend the recurrence of the tremendous moment—twelve minutes past 12. Oh that my sufferings, intense as they have been, might produce some permanent change in my habits of thought! I am horror-struck when I think how little I have done towards the great end I proposed to myself last year; how little habitual the thought of death. I poured out my soul to God upon the bed with an inexpressible intensity of feeling. Oh may this hour be deeply engraven in my thoughts!'

Evening of the same day.

'. . . . O that I could indeed have that "habitual remembrance of the reality and eternal duration of a future state" which I have prayed for every morning. Shall I never, but in exalted moments of love and hope, attain any tolerably practical conviction of the great truth? I feel it almost presumptuous to resolve again. I now resolve to devote a more divided time before going to bed for reading the Bible, which shall include a short, but clear, self-examination. And on every Sunday I will, with devout deliberation, turn my thoughts to this day, to inquire how far its grief and contrition have worked any good. By the renewed assistance of God's grace, I trust in almighty power that my efforts may not continue to be nearly vain.'

For several years the recurrence of this anniversary fills his journals with the most passionate outpourings of grief. No abridgment, nothing but the sight of the journals themselves could give any notion how deep and how enduring that grief was. Strong as were his leanings towards scientific discovery, earnest as his devotion to it was, these his ruling intellectual tendencies were weak to sway him compared with his home affections. The former habits, much as he clung to them, he could suspend whenever duty or a higher impulse seemed calling him to do so. Science was an instrument he could wield, and again lay down; a garment he could put off, when the proper time came. These first affections were himself, his own proper being, and he could no more lay them aside than he could suspend his own identity or the beating of his own heart.

As the sequel of this memoir will be mainly occupied with his doings as a man of science, it is well that this should be understood once for all by those who wish to know what manner of man James Forbes was. In this respect almost all men fall into one of two kinds. In the one the intellectual or professional interest is so paramount that it pervades the man's being to the very core. Withdraw this, and little else is left. The domestic feelings occupy but a small niche apart, and do not really colour and rule the life. In the other kind of man affection is central and paramount, the mental and scientific habits, whatever they be, seem external powers or capacities, clothed as it were upon the deeper affections which are the truer self. To this latter kind Forbes belonged.

Though he had completed the college course at that time usual for Scottish students, he entered on another session of attendance in November 1829; a second course of Natural Philosophy, a second of Chemistry, and a course of Scottish Law. His professors during this year were therefore Leslie, Hope, and Bell. But, while attending these classes, his own private studies and pursuits were coming more and more into prominence. In his astronomical journal the following entry occurs:—

'*December* 31*st*, 1829.—This is the last night of the year, and in revising the fifth year of this work I begin to be sensible that its style has considerably changed. I begin to be sensible that astronomical calculations are, as Dr. Brewster says, a great absorber of time, every spare moment of which I have devoted to original researches during the past year; and the quantity I have written is very great.

'My diminishing periods of leisure must curtail my astronomical calculations, which I must, however, always value for the many happy years they have occupied, and the relief in periods of distress which I hope they will long continue to afford. But I am led to more original fields, and I hope to prosecute practical astronomy with my other pursuits. Having the means of making good observations in geology, I hope to continue those labours which have occupied a good many pages of this work in the two past years. My visit to Dr. Brewster I look upon as one of the happiest points in my life, and the account of it, which I have just read over, as one of the most interesting parts of this work.

'I shall now as formerly carry on the notes of my

printed publications, particularly in Brewster's Journal, the history of which I have always here inserted.

'*April* 1829.—Physical Notices of the Bay of Naples.—No. III.: Pausilippo and Lake Agnano. On the Defects of the Sympesometer as applied to the Measurement of Heights. Physical Notices, No. IV.

'*July* 1829.—On the Solfaterra. Analysis of Schomo's Specimen of Physical Geography.

'*October* 1829.—Physical Notices, No. V.: Temple of Jupiter Serapis.

'*January* 1830.—Description of a new Anemometer. Physical Notices, No. VI.: Bay of Naples, &c.

'No. II. of the Physical Notices, or Herculaneum, was in great part translated in the *Bibliothèque Universelle* for April, and No. I. has, I hear from Prof. Jamieson, recently appeared in German in a periodical of Leonhard. In Jamieson's Journal last year appeared an abstract of observations on the temperature of springs at Colinton: these have been translated in the *Bulletin Universelle* for August 1829. In the same work appeared my paper on registering thermometers, from Brewster's Journal of last year. Jamieson published in his Journal for October a letter of mine to him on a boulder on the Pentlands: a notice of it, and also of my paper on Serapis, appeared in the Edinburgh Journal of Natural and Geographical Science, No. III.'

In his meteorological journal he thus writes:—

'*Friday, January* 1*st*, 1830.—I now enter the fourth year of this work, which is likely to assume daily a more original character, as I intend to make it the receptacle of all my experiments connected with practical meteorology; which, if my present plan holds good, will occupy a good deal of my attention next summer. The inquiries I was led to make in a more close manner than before, when writing the article "Thermometer" for the Edinburgh Encyclopædia, showed me strongly the want of good observations; many fundamental points, such as the application of spirits to thermometers, the expansions of various fluids, hitherto resting on the imperfect experiments of Deluc, to which Biot has absurdly applied arbitrary formulæ; the capacities for heat, the disturbing causes in meteorological observations, of which I have given a sketch in the above-mentioned article, and the Theory of Hygrometry. The method I adopted during the past year of copying the daily observations of the barometer into five columns of hours has succeeded to admiration, and saved me much time and trouble. Since May 7th, I have likewise daily inserted my miscellaneous observations in a regular quarter sheet. I mean, in continuation, to get my rain gauge into better order and placed on the top of the house, and to observe the radiating thermometer more frequently.'

The following extract from his journal expresses the feelings with which he closed what was virtually his last student session at college:—

'FETTERCAIRN, *April 25th*, 1830.

'The College session which has now passed, the last probably which I shall seriously be engaged in, has been less distinguished than the two preceding ones by much excitement. I attended Bell's lectures on "Scot's Law" with extraordinarily little satisfaction, even less than I expected. Dr. Hope I re-attended (along with John Buchanan) with renewed pleasure, and Leslie I also took another course of, though chiefly useful in directing my mind to problems I had not before studied. I easily carried off his prize, and, as I got a medal last year, I am going to have a copy of Newton's "Principia" this time. I cannot look back upon my College education but with peculiar satisfaction, as comprising the happiest period of my life. Towards the University of Edinburgh, and the valuable friends I have acquired in several of its professors, I feel the warmest attachment. Of the professors, I would particularize especially Mr. Wilson and Mr. Jamieson; and as to Professor Leslie, he has been as kind to me, I believe, as his singular nature would permit; and I have had frequent, though slight, proofs of his good opinion. At the Royal Society I have been a regular attender, and made many scientific acquaintances. It was only on Monday last that I fully understood from Mr. Robison that my recommendation as a member of the society was to be received as soon I was twenty-one. He said, 'When are you going to be of age, to come among us?' I answered, "To-morrow"; and it was agreed that the recommendation should be presented at the first meeting of next session,' as that was the last of the past one.

CHAPTER IV.

CHOICE OF A PROFESSION, 1830.

DURING his last session at college, his mind was much exercised with anxious thoughts about his future. The following letter to his uncle, Lord Medwyn, will show the course to which these thoughts were pointing:—

'*January* 30*th*, 1830.

'MY DEAR UNCLE,

'. . . I have in my possession memoranda written about four years ago, which contain the very sentiments I still feel, and which you have begun, the first of my relations, to entertain with almost equal force. Had circumstances been then what they are now, there is little doubt the Cambridge plan would have been followed. But now, though all things are permitted me, still all things are not expedient. I feel that the time has in the fullest sense passed. Perhaps it may be news to you that my earliest wish towards a profession was for the Church of England; when I was eight years old, I began to compose sermons, and long before my thoughts were directly turned to science I was warm on the scheme of the Church. In November 1822, I seriously proposed becoming a clergyman, but was dissuaded from it by my family; and the dislike naturally entertained by one so tenderly nurtured of deserting a beloved home, country, and friends, added to the strenuous advice of my revered father, finally determined me to give it up, though with a gradual struggle of years. Law was the choice of my father, and, on the principle of extinction of others merely, my own. My distaste increased instead of diminishing. . . . Indeed I could not have tolerated the idea of such a struggle of mind, but in the hopes of after a few years of drudgery reaching a sheriffship, which by adding 300*l.* or 400*l.* to my income might enable me to pursue my darling studies at leisure. But my view now is, that it is better to live upon little, than embitter the happiness of life by being more independent in fortune. . . . It is my present object to show that, independent of ties in some respect at home, it would not be for my real happiness or advantage now to go from home. 1st, then, my age is decidedly against me, and it would not be for a long time I could be prepared for such a course of education. 2nd. My attainments are very different from the Cambridge method. Could it be otherwise when I have been entirely my own master? . . . I should literally have to begin from the A, B, C of mathematics, and my preconceived and too advanced ideas must, you will see, be disadvantageous in a regular course of study such as in my youth I never had the advantage of pursuing. 3rdly. Mathematics I have little esteem for, except in their practical application. This is quite otherwise at Cambridge, and some precious years of my life would be spent in the acquisition of theoretical knowledge, instead of practical, which is far more in my line. . . . 4th. My classical learning was always on a very low scale; I have no time to write on this subject, but content myself with saying that from many circumstances, not all at least under my own control, it has been very imperfect; and though, as to enjoyment, I have pleasure in saying that I believe I have extracted more unsullied delight from the perusal of my favourite Latin authors than the nicest grammatical knowledge could have given, yet this is not the point at a University.

'From all these causes, and my anxious attachment to home, and my recollections of past sacrifice, when such a step might have been desirable, I must dismiss for ever from my mind the thoughts of an English University education. . . . You will ask what is to be done. I see but two courses: I conceive in the first place I should pass advocate, and then either to prepare for the Bar and race for a sheriffship; or to set to with a vigour and entireness of application, which to me at present is as a day-dream, to qualify myself for a professor's chair. . . . I should return unnumbered thanks to Heaven on my knees could I engage in so unbroken, so delightful a career, even though the attainment of its end should be distant or uncertain. . . . I assure you my mind was in a very fit state to receive your letter; for I daily became more convinced that I could not go on with this struggle, and that, if I regarded my comfort in life, I ought not to be swayed by the considerations of prudence, which I knew would induce most if not all my friends to set their faces against throwing myself upon all the difficulties of a limited income; and in fact I begin to see that I have hitherto trusted too little to my own judgment, and am tempted to follow the path so distinctly set before me. . . . I may say that the three friends on whom I most firmly rely, and who have done their best to supply the place of my natural protector, are yourself, Mr. Mackenzie, and Sir A. Wood. I do not despair of making them all unanimous, but I am resolved to be greatly guided by the spirited light in which you have taken up the question, corroborated as it is by the conviction I have in the fixedness of my resolution, and the little chance I have of repenting a choice which may go with my heart and deliberate judgment, after having lived to repent so deliberately a choice made chiefly upon principles of duty, which really did violence to my inclinations. . . .'

A few months later he writes to Mr. Colin Mackenzie, asking him for arguments with which he may controvert Dr. Brewster, who alone of all his friends stood out against his giving up the Bar:—

'. . . . You will enter into my anxiety to change the views of the only friend who has opposed them, who, you will perhaps at first be a little surprised to learn, is Dr. Brewster. Do not for a moment suppose that I want to use the strength of your opinion to overweigh his scruples; the fact is that Dr. Brewster and I merely differ upon a point of fact, upon which he seems incapable of being convinced, unless by those who have had more experience in legal life than either he or I have. . . .

'You who know the legal world so well will be amused at the opinions he maintains, and perhaps scarcely think them worth refutation; but you will, I am sure, enter into my desire of obtaining the approbation of one to whom I owe so much, whose advice has ever been so sound and valuable, at this most important crisis of my life, which must materially influence the comfort of all our future connection. I am therefore anxious that you would write me as strong a letter as you conscientiously can, expressing, what I know you must feel, the folly of a young man laying out his future prospects with the sole view of gaining, without labour or application to his profession, a lucrative office, of which the very existence as a sinecure, before he fills it, is eminently doubtful.

'To explain his ideas, I, in the first place, quote the following from the first letter written to me after I mentioned my proposed plans:—

'"I should like to see you measuring the boards of the Parliament House, taking what routine business came in the way, and thus fitting yourself for a sheriffship. During those two years you may devote two-thirds of your time to science, and then you might spend two years abroad acquiring the modern languages and carrying out your scientific pursuits."

'In short, on my return home, I was to find myself Sheriff of Peebleshire!!! much to the discomfort, I should think, of the county. I wrote to Dr. Brewster to convince him that sheriffships were not so easily jumped into; but all the effect it produced was the enclosed, which left me no alternative but to apply for incontrovertible arguments from those whose experience could carry weight along with them. He implores me, too, to pause, as if the past five years had not been one long and anxious pause.'

But whatever arguments Mr. Mackenzie may have supplied him with, they seem to have had no weight with Sir D. Brewster. The following extracts from Brewster's letters to Forbes at this time are interesting, in their bearing both on this subject and on other plans of the young physicist:—

'. . . I have been considerably puzzled how to advise you on the subject of the treatise on meteorology. Considering pecuniary motives as entirely out of the question, I am disposed to dissuade you from your scheme. I wrote a treatise on optics before I was your age, and the principal novelty in it was to be the practical part. I mentioned my scheme to Sir W. Herschel, with whom I had then some correspondence. He advised me against it, and I have since congratulated myself that I did not print such a work. . . .

'A first work clings to an author while he lives, and hence it is of great consequence that it be an original work worthy of the status in science which he is destined to attain.

'I would object less to a system of meteorology than to an elementary treatise. . . . I have no scruple in expressing my convictions that you are destined to do something important in science, and under this conviction I am confident that there is no object of ambition worthy of your pursuit, but that of original discovery.

'My conviction is that the very first thing a scientific man regrets is that of writing a book at the beginning of his career, however good it might be; he sees how much better it would have been had he waited a few years.

'If you should view things in a different light, which I dare say I would do in your circumstances, I shall be most happy to forward your wishes, whatever they may be.

'I would strongly advise you to continue your mathematical studies, but to keep them in due subservience to your physical inquiries. . . . If you should have any spare time, you would oblige me much by a short notice on Vesuvius, and touching very briefly on volcanoes; but if you have not time, do not think of it a second time. . . .'

Again, on February 11th, 1830, Brewster writes :—

'I cannot entirely approve of the line of life which your friends have chalked out for you, though their advice conveys great good sense and great knowledge of the world.

'I would object to your giving up law as a profession, because by a moderate study of it, perfectly compatible with the most ardent prosecution of science, you might qualify yourself for a sheriffship, which you have influence to obtain, and the duties of which in a small county would be an agreeable variation of your pursuits. I cannot tolerate the idea of a Professorship being an object of your ambition, if you mean a Scotch one. There is no profession so incompatible with original inquiry as a Scotch Professorship, where one's income depends on the number of pupils. To young men at College the Professor appears to have a reputation, when he is not known beyond the walls of the University. The Professor can obtain no fame from his teaching powers, and it is only what he does by original investigation that gives him any celebrity ; but all this can be attained without teaching boys and half-men the elements of Euclid, or the principles of chemistry, or natural philosophy, which are often all but the functions of a schoolmaster, and much less useful to society.

'I disapprove also of the idea in your first letter, of making up your income by your pen. I do not object to your making money by your writings, but I am sure that it would be injurious to your happiness to rely on such a source for a permanent portion of your income. The moment you do that you become a professional author, following the worst of all professions.

'You threw out an idea in your first of wishing to remain a bachelor. This I cannot approve of. The married life is more appropriate to a man of science than to any other person ; and though its unavoidable evils may sometimes interrupt the even tenor of a philosophical life, yet these evils, even in their worst form, are useful impediments in our lot, and are incalculably overbalanced, when considered only in their bitterness, by the hourly enjoyments of domestic happiness. . . . But I would advise you not to expect too much happiness, even from the fulfilment of all your wishes. The moment you have distinguished yourself you become an object of envy and malice; men whom you believed to be lovers of knowledge you will then find to be lovers only of fame, and haters of all knowledge that has not come from themselves. You will find that a life of science has in it no superiority to any other, unless it is pursued from a higher principle than the mere ambition of notoriety, and that demagogue or a philosopher differ only in the objects of their selfishness. As you will now have experienced how unsatisfying even the pursuit of knowledge is when insulated from higher objects, I hope, if you have not been fortunate enough to begin the study earlier, that you will devote yourself to the most extraordinary of all subjects, one which infinitely surpasses the mechanism of the heavens or the chemistry of the material world, the revelation of your duty and the destiny of man as contained in the Bible—a book which occupied the best hours of the manhood of Newton, of Locke, and of Euler.'

By March 1830 Forbes's mind was made up. The course on which he had now fixed, with the reasons and views determining him to it, are thus recorded by himself:—

'On the 30th January my uncle, Lord Medwyn, wrote that happy letter which produced all the change in my views from the hopeless slavery of law to the freedom of scientific pursuit, and the approbation of my friends sealed the resolutions which my deliberate choice had so long pointed out. Dr. Brewster alone stands out. The valuable letters of Lord Medwyn, Mr. C. Mackenzie, Sir A. Wood, my brothers John and Charles, John Buchanan, and David Milne, remain in my possession, a convincing body of documents that I have not proceeded in this important matter without due weight of advice; and while I have returned thanks to a gracious God for a happy issue of this important crisis, I have earnestly besought His guidance in the new line of life thus set before me. My friends have all pressed upon me the propriety of assuming the gown, and of going abroad for a considerable time. The former I am now preparing for, and hope to accomplish in June. The latter I think proper, for many reasons, to delay.

'I am now permitted, under the support of those friends whose advice I have most reason to respect and esteem, to follow the bent of my mind, to cast law behind me, and, content with a small competence, to follow science at leisure. I have not patience to repeat all I have thought and said on the subject for the tenth time here. Suffice it to say that my resolution was consummated by John's kind letter of the 2nd March, which greatly affected me. Accepting his splendid offer of 300*l*. a year would indeed have made me rich for life; but I feel too much the tax I should be putting on my friends in order to support me in what the world calls idleness, to entail upon them other disagreements than the nature of my choice necessarily carried with it. I shall enjoy at least the comfort of independence.

'But enough of this. God be praised! my mind is now at ease; and though I have the preliminary drudgery of passing advocate to go through, that is a trifle. In June I hope I shall be free of law for ever. Then for self-cultivation! I mean to study the higher mathematics, and make some original researches connected with the fundamentals of meteorology. If things go well, I intend to go abroad the autumn of next year for eighteen months or two years. But I plan for the future with perfect diffidence. God only knows how far there may be between me and misfortune, disease, or death.'

'COLINTON HOUSE, *April* 20*th*, 1830.

'*On Completing my Twenty-first Year.*

'Most merciful and gracious God, who hast preserved me unto this hour, I most humbly acknowledge Thee as the guide and companion of my youth. Thou hast protected me through the dangers of infancy and childhood, and in my youth Thou didst bless me with the full enjoyment, the happy intimacy, of the best of fathers. Be as gracious and merciful then as Thou hast hitherto been, now that I am about to enter a new stage of existence. Teach me, I beseech Thee, to strengthen in my soul the cultivation of Thy truth, the recollection of the uncertainty of life, the greatness of the objects for which I was created. Revive those delightful religious impressions which in early days I felt more strongly than now; and as Thou hast been pleased lately to permit me to look to a way of life to which formerly I dared not to do, let the leisure I shall enjoy enlarge my warmth of heart towards Thee. Make every branch of study which I may pursue strengthen my confidence in Thy ever-ruling providence, that, undeceived by views of false philosophy, I may ever in singleness of heart elevate my mind from Thy works unto Thy divine essence. Keep from me a vain and overbearing spirit; let me ever have a thorough sense of my own ignorance and weakness; and keep me through all the trials and troubles of a transitory state in body and soul unto everlasting life, for Jesus Christ's sake. Amen.'

The summer of 1830 was spent in preparation for his legal trials, studying the Calculus, and learning German. Early in July he passed advocate, and at once bade adieu to the Bar and its studies for ever.

The autumn months were spent in various excursions: first on a visit to his kindred the Macdonells of Glengarry at Carradale, a beautiful abode on the coast of Kantyre, looking across Kil-Brannan Sound to the peaks of Arran; then with his brother Charles on a tour to the English Lakes, where they visited Professor Wilson, who was then living at his beautiful summer home of Ellery, overlooking Windermere.

'We were on the point of leaving Windermere for ever, when I got a note from the Professor inviting us to come to dinner. We made our arrangements to stay. We found a bachelor party, but the kind Professor was most hospitable and agreeable, and we spent a cheerful and happy evening. It was arranged that next day an excursion through a mountainous tract should be made, and we could not resist so favourable an offer. We accordingly set out, a party of six on horseback. The first part of our way was very good, but at length we came to a mountain pass called Nan-Bild, over which none of the party, and it was alleged no human being, had ever taken a horse. At great risk of the animal's legs we at length disentangled him from a great chaos of loosened, tangled rocks, bogs, and torrents,—in short the "bogs, lakes, fens, caves, dens, and shades of Death," as Milton has it. However we were not a party to stick at trifles, headed by so manful a mountain general as Mr. Wilson; and we reached Mardale, a sequestered valley above Hawes Water, in safety, where after a due stay we returned home another road. All declared that nothing should induce us to take horse again over Nan-Bild. Our ride was nearly thirty miles. Next day was to be the Windermere Regatta; so of course we were tied down to remaining to it: but what a day! We went down to the ground with the Professor, where the Regatta was agreed to be postponed, and a dinner got up among about twenty there. I had much conversation with Professor Wilson, and rode and walked much beside him, and was certainly anew delighted. He dotes upon the country, and knows every inch of it; so I was very glad to make a mountainous excursion with him.'

In October Forbes returned to Colinton House; and the following extracts give the views and interests with which he met the approaching winter:—

'*October* 24*th*, 1830.—Almost had this day passed over my head without a due remembrance. Returned only yesterday from a tour to England and Wales, and the accident of being obliged to date a note brought before my eyes these deeply associated characters, 24th October. By a curious enough accident I had been reading some notes of my grandmother's in her old Bible connected with this day. I have gone through the whole of the prayers intended for the past week written last year, and with great advantage, for the carelessness incident upon travelling had made me forget much of the duty I owed to the kind Protector of my often dangerous path. My self-examination had been degenerating, my serious reading grown more scanty.

'Soon our parting from Colinton House will be among past events—God strengthen me!'

'*November* 14*th*, 1830.—Kind, excellent Mr. Mackenzie died on the 16th September. How much I owe to that excellent man! How sincerely I loved him God only knows. Eulogy were vain!

'Entering now upon another winter, and relaxed by two months of varied excursion, I have commenced the delightful and engrossing studies which have now, blessed be God! become my principal and legitimate employment, untrammelled by jarring occupations and conscientious scruples.

'I have recommenced the study of higher analysis, and have far advanced with Boucharlat's "Integral Calculus." Feeling the necessity of gaining a more practical knowledge of what I have gone over, I have commenced the Differential in Lardner's work. I have begun experiments on heat, which occupy a good deal of my time and thought; studying and analysing Leslie on that subject, and reading Thomson, Prevost, Pictet, &c., on the same.

'A paper for Brewster's Journal on the sympiesometer has occupied, and will continue to do so, a good deal of time and study.

'When I view my situation at this moment, it is one of great comfort and satisfaction, what a year or two ago I could not have dreamt of—relieved from all but the toils I delight in, receiving frequent assurances of the goodwill and support of those most in my own line of study. The fears are for my simplicity and steadiness of character. I am ever too apt to forget that my youth is the principal cause of the attention and approbation I receive. I shall soon be admitted to the Royal Society, with the principal members of which I have the happiness of being on good terms. Mr. Robison in particular has been very kind, and in a note written this summer expressed his wish that I should be some time in the Society before an opening occurred in the Secretariat. But a more remarkable, and to me very surprising, communication reached me the other day; which I confess astonished me much; and I am afraid lest it should influence too strongly my views and hopes. On Wednesday last I went with Charles to hear Professor Leslie's first lecture. He sent for me after the class, and after apologizing for not answering my letter, said he was glad to hear I was pursuing my studies, but recommended me not to give up the Bar. He then very explicitly informed me that when he proposed going to the East, last summer, he had thought of getting me to officiate for him, but was afraid the public might think me too young. He then broke off abruptly. Such a declaration was to me a matter of considerable wonder. That he should have pitched on me, whom he could have no interest to serve, was equally flattering and unexpected; especially as I had never done anything to induce him to make such a declaration, never cringed to his authority or opinion. Ever since Wednesday my imagination has been perpetually building castles in the air upon this declaration of Professor Leslie's.'

The proposal alluded to in the last sentence brought out into definite relief a prospect which, though he may hardly have named it to himself, had probably been slumbering in his thoughts for long undefined. Notwithstanding Forbes's high promise and congenial interests, Sir John Leslie does not seem, up to this time, to have shown him any favour, or invited his confidence. That he should now suddenly let drop such a hint must have appeared all the more marked, from his previous reserve. But whatever thoughts this remark of the

Professor's may have awakened in young Forbes, there was nothing to be done, but to keep working steadily on the line of self-improvement he had planned. During the winter, while carrying on his physical researches, making barometrical measurements, and reducing his barometrical observations made at Colinton, he attended Dr. Chalmers' Divinity Lectures and Dr. Reid's class of Practical Chemistry. With his experience of both of these classes he expresses much satisfaction:—

'I am attending two classes at present, with both of which I am delighted—Dr. Chalmers' First Theology, and Dr. Reid's, Dr. Hope's assistant, Practical Chemistry. I am strongly of opinion that to hear such masterly lectures by Chalmers upon Natural Theology and the Evidences is a most fitting conclusion to a course of liberal education, and singularly well calculated to prevent injury from the sceptical insinuations of Laplace and other modern philosophers, whose works are likely to become oracles with those treading the path of exact science. With Reid I am almost equally delighted, though in a different way.' Besides this he was continuing his study of German, and taking lessons in Elocution.

The month of December 1830 saw the break-up of the old family home at Colinton. To this he had looked forward ever since his father's death. And now, when it came, he had to go through that pain which has been the experience of so many in every generation.

> 'We leave the well-beloved place,
> Where first we gazed upon the sky;
> The roofs that heard our earliest cry
> Will shelter one of stranger race.'

What his feelings were Forbes thus records:—

'We are about to leave this delightful and endeared spot, endeared by its beauties and comfort, by long habit, by the associations of childhood and youth, and by the tenderest recollections of riper years. So deeply and heavily has this event pressed upon me ever since I saw its necessity, that I am almost ashamed to confess the weakness of my feelings connected with inanimate objects.'

Quitting Colinton he removed with his two sisters and his brothers, Sir John and Charles, to Greenhill House, on the south side of Bruntsfield Links, a pleasant abode, belonging to his family, in which he lived until Sir John's marriage. Here are the thoughts with which he closed the year 1830 and entered on 1831 :—

'GREENHILL, *Sunday, January 2nd*, 1831.

'Farewell to the past year! momentous in its results. How my views have changed since this time in 1830! How kind were the friends who have left this world of care and trial since I then wrote. With what utter blindness can I look towards the characterless scroll of time stretched out anew before me, and soon to be imprinted with records to me I know not how important. Much, much has there been in 1830 to call forth my warmest thanks to God; much to accuse myself of in the neglect of my plighted adherence to resolutions of amendment. I this day received the Sacrament. I reviewed my resolutions in August, and more particularly adopted the following as being least fulfilled. 1st, Resolution for self-examination and religious thought. 2nd, For dependence in God. 3rd, For social love and confidence in my own family. 4th, For humility.'

This winter seems to have been an eminently happy one, in the enjoyment of newly-won liberty and unimpeded devotion to his favourite pursuits. It is thus he speaks of himself at this time in a letter to his uncle, Lord Medwyn :—

'COLINTON, *November 28th*, 1830.

'. . . Every day has convinced me more strongly that a distinct election was necessary; and each day has likewise convinced me that it was properly made. . . . A summer of steady, and I hope not unprofitable, labour has only strengthened my ardour of pursuit, and refreshed by an ample period of relaxation I have renewed with zeal my studies, which were never wholly interrupted, and at this moment feel myself so entirely happy, alike removed from turbid excitement and monotonous dulness, that I should be ungrateful to God and to my kind friends, and a traitor to myself, did I not acknowledge myself so. . . . I have mastered about two-thirds of Boucharlat's Differential and Integral Calculus; and the study of some parts of this most amazing branch of human inquiry has, I confess, astonished and delighted me, and given me new views of the wonderful powers which have been confided to man. Of nothing am I more assured at present than this, that a suitable acquaintance with the higher analysis is the strict basis of real scientific inquiry in the present day; and when we see everything as we do reduced to the popular scale, knowledge diffused but not deepened, and all severe mental labour received with disgust, this is the time, if any, to lay deep the foundations of those acquirements to which there opens no royal road. Both from inclination and expediency I therefore resolve to pursue my mathematical studies, always keeping the application strictly in view, and acting upon them in the course of my other pursuits.

'It has long been Dr. Brewster's particular desire to engage me in some field of original research, to which I might devote my whole attention; and that which is every way most agreeable to me, and which seems most fitted for my exertions, is Heat, in the widest sense of the word, opening a field of the widest interest. . . .'

This winter too saw his entry into the Royal Society of Edinburgh, from which only his youth had kept him hitherto, and of which for more than twenty years he was a main stay.

With the coming of spring 1831 he left Edinburgh with his sisters on a visit to London. Being well furnished with letters and introductions, he at once made acquaintance with the chief scientific celebrities of the time, and was by them warmly welcomed. Drs. Murchison and Babbage, and Mrs. Somerville, he mentions as having received him with special kindness in London.

At Cambridge he spent one of the happiest weeks of his life in the society of Whewell, Sedgwick, Airy, and Peacock, from whom he received much sympathy and kindness. Oxford he visited, and was present at Commemoration. Buckland did the honours of the place; but Oxford of course had fewer intellectual attractions for a young physicist than Cambridge. With Drs. Buckland, Conybeare, and Phillips he went to Shotover to see the upper strata of the lias exposed, and afterwards was present in the theatre when Washington Irving and others received degrees.

'LONDON, *May* 1831.

'Only arrived, when Sir A. Wood came in and told us that the king was on the point of going down to prorogue Parliament in person, preparatory to its dissolution. We were from accidental circumstances too late to see him go down, but saw him leave the door of the House in state. King looking strange and nervous. Excitement in London very great. Turbulence in both Houses very great. Went into the House of Commons, and saw the members coming out of it, many of them trembling for their seats. Afterwards saw Lords Grey, Durham, and Brougham in carriages, and Dan O'Connell walking.

'Breakfasted with Mr. Lyell. Met Mr. Babbage, whom I had long wished to see: very much interested by my account of Brewster's analysis of the spectrum, and with regard to Fraunhofer's rings. Spoke of his machine as a work of great anxiety.'

'*May* 14*th*.—Breakfasted with Mr. Babbage; much pleased and interested by him: present, Mr. Lyell, Mr. Harris, Mr. Drinkwater, Captain Drinkwater. Mr. Babbage showed to me some of his travelling instruments, Kater's circle, horizons, &c.; and some curious notes, especially on the temple of Jupiter Serapis. He has a very fine telescope. Saw his mother and daughter. Got a letter to Professor Whewell of Cambridge.

'*June 5th.*—Had an appointment at one o'clock to meet Mr. Babbage at his house. Thence we went to see his calculating machine with Mr. Harris, at the place where the machine is making, near the Bethlehem Hospital, St. George's Road. A man of the name of Clement has the entire direction of it, and the mechanism is most splendid: it is almost entirely made by the turning lathe, which he has in high perfection. . . . Mr. Babbage asked me to go and dine with him. I had occasion to see a good deal of his character, which is very peculiar, and particularly interested me. . . . It was with the greatest difficulty that I escaped from him at two in the morning after a most delightful evening.'

The following are some notices of the visit to Cambridge:—

'*May 17th.*—Set off for Cambridge. Called on Professors Sedgwick and Whewell, and Mr. Ramsay. Walked about the charming grounds behind Trinity and King's College. The day delightful, and the groves looking truly academic. Under the circumstances could not fail to muse a little upon the knife edges upon which a man's fate in life may turn at a critical period.

'Strolling through Trinity Court met Sedgwick, and went with him to dine in hall, when I met Whewell. Two very interesting, indeed fascinating men. Took wine in the Combination room. Had a remarkable proof of Sedgwick's temper, kind feeling, and personal forbearance in a miserable juggler's exhibition. Strolled through King's College by moonlight, and returned with Sedgwick to his rooms, and with difficulty escaped from his delightful conversation after midnight some time. He is a man of great and varied talent, of true self-estimation, of most liberal spirit to his contemporaries, of most kind and conciliatory manners.

'*May 18th.*—Breakfasted with Mr. Peacock, tutor of

Trinity, a very able mathematician: present, Messrs. Hawkin, late M.P., Sedgwick, Clavering, &c. &c. Delightful rooms, looking into Trinity grounds. Went with Mr. Whewell to Mr. Airy's lecture. These lectures by the Plumian Professor, and at present probably the ablest man in the University, are in the highest estimation. With great talents for perspicuous though unadorned explanation, he is able to carry his class through propositions, especially in physical optics, of the highest profundity, and by his singular ingenuity to illustrate some of the finest and most delicate experimental truths on a most magnificent scale. The lecture was upon mechanics. . . . After it was over he exhibited a magnificent series of experiments on polarization with apparatus which he had taken the trouble to prepare, hearing that I was to be there. . . . Drank wine with Professor Whewell in his rooms, and had a most delightful party, including Airy, Sedgwick, Jarrett, Challis, and Bothman. Much delightful and brilliant conversation. Went at seven to see a boat-race of the University; a beautiful sight, great crowds assembled, particularly of oarsmen. Went with Sedgwick and Whewell. Charming evening, and enjoyed it much. Professor Sedgwick came and drank tea with me.

'Dined in Trinity Hall. Took wine in Sedgwick's room: present, Whewell, Thirlwall, Hare, Bothman, &c. Peacock came in and had some very pleasant talk about Airy, Herschel, &c. Memorandum—Herschel in his fortieth year; Peacock the same; Whewell, 37; Airy only 30—inimitable man! Average age of taking degree 22nd year. Separated pretty late.

'Went to Professor Airy's lecture on mechanics. Beautiful extempore explanation of the three axes of rotation and the cause of precession. After lecture he had the kindness to show me some experiments on interferences of light. . . . Mr. Airy next showed me some beautiful experiments with coloured shadows of bodies thrown from a minute point of light, which he views with eye-glasses.

'Mr. Airy took me over to the Observatory, where he lives. Saw a very pretty polarizing apparatus on a moderate scale. The only considerable instrument is the transit—a fine ten-feet, and perhaps the best worked instrument in Britain. Airy is a wonderful man. His observations hitherto published have all been made with this single instrument, and contain a great mass of good matter, all beautifully reduced. He had the kindness to present me with the three volumes of his observations now published.

'Dined with Mr. Ramsay at Jesus College; very pleasant party, including Peacock, Whewell, Henslow, and Bowstead. Delighted with Whewell.

'Professor Whewell sent me his paper on mathematics applied to Ricardo's political economy, which I employed myself a good deal in reading. Breakfasted with Mr. Ramsay. Saw Jesus College Chapel; pretty Gothic. Tomb with 1263 on it in Arabic numerals. Whewell doubts its genuineness.

'Sedgwick called on me, and I went with him to morning prayers in Trinity Chapel, being Whit Sunday. Heard Mr. Blunt preach before the University in St. Mary's Church. Dined in Trinity Hall, which was in greater style than usual, being a Feast day. Several strangers: James Hall and I went to Trinity Chapel in the evening. Delighted with the service and the appearance of the chapel, which was well filled with surpliced students, and the music excellent. Effect truly grand; would on no account have missed spending a Sunday here. Took tea with Sedgwick, in company with Whewell, Romilly, Hon. — Grey (Premier's son), and Mr. Sheepshanks, a distinguished man and an eminent member of the Astronomical Society. Charmed more than ever with Whewell. His notions of the prosecution of science liberal, on the scale of his vast attainments.

'*May 23rd.*—Breakfasted at Prof. Sedgwick's, with James Hall, Mr. Thorp, a junior Fellow of Trinity, with whom I only became acquainted on Sunday; and spent half an hour with Mr. Whewell in his rooms, and saw some of his minerals. Astonished at the vast range of his library in subjects and language. Took a kind farewell of him and Sedgwick, and went to hear Airy's lecture before setting out. It was upon hydrostatics and pneumatics. Took leave of Mr. Peacock, Mr. Ramsay, and Mr. Thorp, and proceeded by coach to London more than delighted by my stay at Cambridge.

'I omitted to mention one of the finest mechanical illustrations of an abstruse theory that I ever saw, at the Observatory, contrived by Professor Airy. It was to explain how, on the theory of undulations, the coincidence of two undulations will polarize light in one plane, the semi-coincidence in a plane at right angles to the former, and by a difference of a quarter of an undulation polarize it circularly. . . .

'Persons to whom I was introduced at Cambridge :—Professors Sedgwick, Whewell, Airy, Henslow ; Mr. Peacock, Thorp ; Dr. Ramsay, Vice-Master of Trinity ; Mr. Jarrett, Romilly, Bowstead, Ash, Turner, Bothman, Thirlwall, Hare, Sheepshanks.'

A visit to Mr., afterwards Sir John Herschel, is thus recorded :—

'*May* 27*th*.—Went by appointment to Slough to see Mr. Herschel. Met Mr. Beaumont, who is setting up astronomer. Herschel was engaged most of the afternoon in putting his twenty-feet reflector in order for the evening, in case we should have the good fortune to have any observations. Mrs. Herschel was unwell, and I did not see her. We dined, and the sky began to look favourable after a cloudy day. As soon as it was dark, we went to the telescope, just furnished with a beautiful new polished speculum, which had not been used before. We got hold of Saturn, and, notwithstanding a very indifferent sky, saw him in a manner which was very delightful to me. His body was but indifferently defined. I saw, however, distinctly the shadow of his ring and five satellites, which is all that have ever been seen with the twenty-feet telescope. Speedily, alas! the sky clouded over completely, and we were obliged to give up the night, which continued impenetrably thick with an east wind.

'*May* 28*th.*—Mr. Herschel showed me his apparatus for grinding and polishing specula, which he does himself, and with great success. . . . I spoke to Mr. Herschel about a course of reading in analytics, expressing my conviction of the necessity of a good foundation in the highest mathematics. He considers Lacroix' Differential and Integral Calculus indispensable, but to be read not through but with selection. . . . . Mr. Herschel drove me down in his phaeton to Stoke Church, the scene of Gray's Elegy; a beautiful and a most sequestered spot it is. The poet is buried under a black slab elevated above the ground by brickwork at the east end of the church. His mother's name is inserted, not his own.'

On his way northward Forbes stopped at Manchester, and this is the most noteworthy memorial of his stay here:—

'The most extraordinary man I met is John Dalton, whose name is better known in almost any country of Europe than his own, and in any town than in Manchester. He is generally styled by continental writers the Father of Modern Chemistry, and is one of the eight associates of the Institute. Yet this man between sixty and seventy is earning, as I had a peculiar satisfaction in seeing with my own eyes, a penurious existence by teaching boys the elements of mathematics, with which he is so totally occupied, that he can hardly snatch a moment for the prosecution of discoveries which have already put his name on a level with the courtly and courted Davy. But the remarkable thing is that this simple and firm-minded man preserves all the original simplicity and equanimity of his mind, and calmly leaves his fame, like Bacon, to other nations and future ages.'

The hearty reception he had met with from the scientific brotherhood in the south sent him back to Scotland with spirits braced for new exertions. And well it might: to be thus welcomed by the chief men of the scientific world, while still a mere youth, is not the lot of most aspirants. It seemed, as if passing over the long obscurity through which the early manhood even of the most gifted has to struggle, he had at one step taken his place beside the foremost men of the time, if not yet as their equal, at least as one who was soon to become so.

This autumn saw the first beginnings of what has since become a world-wide Institution, the well-known British Association. In the foundation of it Forbes bore a part which in a youth of two-and-twenty seems wonderful. The real father of it was Sir David Brewster. By him the idea of such a congress was first conceived, and its realization was mainly due to his enthusiastic and untiring efforts. Everyone who ever listened to his conversation or read his more popular writings will remember how unweariedly he dwelt on the way this country neglects its scientific benefactors, how earnestly he pleaded their claims to fuller honours and larger salaries. It occurred to him that if the scattered scientific intellect of the country were gathered into one spot, and presented in bodily mass and weight to the eye and ear of men, the crass public might be roused to recognize its existence and value, as they never would do while it appealed only to the pure intellect. The Association which he originated has grown to a size and importance which Brewster perhaps little dreamt of. But it still combines and represents what were the two master tendencies of Brewster's life—his love of science and his love of sociality.

The first hint of it is contained in the following letter from Brewster to Professor Phillips at York, dated from Allerly, 22nd February, 1831:—'It is proposed to establish a British Association of men of science similar to that which has existed for eight years in Germany, and which is now patronized by the most powerful sovereigns of that part of Europe. The arrangements for the first meeting are in progress; and it is contemplated that it shall be held in York, as the most central city of the three kingdoms.'

Nineteen years later Sir David, speaking from the President's chair to the Association assembled in Edinburgh, said, 'On the return of the British Association to the metropolis of Scotland, I am naturally reminded of the small band of pilgrims who carried the seeds of this Institution into the more genial soil of our sister land. . . . Sir John Robison, Professor Johnston, and Professor J. D. Forbes were the earliest friends and promoters of the British Association. They went to York to assist in its establishment, and they found there the very men who were qualified to foster and organize it. The Rev. Vernon Harcourt, whose name cannot be mentioned here without gratitude, had provided laws for its government, and along with Mr. Phillips, the oldest and most valuable of its office-bearers, had made all those arrangements by which its success was ensured. Headed by Sir Roderick Murchison, one of the very earliest and most active advocates of the Association, there assembled in York about two hundred of the friends of science.'

In this account of the founders of the Institution Sir David omits only one, but that the chief name—his own.

If to these words of Sir David Brewster we add those of Forbes, when in 1866 he spoke at St. Andrews in favour of the Association, then about to meet for the first time at Dundee, enough will have been said to show with what feelings that earliest meeting at York in 1831 was long after remembered by two who had taken the chief part in it :—

'. . . Mr. James Johnston, a chemist of repute and a friend and correspondent of Sir David Brewster, communicated to his scientific journal in 1830 details of the proceedings of the German Association, and their results. Sir David Brewster warmly took up the idea of introducing into Great Britain a similar institution, which occurred the more opportunely that about that time a controversy took place, in which he bore an active part, as to whether the science of England, as compared with that of continental countries, was or was not in a condition of decline. Sir David, following Mr. Babbage, espoused the less flattering side of the debate, and attributing the alleged decline of science to defective organization and the want of support of the State, he naturally considered that an annual congress of scientific men would afford the occasion which he desired of stimulating their energies in a common pursuit, and of impressing upon the attention of Government any disadvantages of a public nature under which the science of Great Britain might be thought to labour.

'Sir David Brewster brought his proposal under the notice of his scientific friends in London, and being fortunately in communication with Mr. Phillips of York, now Professor of Geology at Oxford, the city of York, the seat of one of the most flourishing of the provincial societies, was fixed on for what was termed in the original circular "a meeting of the friends of science," which should take place in September 1831. The time appeared in one respect unfortunately chosen. The excitement of public feeling incident to the discussions on the Reform Bill was so great, that the postponement of the meeting was at one time contemplated. It however took place without the smallest infusion of political bitterness.

'In an address, distinguished by thoughtful elaboration, Mr. Harcourt propounded to his select auditory at York a scheme for a British Association for the Advancement of Science, of which he described the aims and the working details with a completeness which took his hearers somewhat by surprise, but in which they found little to alter or amend; and the constitution proposed by Mr. Harcourt remains in all its important details the working code of the Association to this day. "I propose to you," he said, "to found an Association including all the scientific strength of Great Britain, which shall employ a short time every year in pointing out the lines of direction in which the researches of science should move, in indicating the particulars which most immediately demand investigation, in stating problems to be solved and data to be fixed, in assigning to every class of mind a definite task, and suggesting to the members that there is here a shore of which the soundings are to be more definitely taken, and there a line of coast along which a voyage of discovery should be made. . . . I am not aware that in executing such a plan we should intrude on the province of any other institution. Consider the difference between the limited circle of any of our scientific councils . . . and a meeting at which all the science of these kingdoms should be convened, which should be attended by deputations from every other society, and in which foreign talent and character should be tempted to mingle with our own. With what a momentum would such an Association urge on its purpose; what activity would it be capable of exciting; how powerfully would it attract and stimulate those minds which either thirst for reputation or rejoice in the sunshine of truth!"

'Blended with such stirring appeals did Mr. Harcourt unfold his intended constitution of this new "Parliament of Science," as it has since been happily termed. Their echoes seem still to vibrate through the long interval of five-and-thirty years, the interval of one entire generation of man. Yes, gentlemen, they still seem to vibrate, for I myself, then little older than some of the senior students whom I now address, was an attendant at that meeting, and a profoundly interested auditor of this inaugural discourse. Year after year without intermission, from 1831 to 1866, has the British Association held on its course, visiting town after town, university after university, from Oxford in the south to Aberdeen in the north, and from Cork in the west to Newcastle in the east, carrying everywhere with it the prestige due to its founders and supporters, and developing in comparatively remote districts local talent and enthusiasm, which the advent of so renowned and comprehensive an Association is sure to excite.'

To this first meeting at York, Scotland sent a numerous and powerful contingent. Forbes set out in September 1831, along with Sir D. Brewster, Sir John Robison, and others. The success of that meeting more than fulfilled the expectation of its founders. During the immediately following years, while it was still in its infancy, the Association had to run the gauntlet of not a little ridicule. It was a butt for the shafts of the 'British Critic,' the chief organ of the Oxford movement then beginning, which attacked it with no sparing satire. And though no doubt the gala days of the *savants*, enlivened by good dinners, railway excursions, and abundant talk, may still provoke a smile, those who know best the history of these meetings seem most assured that they have really been fruitful of solid results. Of these services to science one of the most tangible is the establishment and maintenance by the Association of the Observatory at Kew, for observing the phenomena of meteorology and terrestrial magnetism. During all his active life Forbes maintained the same interest in promoting the work of the Association which he had shown in its first foundation.

The winter of 1832 was spent in Edinburgh, carrying on the work of self-education in those scientific subjects to which he had now entirely devoted himself. Attendance on classes and instruction from others were now past. It was by private reading and by making experiments for himself that the work of improvement must henceforth be carried on.

The following fragments from letters and journals show how this winter was spent, and with what thoughts it was engaged, till in spring he went again to London :—

To Sir John Robison, February 10, 1832.

'It is my intention, should circumstances permit, on Professor Leslie's resignation to offer myself as a candidate for the Natural Philosophy chair. I am aware of the chances of disappointment, and that, if once filled up, it might not again be open during my life ; but, if circumstances should render my chance a good one, I mean to put myself in the way of it, without at all compromising my happiness by a sanguine desire of success. I feel the boldness of such a project, and am humbled by comparison of those who have filled that chair with myself; but I believe that under the existing state of science in Scotland—I say it without any emotion of vanity—there are few who have had the means and inclination combined to pursue the course which I have done. You must be aware of the delicacy I feel in writing of the chair Professor Robison once occupied; but, Sir, "there were giants in those days," and I fear we are now reduced but to a dwarfish state of science in Scotland:—this alone is the footing on which I can place it.'

This is his chronicle of the first half of the year 1832:—

'*January* 1832.—Living at this time with John, Charles, and my sisters, and Miss Ballingall at Greenhill. This month made experiments with Dr. Christison at his house, on comparative readings of Royal Society's thermometers. Results in my pocket-book of that date.

'This winter saw a good deal of the amiable Professor Louis Neckar of Geneva, living in Edinburgh.

'*Feb.*—Making experiments of conducting power of metals, with Fourier's thermometer of contact. The notes in pocket-book. Also on magnetic intensity with Hansteen's apparatus.

'*Feb.* 26.—Reached London; Sir D. Brewster there. Saw much of him, Babbage, Captain King, Neckar, and others. Occupied in ordering instruments for foreign tour.

'*April* 4.—At Allerly with Sir D. Brewster.

'Obtained spark from magnet.

'*May* 17.—At Cambridge.

'*April, May, June.*—Engaged with report on meteorology for British Association and preparations for tour.

'*June.*—Association meeting at Oxford. Read report on meteorology.

'*June* 7.—Elected into Royal Society of London.'

These early years I have described more at length, and illustrated with fuller details from his own journals, than I shall think it necessary to do in the sequel. I have done so from the conviction that of all men's lives the opening years most repay close attention, because it is then that the master tendencies first show themselves, that the materials are gathered in, which pass most deeply into the being, that the hues are laid, which colour the whole life till the end. But if this is true of all men, it is especially true of strong intense natures which take their line early, and keep it unfalteringly.

That saying, 'The child is father of the man,' though perhaps first uttered by Wordsworth, sounds like a world-old proverb. In no man was this more distinctly seen than in James Forbes. The moral habits and mental pursuits, begun in the nursery at Colinton, remained with him all life through. But there is another circumstance that makes those first years instructive, especially to those who have still to shape their course. Forbes may be said to have been a self-educated man. Till his sixteenth year, when he entered college, the instruction he had was of the slightest and most desultory kind. No doubt he was surrounded by a refined and intelligent atmosphere, one in which the tone was more than usually pure and serious. But the mental habits and the definite information which he took with him to college were not put in him by the spoon; the habits were formed by his own self-discipline, the knowledge gathered by his own unwearied seeking; and even during his college years this continued. The main thing was not the information furnished, nor the stimulus given by the professors' lectures, but the widening of his mental horizon by the action of his own swift intelligence and persistent effort. The journals and observations which he had begun years before he entered college continued to be his most cherished employment and his chief means of education throughout his college course. That one year spent in Italy,—we have seen what an impulse it gave to all his thoughts and early predilections. His after labours in the Alps were only continuations of habits and processes of observation then begun. To us now looking back it is clear that physical science was his natural element. Towards it he was borne by a tendency as strong as that which leads the duck to swim or the hawk to fly. But between these unreasoning instincts and the rational bias of man there is one great difference. The lower creature has the instinct, and the opportunity to act on it is never wanting. But men, especially in youth, before they know either themselves or the world, feel many an impulse which they cannot explain 'to themselves, much less to others,' many a natural tendency for which there seems no outlet, against which all outward circumstances,—and these are one large part of the guidings of Providence—seem to reclaim. In such cases it is often very hard for a young man to know whether he should cling to his natural tendency at all hazards, or yield to the heavy necessities which seem warning him to abandon it. It is many a time a great and sore perplexity. The inward bias and the outward surroundings are, we believe, both from God—which of the two is a man to follow? No definite rules can be laid down. But something may be learnt from the case of young Forbes, and of others like him, by whom the difficulty has been deeply felt and fairly met. Before he could make science the principal and legitimate object of his life, untrammelled by jarring occupations and conscientious scruples, there are proofs enough to show that he had many an hour of anxious self-scrutiny. He pondered well all the circumstances, he sought the advice of those most entitled to counsel him, he took his motives and aims honestly into the light of conscience and of God, and he sought guidance from above. In due time the way was opened, and the result we are now to see.

His period of exclusive self-education and preparation for the active duties of life was to be closed by a tour on the Continent, which he had long planned, and in which it was his intention to visit those places, in which either by converse with scientific men, or by observing striking aspects of nature, he might enrich his scientific resources. It probably was his intention to prolong this tour for some years, and not to return till he had laid up a store of facts and observations, which he might give to the world in some mature work. At the beginning of July 1832 he left London with his brother Charles. Their course lay through France and by the Rhine on to Switzerland. By the middle of November he was at Geneva. Thence he was suddenly recalled to Scotland by the news of the death of Sir John Leslie, and the consequent vacancy in the chair of Natural Philosophy in the University of Edinburgh. It was exactly two years before this that Sir John had made to his young pupil the announcement I have already noted, which, though it took him by surprise, must have helped to define his views for his future course.

When James Forbes went to the Continent he had left instructions with his brother Charles, that if anything should befall Sir John Leslie during his absence he wished to be put in nomination for the chair. Returning rapidly, as he passed through Cambridge on his way to Scotland, he found that his brothers and his uncle, Mr. George Forbes, having already entered his name as a candidate for the vacant chair, were warmly prosecuting the canvass, and had procured testimonials from all quarters.

Besides James Forbes there were four other candidates in the field—Dr. Ritchie, Professor of Natural Philosophy in the London University; Mr. Thomas Galloway, Professor of Mathematics in Sandhurst College; Professor Stevelley, Professor of Natural Philosophy in Belfast; and, last and greatest, his firm friend and adviser, Dr. afterwards Sir David Brewster. But young Forbes was in the field and had fairly committed himself, before he knew that Sir David was to come forward. The tone which Brewster had always held about the work of professorships, as fatal to original investigation—an opinion which is expressed in a letter already given—led Forbes to believe that he would not now think of the chair. It was therefore with surprise, but not dismay, that some time after he had returned from the Continent he learnt that Brewster was really to be a candidate. To have such a rival as Dr. Brewster, his elder in years, his superior in reputation, one whose achievements in science he so greatly admired, to whose friendship he owed so much, must have been a severe trial to Forbes, and would have made many men shrink from the contest. But while he admired the performances of his elder friend, he had confidence in his own powers, and was not a man to be daunted by any reputation however great. I have before me a volume in which Forbes collected the printed testimonials of all the candidates. It is interesting now to read over some of these, and to observe the way in which the same Cambridge magnates testify to the performance of Brewster and to the promise of Forbes. Professor Airy, the Astronomer Royal, in writing of Sir David expresses his conviction 'that there is no person of the present day to whom his (Brewster's) country owes so much for its scientific character as himself.' He adds, that if the choice of the electors is to be determined by an estimation of past services, it must fall on Brewster. He consoles himself in the end with the thought that, 'whatever the decision of the electors between Brewster and Mr. Forbes may be, it cannot be unfavourable to the University.'

Dr. Whewell said of Sir David that 'in the department of optical science he had brought to light more new facts and new principles than any other person in any country, and he might almost say than all other observers together.'

Nothing could well be stronger than these testimonies. Comparing them with the language employed by the same men when speaking of Forbes, we find them testifying that Forbes's performances were already important in themselves and very remarkable as coming from one so young, and they drew from these and from their knowledge of himself an augury of a brilliant future.

From among the numerous testimonials which Forbes produced from men of the highest scientific name, Whewell, Airy, Peacock, Buckland, Vernon Harcourt, Chalmers, Sir William Hamilton of Dublin, and many more all speaking one language, the following words of

Sir John Herschel need alone be quoted. He speaks of Forbes 'as marked by Nature for scientific distinction, if he should continue to aim at its attainment.' 'I adhere,' he adds, 'to the expression; and having, in consequence of the interest attaching to this matter, been led to an attentive re-perusal of his meteorological and magnetic researches, as well as of many of his earlier papers, I must say that I find this impression greatly strengthened and confirmed by the evidence they afford of a most valuable union of careful diligence in the observation of facts, and just philosophic views in combining and reasoning on them, together with a remarkably extensive knowledge of the investigations of predecessors and contemporaries in a great variety of different branches of inquiry. . . . . It would be the height of absurdity to think of raising any objection on the score of standing, to one who has already brilliantly distinguished himself, and whose talents and application can only be rendered more precious by the vigour of age—youth he means—to which they are attached.'

Excessive youth was one of the main objections urged against Forbes. These words of Herschel must have done something to repel this foolish charge; for in truth there can be no more foolish objection urged against a candidate, who is otherwise well qualified. Where character and genius are found, as they sometimes are, combined in a very young man, he cannot, when once he has attained his majority, too soon be placed in the chair for which his gifts have evidently designed him.

It cannot, however, be doubted that, if the election had been decided solely by achieved results, Brewster must have been successful. I have heard it said, however, that one element which does not appear in the testimonials weighed with the electors—a very grave doubt whether Sir David had the power of lecturing before a large audience. It is well known that he never could venture to speak without manuscript at any length in public. It may be that the same nervousness would have unfitted him from being a successful lecturer. There was still another element seldom wanting in such cases, and in this contest more than usually prominent. It was the day of the Reform Bill, and the spirit of party was then running very high in Edinburgh as elsewhere. One still more embittering ingredient, however, which has since then been added to these professorial elections, church partisanship, was not then present. It might then have had some effect on the decision whether a candidate was a Tory or a Whig; it was not then inquired whether he belonged to the Episcopalian or Established Church, whether he was a Free Churchman or a United Presbyterian. How far political feeling may have influenced the choice which the electors made I cannot say. It was the last of the unreformed Town Councils of Edinburgh with whom the election lay, and it was the last time in which they had to exercise their prerogative.

After a more than usually excited contest the election took place on the 30th January, 1833, when the choice fell on James Forbes by a majority of twenty-seven to nine.

Likely enough the electors may have been glad to see in the field a young candidate so highly gifted, sprung from an ancient family, long and honourably identified with the Tory party in Scotland. But, politics apart, of the many qualities which may be taken into account in the choice of a professor, though they cannot figure in testimonials, few candidates ever had a larger share than James Forbes. If, in addition to high scientific genius, a finely cultivated literary taste and style, natural powers of eloquence perfected by the best aids of art, a dignified and commanding presence, gentle and refined manners, and these all wielded by a will of rare strength, purity, and elevation—if gifts such as these enhance a man's claim to such a chair, then the claim of James Forbes was of the highest order.

Whatever reflections may at the time have been made on this preference of the young man's promise to the performances of the veteran in science, the result well vindicated the wisdom of the choice. And no doubt the sense of his own exceeding youthfulness in years if not in learning, and the knowledge that he himself and his supporters were on their trial, acting on his tensely-strung and resolute spirit, urged him to put forth his utmost strength, and to silence all murmurs of detraction. But though such feelings may naturally have had some place in his mind, it was from higher and more stable sources that his habitual strength was drawn. The following extracts show with what aims and motives he entered on his new career :—

'GREENHILL, *3rd February*, 1833.

'God of His great mercy has brought me safely through another year: He has directed my steps abroad, He has watched me there, He has given me much richly to enjoy, He has called me home at an earlier period than I proposed, He has fulfilled the object which was the immediate cause of that return. In all these proceedings I see His hand. If ever I felt disposed to say, "To-day or to-morrow I shall go to such a city, and stay there a year," He has convinced me of the uncertainty of human affairs, and, by analogy, of my own life. "His ways are in the sea, and His path in the great waters, and His footsteps are not known." By the fulfilment of the object of my return to this country, I find myself at an earlier period than I proposed in the position of Professor of Natural Philosophy in Edinburgh, to which I was elected on the 30th ult. May it be for my own welfare, temporal and eternal. It is under this condition alone that I have prayed for it. To-day I received the Communion, and expressed my wants and wishes in such words as these:—O most powerful and gracious God, look down with mercy on the infirmities of Thy servant. Amid the changing scenes of life give me firmness to keep my eye steadily fixed on the great object of that life which Thou hast given me. I trace the workings of providence in Thy recent dealings with me. Let my object be so to use the endowments and temporal advantages wherewith Thou hast blessed me, as may most redound to Thy glory, and to my own spiritual and eternal welfare. Take from me all fear of the world's frowns, all elation at its smiles. Enable me to fulfil with integrity the new duties placed before me, to act becomingly in the wider sphere to which Thy goodness has extended my exertions. Aid me to cultivate Christian charity, to lay aside all feelings of animosity, and to cherish a principle of universal benevolence. I beseech Thee to preserve me from an undue anxiety in temporal matters, to which I am so prone, and to strengthen my faltering confidence in Thee, which I acknowledge has been sinfully and ungratefully imperfect. Grant me, I pray, health for the pursuit of my undertakings, yet not a slavish love of life. Grant that I may take a fearless view of my latter end. If I be not so grossly forgetful of the uncertainty of life, let me not mistake the very fault of too much thoughtfulness of the morrow for the saving conviction of the thought of a last judgment. Teach me not merely to number my days, "but so to number them that I may apply my heart unto wisdom." Further all my good resolutions, however imperfect, and give effect to the weakness which, unassisted, would render them unavailing. This, and all I ask is through the merits of Jesus Christ. Amen.'

Our senses are impressed when we see high and energetic character in men, and we say that nature has made them vigorous or persistent or conscientious. We say so, and in part truly. But there is another and deeper side which we do not see. The inward struggles, the self-scrutiny of motives, the self-dedication, the casting of self back on a higher strength, these are things hidden from all eyes. Only when, as here, some secret record long after comes to light, do we get a glimpse of the inner springs whence that strength was fed, whose outward results men saw for a time, and approved.

This chapter, which attempts to sketch the growth of Forbes's mind and character from childhood to manhood, may perhaps be fitly closed by the following autobiographic letter, written at a later date than we have yet arrived at. One thing only must be noted. While touching on some peculiar circumstances of his youth, already alluded to, he gives an estimate of his own character, which, if it have a side of truth, seems certainly drawn with too severe a hand :—

'SUNDAY, *November 3rd*, 1839.

'I was born 20th April, 1809, whilst my mother was in very delicate health. She went to Devonshire, taking me with her, and died there the following year. My poor father, as I have reason to know indirectly, was almost distracted by his loss: a man of the most virtuous, amiable, high-minded, and singularly unobtrusive disposition, he was evidently formed for the complete enjoyment of domestic happiness. He had, it appears, so concentrated his affections on my mother, that with her loss he was a changed man: he lived as a Christian ought to do, striving to fulfil his duty to his family and to mankind by the most active but generally secret benevolence; but from the time of my mother's death I suppose no one shared his entire confidence. To his family the most affectionate, considerate, and uniformly indulgent parent, he yet spent but little time in their society, though he lived always at home, never, I believe for years together even dining out, until the increasing age of my sisters induced him to mix in society for their sake, and live part of the year in Edinburgh. Till then we always lived at Colinton in absolute seclusion. On good terms with everybody, he had no intimacy with any person in the neighbourhood, and consequently we had none. In all the time I knew him he never mentioned my mother's name in my hearing but once or twice at the most. I, the youngest and most delicate of the family, was his peculiar favourite, and never being sent to school or anywhere else, was scarcely separated from him till the day of his death. You must not suppose him sad or moody—it was not so; he lived in the very hearts of his children, and it is now, as I look back, that I clearly perceive that as his treasure was not in this world, so neither was his heart. I was educated with my brother Charles at home, my two elder brothers went to school in England. Even when living in Edinburgh we rode out every day to the schoolmaster at Colinton for private lessons, so that we formed no acquaintances even amongst our nearest relatives until we were almost grown up, and literally knew nobody beyond first cousins, and few of them. This was no doubt a great mistake, and you cannot fail to have observed the effects of it in my brothers and sisters as well as myself. In due time I went to College, having been abroad for a year, but my habits were by this time too much formed voluntarily to seek for acquaintances. My single playfellow, Charles, was withdrawn to other employments, and my habits, which were always solitary, became confirmedly so. About my thirteenth year tastes for reading and inquiry had spontaneously developed themselves, without altogether the means of gratification, for our reading was very rigorously regulated, and my first essays in science, which were soon after, were indulged with almost criminal secrecy, and of course, if discovered, shared the usual fate of being laughed at by my brothers. This still more increased my reserve. One friend I took a fancy to at College, but it was not altogether happy; my reserved manners told too truly then what they have done ever since; he is now alive and well, but he never knew what I felt for him. When seventeen, I commenced writing anonymously in Brewster's Journal, and soon after went to Italy. My taste for science always grew, but my shyness made me conceal it still more sedulously. For years I made observations every hour of the night on two days without anybody in the house suspecting it, except my brother who slept in the same room; even he did not know of my printed Essays. I am not going to give you a history of my progress, because that is not to my present point, except in so far as it illustrates the system of education under which I was brought up. My eldest brother died in 1826, whilst we were abroad, and my poor father never had a day of health after. I need not dwell on the two years that he lingered, rendering himself more and more beloved by his family till it pleased God to remove him just eleven years ago. I need not say that we lived secluded these two years and for long after, living entirely at Colinton. For myself, being quite keen on my studies, I never missed society, and but for the catastrophe of my father's death, I could spend those meditative years with delight again. At twenty-one I knew almost nobody, and was oppressed by dyspepsia, against which I had never been warned, and had carried the seeds from infancy. Though I had long enjoyed independent thought, I had but little notion of independent action, and felt an awkwardness and diffidence in my dealings with people in general, and women in particular, of which no doubt you have remarked the traces. This diffidence did not extend to intellectual society, and I often wonder at myself when I think of the memorable epoch in 1831, when just twenty-two, and furnished with a great number of letters, chiefly from Dr. Brewster and others, whose acquaintance I had made for myself, I made my *début* in London, Oxford, and Cambridge. I wonder, I say, at the self-possession I then felt, at the intense enjoyment I felt at being able to communicate on subjects bottled up in me so many years, and I was immediately a new man. I had thus, you will see, a world of thought and of intellectual intercourse opened up to me which became my world: my familiar world was not altered. My friends, and they soon proved themselves really such, were generally double my age. Whilst my, now, eldest brother was moving at the same time in the best circles in London, I had no wish to accompany him; it required many years of the intellectual society into which I had literally worked myself to give me leisure to think of any other. I fully indulged, and secretly prided myself on my independence of society. In Edinburgh I could not find what I had in London and Cambridge, and therefore I had little or none. In 1832, I travelled alone abroad, and fully worked out a taste, which you will see was natural to my then state of mind, for solitary travelling, and the independence which it confers and indulges. These unsocial journeys, which you profess to like me to make, are the very evidence and food of the reserved temper in ordinary society which, because you observe it most where you feel personally interested, you sometimes lament. Since I have acted for myself I have ventured to do so very independently on the subject of associates. I have formed my own acquaintances, maintained them without consulting other people's opinions or prejudices; and on the other hand, I have treated with mere civility those accidentally thrown in my way. The consequence is that if I have offended a few, I have gained much time and a solid phalanx of useful allies, who like me and I like them simply because we have something in common, however we may differ on other points. General mixed society never had charms for me. As I ask no homage from it, I have nothing to stoop for, and it asks as little from me. . . .

'I should add that in childhood, partly from constitution and partly from indulgence, I was in a high degree timid, excitable, and nervous; that as a more independent spirit came over me I became energetic and ambitious; this stage continued from my seventeenth to my twenty-seventh year; since that some mellowing ingredients have softened some of the asperities of my character, and the providential goodness of God which made me acquainted with you made that friendship in many senses blessed, for which I desire to thank Him as I ought.'

CHAPTER V.

PROFESSORIAL LIFE.

JAMES FORBES'S election to the Chair of Natural Philosophy took place on the 30th January, 1833, and as he had not to enter on its active duties till next November, he had nine months to prepare for the new task that awaited him. When he was elected he was under twenty-four, and he had to take the place of a professor who had made himself a name in his generation. But to the work before him he girt himself with all the energy and courage of opening manhood, and with a methodical and sober wisdom beyond his years. His feelings of satisfaction at success were well tempered by thoughts which do not often occur to one so young.

The way in which he viewed his then situation is seen in a confidential letter to his sister, written in the May after the election:—

'PARIS, *May 26th*, 1833.

'DEAR SISTER,

'. . . Were I asked whether I derived more unmixed pleasure from my recent appointment, or from my tour in Switzerland, I must answer, from the latter; for however intense and piquant the pleasures derived from one's position in society may be, they undoubtedly fall far short in singleness, purity, and charm in retrospect to intercourse with nature—where we have nothing to regret,—where the pursuit is the pleasure, and the delightful elasticity of mind and body contrasts strongly with the opposites which almost inevitably accompany the pursuit of the other. I feel assured that my recent success will be beneficial to me in a way which, on a superficial view, may not be apparent. . . . I have now the arduous task of fulfilling expectations before me, which, while it is humbler, is more wholesome than that of creating them. The chief danger of courting praise or gratifying ambition is either when a man is mean enough to wish to have more than his due, or unhappy enough to combine a feverish love of ambition with an adverse fortune which makes his whole life a struggle to obtain his due, and upon which he therefore naturally puts a fancy price. Let a man take once his true level, and it is not likely, other things being equal, that either the prosperous gale or storms of the world's ways will readily overset him. At least this is my view of the subject, and I feel that Providence has put me in a situation less tempting to the sins that most easily beset me, than if it had left me to buffet my way against opposing circumstances.'

While he was thus girding himself for his work, and cheered by the congratulations of his friends, it must have been felt as not a little painful that there was one of the oldest and most hearty of his friends who could not join in these congratulations. Contests for professorships, as for higher posts, sometimes bring with them not painless collisions. But seldom has the irony of destiny been more conspicuous than in the circumstances which pitted James Forbes and Dr. Brewster against each other as rival candidates for Sir John Leslie's chair. It was not in ordinary human nature but that the success of the younger candidate should cause some soreness in the elder. At the height of the canvass, in anticipation of such a result, Forbes had written to his uncle: 'With Sir D. Brewster I am resolved not to quarrel. With him I believe it is difficult to quarrel, except by one's own fault. I hope and expect that, however this business may turn out, I shall find myself on the same footing as ever with him.'

If this amiable expectation was not exactly fulfilled, it is pleasing to know that after a few years the old intimacy was resumed, their correspondence renewed, and that their friendship continued steadfast till the close of their lives.

Meanwhile, when, during the preparation for his first year's course, he found himself in need of a counsellor, he was obliged to turn, not as of old to Brewster, but to some one of his many scientific friends at Cambridge or in London. Chief among these was Dr. Whewell.

Forbes had hardly set himself to write his lectures before he found the urgent want of simple text-books in some of the most important departments of Physics. This want was the more felt, because he knew that text-books used at Cambridge would be useless for his class at Edinburgh, owing to the then low state of mathematical knowledge among Scottish students. In this strait, he turned at once to invoke the aid of Dr. Whewell:—

'*March* 31*st*, 1833.

'. . . It is most urgently pressed upon me during my present laborious task of writing lectures, upon which I have been six weeks at work, that the difficulty lies, not so much in that of the subject, as in the very elementary manner in which it must be taught, the state of preparation here being low to a degree which, with your high academic notions, fostered by the spirit of your noble university, must appear almost incredible. From the moment of starting for the Chair, I resolved that, should I be successful, I should make a sacrifice, at least a probable sacrifice, of my popularity, to an endeavour to raise the standard of science, and to rescue the noblest walks of learning from the exclusive dominion of the penny literature. . . . In introducing into my lectures a cautious mixture of pure demonstration with experiment and collateral illustration, I have felt that all my labours are likely to be rendered useless for want a fit text-book—I mean for theoretical mechanics. Notwithstanding the number of works on the subject, without, I hope, being fastidious, which my qualifications do not entitle me to be, I have found no work to my mind. . . . In writing on a subject so vitally important, I need not assure you that compliment is wholly foreign to my thoughts, when I state that your Mechanics has appeared to me far the best book I have met with for teaching from, both from its admirable arrangement, which I have hitherto almost strictly followed in writing my lectures, its mixture of geometry and analysis, its subdivision into clear propositions and illustrations. But for my purpose it is too long: it is, on the whole, rather too difficult.

'What I want you to be prevailed upon to do is to publish a sort of abridgment of the work requiring a grade less mathematics, and introducing into the dynamical part problems from your new Introduction to Dynamics, which, by the way, I mean to teach at a separate hour to more advanced students. I would have nothing farther than the very elements of the calculus, and that part printed in a smaller type. . . . Your labours are most properly devoted to your own noble institutions; nor would I venture to ask you to do anything which would employ much time. Such a text-book as I want would be almost composed by the easy art of clipping from your own writings, and if you would add a short system of Hydrodynamics, it would be an important addition. The whole should not exceed 300 pages octavo, as I can hardly devote three months of my course to it.'

One cannot but wonder at the boldness with which this youth of three-and-twenty approaches the great Cambridge Don, one of the chief scientific celebrities of the time, to lay upon him so exacting a demand. It is still more surprising to find the Master of Trinity quietly submitting, and setting himself to execute the imposition.

From time to time, during the preparation of his

lectures, as difficulties arose, or interesting questions suggested themselves, Forbes gave his thoughts to Dr. Whewell, as in the following letter :—

'*July 20th*, 1833.

'. . . Since I came home I have been at work upon the fundamental principles of Dynamics. I confess I am half afraid to write to you upon your hobby, but I do want to put one or two questions to you about Newton's laws. . . . I have read, I believe, all that you have written on the subject, which is saying something, seeing that it is scattered through seven volumes, and I feel convinced with you, that the composition of velocities and the proportionality of velocity to pressure must be separately proved. What appears to me is this, that your three laws do not correspond to Newton's three laws, though you lead one to suppose so. My notion is that Newton's third law is not included in yours ; that he speaks simply of action and reaction in the ordinary, accurate, and statical sense of the word. No one can, I think, read his second law without seeing that it contains two distinct propositions. . . . You will find, I think, that all who follow Newton closely, find this double meaning in this second law, and reduce the third to the literal statement of action and reaction. I have no doubt that but for a little love of the numbers three and seven, Newton would neither have so divided his laws of motion, nor his spectrum. According to my view his laws of motion would have been four. 1. Inertia ; 2. Composition of Velocities ; 3. Proportionality of Force to Velocity ; 4. Equality of Action and Reaction. The French, like you, consider the last a necessary truth, and reduce the first principles to Newton's first laws ; you do the same, but break them up into three. I am not quite convinced that action and reaction can be argued à priori satisfactorily, at least human minds are by no means clear upon it, and Robison well observes that no one before Gilbert seems to have thought that a magnet was attracted by iron, as much as iron by the magnet. . . . I cannot tell you what a weight is off my mind since you agreed to modify your Mechanics. I am so fully convinced that no inexperienced person can undertake to write a good systematic text work, that I had resolved that nothing should induce me to do it at present. The same reason makes me very cautious in offering any suggestions, which I might, on mature reflection, repent of; but if you will permit me, I should like to send you my copy of your last edition, in which I have marked in pencil on the margin the portions which occur to me as suitable for the first volume of the two into which you mean to divide it, according to the ideas I have of utility in my own department. I trust you will make every exertion to have the first volume out by the middle of October; and I hope you will induce your bookseller to make the price as low as possible, which would much increase its utility.'

'*August 8th,* 1833.

'. . . Any doubt as to the propriety of viewing mixed mathematics as belonging to a natural philosophy class is at this moment peculiarly untenable; for the whole progress of general physics is happily so fast tending to a subjection to mathematical laws of that department of science, that in no very long time magnetism, electricity, and light may be expected to be as fully the objects of dynamical reasoning as gravitation is at this present moment.'

But while he was giving his main strength to these weightier matters, which are fundamental, he did not neglect the lighter accomplishments which tend to make the perfect lecturer. When in London in May he took lessons in elocution from Mrs. Siddons. At an earlier date he had studied this art under other teaching, but from the great actress he received, we may believe, some finishing touches in the art of speaking. In the case of many men what they learn from such instructors seems to be a very doubtful gain. The clearer pronunciation and more effective delivery thus acquired are bought at too high a price in the loss of naturalness. But nothing of this kind was to be seen in Forbes. His nature was too strong and serious to be open to the inroads of affectation. To the last his pronunciation and tone of voice continued to be, what it had always been, unmistakably that of a refined, but entirely Scottish gentleman. But he had absorbed into his own nature whatever the rules of art had taught him; and the result was that clear and well-articulated utterance in which no syllable was lost, and that impressive and graceful delivery which none who have heard it will forget, and which made him one of the most winning and effective lecturers of his time.

In the midst of preparation he found time to turn his thoughts to other interests of science than those connected with his future class-work. The British Association was to meet in the summer of 1833 at Cambridge, and before this meeting took place it was necessary to fix on a place of meeting for the ensuing year. Dublin was putting in strong claims; but to Forbes much stronger seemed the claims of Edinburgh, which he urged, not in vain, in the following letters written in the spring of 1833:—

'My dear Sir Thomas Brisbane,

'. . . It appears to me that it is little short of the duty of the Royal Society to convey to the meeting of the British Association an invitation to make Edinburgh the place of their next meeting. Since you, as well as Sir D. Brewster and myself, have been both at Oxford and at York, you will not wonder that I feel this strongly. It has always been considered that the origin of the Society was in a great measure Scottish, and Sir D. Brewster, Mr. Robison, and yourself have always been looked upon as among its founders. Consequently, at both the late meetings a very strong feeling has been expressed that Edinburgh should have the honour of an early visit. Did it lie entirely with the Council of the Association, we should naturally wait till an offer so advantageous to Scotland was made to us. Circumstances which took place at Oxford will, however, at once occur to you as showing that a more decided part is necessary. With a view to put another place between the two Universities, Liverpool, certainly the best place in Britain, was suggested. When, however, it was found that no offers had been made—no representation from the existing provincial Society,—the idea was abandoned. Since the place and time must be fixed within a few days after the meeting of the body, there is no leisure for preliminary inquiry, and the promptitude of the Provincial Societies, to which the Association chiefly looks for support, must decide the matter. You may remember at Oxford a most pressing application on behalf of Ireland was made by the Astronomer Royal of Dublin, but was set aside on the feeling that Great Britain should first be visited, and, most specially, that Edinburgh had a prior claim to Dublin. When I think of all this, when I remember the strongly expressed desire of some of the first scientific men in England to visit or re-visit Edinburgh, and when I think of the immense advantage which such a *réunion* could not fail to produce upon scientific men here—an advantage which but a year or two ago might have appeared chimerical—I am convinced that you will join with me in thinking that we are in every point of view bound to send a warm invitation to the Cambridge meeting. From the feeling which was shown towards Scotland at both the preceding meetings, I have no doubt that the offer would be accepted, though I am sure it will be met by others from various parts of the kingdom, and particularly from Dublin, the Association now enrolling among its number the most distinguished members of the Royal Irish Academy.

'The Association is above being patronized now: the honour conferred is on their side, the advantage on that of the places of its annual migration. Both individuals and societies will now be ready enough to give the aid which is no longer required, and which was withheld whilst the success of the Association was problematical. . . . . .

'At all events of this I feel perfectly assured, that the Council of the Royal Society, as representing the interest of science in Scotland, ought to offer such accommodations as it may be in their power to procure, if the Association will honour Edinburgh by making it the place of meeting for 1834. . . .'

On the same subject he makes this energetic appeal to Mr., afterwards Sir Roderick Murchison, who seems to have favoured the claims of Dublin in preference to those of Edinburgh :—

'My dear Murchison,

'I cannot delay an hour in writing to you about the Association, having taken the deepest interest in its coming here next year, and being horrified at your proposal to put it off for three years. I entreat you as a personal favour to keep the matter open, and in the mean time I can prove to demonstration that your reasons are null and void.

'1st. You say that Dublin has secured a prior claim to Edinburgh. This I positively deny. It was specifically understood, both at York and Oxford, that Edinburgh, from having to a great extent originated the meeting at York, should have the first visit; this, you will see, I have distinctly expressed in the enclosed letter to Sir T. Brisbane, which I send for your private perusal.

'2nd. Then as to Bristol, the idea is a new one. Liverpool was spoken of, but, as far as I recollect, not the other, nor do I think it a good position. But putting this out of the question, what I object to is your calling Edinburgh a University town, and therefore that it ought not to follow Cambridge. This is quite a mistake. The University gives no character to Edinburgh, and I fear will give little to the meeting. You must be perfectly aware that it is not an academical place, and that the University has nothing to offer. It has no status, no funds, no power. In short, you must never think of the University when you come here, nor compare it in the remotest degree with Oxford and Cambridge. I assure you you are proceeding on a fallacy. . . . My dear friend you are a Scotsman, and though a deserter, you should not quite desert what is due to your country. Only look back and remember what Scotland did for the Association. Was there any talk of Dublin then? My dear Murchison, do not commit yourself. I daresay you think I am mad. . . .'

A few months later we find him writing, through M. de la Rive, of Geneva, the following invitation to his foreign friends to attend next year's meeting, which had now been secured for Edinburgh:—

'*August 26th*, 1833.

'. . . The scientific meeting is to take place next year at Edinburgh, and I do earnestly hope that my Swiss friends will come en masse in September 1834. Pray present my warmest invitation to the whole of them. In particular my best respects and grateful recollections to M. de la Rive-Boissier, and to M. Gautier, who, I hope, are quite well.

'I am exceedingly occupied with my preparation for my winter's course, which must be my excuse for this short letter. I have been almost obliged to give up correspondence altogether. . . .'

Even in the press of immediate work his thoughts turned at times to the scenes of his foreign travels, as is seen by the following letter addressed to Mr. William Burr:—

'*September 4th*, 1833.

'I think it would really be most interesting if you could prepare a correct sectional plan of the Cloaca Maxima of Ancient Rome. As you know, I have in vain sought in different books for it; and now the more regret that when in Rome I omitted what I intended to do, namely, to make a sketch by actual measurement.

Most authors treat of the arch as if it were circular. My recollection leads me to think that it is oval, as I remember that I intended to determine its mathematical figure by offsets from a vertical line. The place where I should wish it particularly examined is where it appears at the Arch of Janus, not far from the Forum. Its indubitable antiquity—Pliny says it existed 800 years before his time—gives it a very high interest—merely antiquarian; but my present inquiries are rather connected with the history of the Arch as founded on scientific principles, and it would be difficult to find in any age or country a more striking work of its kind than the massive triple arch of the Cloaca Maxima.'

Of the rest of the summer of 1833, the following short entries in a diary of retrospect are the only record:—

'In May and June I went to London and Paris. Very kindly received in Paris, more so than even in June at Cambridge. Meeting of British Association. Worked hard during summer and autumn writing lectures at Greenhill. Elected to Athenæum Club by Committee, July 5th. My brother Charles married at Blair Vadock on Gareloch. He then went to live at Hermanstone, and for some years I paid him many pleasant visits there. In September to Fettercairn. Back by St. Andrew's, and visited Dr. Jackson there.'

This was probably the first time he had seen St. Andrew's. Dr. Jackson here mentioned was at that time the not undistinguished Professor of Natural Philosophy in that University.

When the college session 1833-34 opened at the beginning of November, Forbes made his first appearance as a Professor. As is usual on such occasions in our Scottish universities, he had to deliver an inaugural lecture. Such lectures, as they are supposed to strike the key-note of the new Professor's course, are occasions of no little interest to academic persons, and of some anxiety to a young Professor. As he succeeds in interesting the audience or not, men are apt to draw auguries of his future career. To Forbes's first appearance circumstances added something more than common interest. He was succeeding to one distinguished man; he had been preferred to another equally or more distinguished. His youth, reckoned by some a fault, was to all an attraction; his promise was high, greater than that of any other young Scotsman of the time. He was sprung from a race well known and highly esteemed in Edinburgh and throughout Scotland; he was of a tall commanding figure, of a fine and impressive countenance, his delivery was clear and resolute, yet conciliatory. Altogether, it is not often that a man, who combines with so many inward gifts such outward attractiveness, addresses an audience from a Professor's chair. The impression left on all who heard that first lecture was, I believe, that he was in the place where he had a right to be.

The following is the short notice of it in his own diary:—

'*November.*—Delivered my first lecture. Owing to the struggle about the election, considerable excitement, excessive crowd. Passed off well. The attendance at my subsequent Monday lectures was also distinguished. Well pleased with my success.'

A friend of his, the venerable Archdeacon Sinclair, who heard that first lecture, still survives to write of it thus:—'I remember well his appointment to the chair, and I attended his inaugural lecture. My expectations were high, but he surpassed them. I can still with pleasure call to mind his tall, thin, elastic figure, with a long wand in his hand, pointing to various diagrams, and pouring forth interesting information in clear language and in a condensed form, with as much facility as if he had been lecturing for years.'

Within a fortnight from the beginning of this his first course, we find him writing thus to Dr. Whewell:—

'*November 15th*, 1833.

'I am giving weekly lectures, besides the usual ones, on the study of Natural Philosophy, in relation to its present advanced condition, which are largely attended by non-professionals, and I hope that the success which has attended them may in some small degree be of use in retarding, for we can do no more, the downfall of solid literature.

'I was much interested by your account of the Poissonian demonstration. I knew it was remodelled, but have not examined the book, though I have it. I, too, have been dabbling lately in rotation, and a confounded subject it is: though I have only studied it elementarily: I shall be very glad to see what light you throw upon it. I discovered a most notable error, which my learned predecessor annually inculcated, on rolling bodies; and as I learned from his assistant, annually tried to illustrate by experiments, and fancied he succeeded!'

Not long before the close of his first session, he thus again writes to the same correspondent:—

'*March* 29*th*, 1834.

'I find the greatest advantage from having been obliged to study these subjects in a way necessary to convey a precise idea of them to others; which I feel that almost no other circumstance would have induced me to spend so much labour upon. And I find what is natural enough, that in the course of last summer, when I worked very hard, I had prepared such a treatise on Mechanics as would almost have required a six months' course to go through, without anything else. . . . But to return to your paper. In general, I think we are much at one. I am still disposed to adhere much to the sketch I sent you in a letter last year, and which I employed in expounding the subject to my more advanced students. So far you accede to my views in what I think an important point, the derivation of the proportionality of Velocity to Force, that force being measured by Statical methods, with which the student is presumed to be familiar; which gave what I called the measure of a 'definite amount of Force,' and which expression you may remember you objected to. . . . A month hence, I shall have finished my course, and then propose to escape for a little relaxation. I shall probably go to London, and hope to see you. I am certainly relieved at having got well through so much of my course. The responsibility I felt was oppressive. But my labours have been more than rewarded by the efforts of my pupils, and the obvious improvement in the method and degree of study which has been the consequence. I have given about twenty lectures to the more advanced, going as far as 'Poisson's Demonstration of the Direct Problem of Central Forces,' which, humble as it may appear to you, is a step among us 'hyperborean sages.'

He might well feel a sense of release and exhilaration at the prospect of the close of his first session as a Professor. In any case, it is something to feel that the strain of six months' continuous lecturing is nearly at an end. But to feel that his first year's course was achieved with satisfaction to himself, with benefit to his pupils, and with the approval of all, was more than a common joy. Whatever misgivings he may have had as he entered the Natural Philosophy Class-room in November were now well over. Whatever doubts as to the fitness of his appointment others may have entertained on the ground of his youth and inexperience were silenced by the test of fact. Henceforth he must have felt he was fully 'master of the situation,' and, as long as health lasted, his work there was not an oppressive burden, but an ample field for his abundant energy. It was not, however, self-complacency, but feelings of quite another kind that the retrospect of the session so successfully closed called up within him. This is the entry of his private journal :—

'*Greenhill, April 20th,* 1834.—I will not let this day pass without recording my deep and heartfelt sense of the special goodness and protecting power of God, which I have experienced during the past fifteen months of almost incessant labour, a share of bodily and mental vigour which has enabled me to pass, with comparative ease and credit, through the first session of my professorial career. I have had the happiness to satisfy my friends, and to have my toils more than rewarded by the zeal and application and gratitude of my pupils. I desire to acknowledge God as the source of all this good fortune, and to bless Him that I am yet spared to pursue the course in which He has so parentally guided and strengthened my exertions. Having attained the summit of my wishes in a worldly point of view, I hope that I could now, with a reasonable share of resignation, say if required, Lord, now lettest thou thy servant depart in peace.'

The life of most Scottish Professors was then as now divided into six months of unbroken work in College and six months of vacation. To strangers unacquainted with the ways of Scotland and the habits of its students, so long a vacation appears a strange anomaly. But there are reasons enough grounded in our social facts and habits which have justified it for generations, and which satisfied the late University Commissioners when they carefully inquired into all the bearings of this question. It must not be supposed that these six months are either to student or Professor times of idleness. The former is often employed in some useful work for self-support, as well as in carrying on his College studies. The latter, when he has recruited himself after the toils of the session, finds full employment in preparing new lectures or recasting old ones for the approaching session. Besides this, whatever Scottish Professors have done for Science, Philosophy, or Literature, has been the fruit of their summer leisure.

No man ever employed his summers more methodically and energetically than Professor Forbes. Indeed, it is probable that the world has received fully as much advantage from what he achieved during his summers as from his regular winter labours.

In May 1834, we find him spending three weeks in London, accompanying his sisters to Oxford, and himself visiting Cambridge, and there attending Airy's optical lectures.

In June, he went by himself for a tour through the south of England, and had his first interest in Gothic architecture awakened.

'NEWPORT, ISLE OF WIGHT, *June 9th*, 1834.

'. . . I have grown very architectural in these southern counties, in proof of which I travelled sixteen miles yesterday to see a country church, Romsey, which I had reason to suspect was curious. I was amply rewarded : an almost perfect church, said to be of the time of Edgar, scarcely polluted with Gothic, and rich in horseshoes. Being quite innocent of theory on the subject, or of having read Rickman or Milner or Whewell, I affect to become knowing on facts only, and if I form any theory, I shall of course be very obstinate. Though tempted to be disappointed at first with Salisbury, when seen under a rainy sky, it constantly grew upon me during the two pleasant days I spent in the neighbourhood. I examined it internally and externally with some care, and left it full of unalloyed admiration. I next went to Winchester, which has much greater variety of style, and interested me very much. But what particularly gratified me was the small church of St. Cross, a mile off, which I accidentally heard of, and immediately went to. It appears to me most interesting, as pointing out the tendency to transition in Stephen's reign, from the round zigzag style to the Gothic. There are interesting arches, lancet windows, and actually Gothic arches with the zigzag ornament, formed by the crossing of two circular arches, placed obliquely to one another, in the roof of the aisle. Connected with this remarkable church, I found a most interesting establishment kept up by funds given by Cardinal Beaufort. It supports thirteen decayed tradesmen, who live in the most charming range of houses connected with the church. They have a common hall, common laws, have a nominal master, Lord Guildford, a resident chaplain who reads prayers daily, and they are obliged to distribute for ever bread and beer gratuitously at their gate, of which I partook.'

Although he never had leisure to devote himself earnestly to the study of architecture, yet interest in old historic buildings, and love for all that they contained of ancient and beautiful, continued strong in him to the end. In this journey he turned towards Lympston, which had for him another and more sacred interest. It was there his mother died when he was yet only a year old. There he found an old sexton who remembered his father and mother. 'Some particulars he told me, calmly enough, thrilled me.' So he writes years afterwards. It is thus he describes this visit in a letter to his sister at the time:—

'EXETER, *June 22nd*, 1834.

'. . . You may conceive, better than I can describe, my feelings at that moment. For some days my sympathies had gradually been warming towards a point in which I felt so deep and melancholy, and at the same time so romantic an interest. It is but rarely that in the life of a sober man this species of interest can be so fully worked up as in my case. I left Exmouth early in the morning, and sauntered up the bank of the Exe, under a melancholy sky, and with somewhat excited feelings, determined not easily to forego the object of finding the spot where my parents lived, though by no means sure how I should easily effect it. I walked almost through the village without seeing anything like a communicative face. I almost fancied I should have been received as a known face, instead of being an object of not a little curiosity. At length I inquired for the church, which is not very conspicuous, and, as I expected, found an old sexton who, I thought, might be a chronicler of more than twenty-four years back. I soon entered into conversation with him, found that he recollected the individuals in question; nay, the first thing he told me was that he had with his own hands tolled the bell at four in the morning on our poor mother's death. Like most of his trade, he was not of very fine feelings, but he was very communicative, and you may imagine the impression which his little anecdotes, as fresh as if of yesterday, made upon me. I made him take me to the top of the church tower, and show me the house where they lived, and those of the neighbours, about the fate of some of whom I inquired. I made him show me where they sat in church, and I walked from the church to the house by the very path leading also to the Parsonage, which they had often trod. It seemed a sort of hallowed ground. The house has nothing very prepossessing in its exterior, except in plain neatness. I wished I could have had my mother's letters to collate them on the spot. You may recollect the interest I took in them, when you were so good as to send them to me. I copied them with such a species of veneration, that I retained not merely word for word, but page for page and line for line. Of course I felt a peculiar and additional interest in every thing on the spot, considering that I was not visiting these places for the first time, that I had been the almost unconscious inhabitant of this very house; nor could I help speculating upon what identity subsisted between my then and my present state of existence. The old sexton recollected the baby, and certainly was somewhat moved when I said, 'I am that baby.' I strolled from the church to the house, and from the house to the church, and could hardly tear myself away, which I did too late to avoid an impending thunder-storm. It may not be amiss if I join a sketch of the position, in case, when the next person goes on the same errand, the old sexton should be dead, and other witnesses too.'

In September 1834, the British Association met at Edinburgh. It was in a great measure owing to Forbes' exertions that the meeting took place there. During its stay he entertained in his house at Greenhill, Dr. Whewell, Mr. Peacock, and Mr. Vernon Harcourt. Of the latter he says in a letter to Professor Phillips about this time: 'I learn every year to look with more admiration and affection on that remarkable man; nor shall I ever cease to look back with peculiar satisfaction on that meeting at York which brought me first into connection with him and with yourself.' With the success of the meeting Forbes was well pleased. The Association had now quite established itself as a national institution, and the gathering at Edinburgh, which was the third since its origin, attracted foreigners of distinction, among whom was M. Arago. Of this eminent man Forbes writes: 'The impression he has left here cannot be forgotten, and I look upon my improved acquaintance with him as a very happy event in my life. I trust before very long to extend it on the other side of the Channel.'

CHAPTER VI.

PROFESSORIAL LIFE (continued).

HAVING so entirely succeeded in his first year's course, Forbes no doubt entered on his second (1834-35) with full confidence that like success would attend his future efforts. Henceforth, session followed session with that uniform success and that energy and devotion to his work, which never flagged, as long as health lasted. Nothing can be more uniform than a Professor's winter's course—so uniform, that to lookers-on from without it may appear monotonous. But from this it is saved, at least in the case of a vigorous and advancing teacher, by the deeper insight and wider range which he is year by year obtaining in his own field of inquiry. And the sense that the teacher is one who is not merely retailing old knowledge taken from books, but by dint of mature reflection or original research is opening up fresh fields, and adding something to the store of human thought or knowledge, adds a wonderful charm to all his intercourse with his students. Such a charm generation after generation of students felt, at least those who could appreciate these things, as they sat in the Natural Philosophy Class Room, while Forbes was in his prime. But though such impressions, received in youth from some master of thought or science, are among the most lasting and delightful which men ever partake, they furnish little that can be told in narrative. What these impressions were I shall have occasion to show in the words of those who once felt and still vividly remember them. But as one session in outward appearance was much like another, as the incidents in them were few, and the adventures none, I shall not attempt to describe them in detail. The main thing to notice will be the course which Forbes' own investigations and thoughts during each winter were pursuing, and this will be best seen in selections from his wide correspondence.

What events or adventures his life contained were reserved for his summers. Of these, I shall notice in this part of the narrative those which he spent in Great Britain. His foreign tours and Alpine explorations will be given in some chapters devoted to themselves.

In November 1834, Forbes left his second home at Greenhill, and went with his sisters to live in Melville Street. His two brothers, each on his marriage, had before then quitted the family home: his brother Charles had married in July 1833, and taken up his abode at Hermanstone; Sir John, in August 1834, had gone to live at his country seat, Fettercairn. James lived in Melville Street only one winter. In May 1835 he and his sisters removed to Dean House. This, which was then an old country house, has since been removed to make way for the beautiful cemetery of Dean.

The following letters will show what subjects were engaging Forbes' thoughts during the winter session of 1834-35:—

To Professor Powell.

'*December* 1*st*, 1834.

'I write to tell you, as you particularly interest yourself in the affair of radiant heat, that I have worked a great deal since I saw you, with Melloni's Multiplier, and with entire success. It is a most manageable, comparable, and satisfactory instrument. I have succeeded in repeating some of his more delicate experiments, such as the refraction of the heat of boiling water, with the most ample success, and have shown many of them to other Professors here.

'But what I principally write about is to tell you that I have proved to demonstration, by its means, the polarization of non-luminous heat, and have now to look to further anxious results connected with it. So that I consider the question for the first time decided. . . .'

To M. QUETELET, *Brussels.*

'EDINBURGH, *December* 5*th*, 1834.

'. . . I have recently been experimenting with Melloni's Thermo-multiplier, and have been much delighted with it. Very lately I have been enabled to establish beyond a doubt the polarization of non-luminous heat; and have verified Melloni's experiment of the refraction of the heat of boiling water.

'To-day I commenced a register with a particular view to you. I have got an apparatus for weighing and measuring men, and shall collect annually as many results from the students of my class as possible, and also their strength by Regnier's Dynamometer. I distinguish their age and native country. . . . Amongst my many other pursuits, as I mean to begin on optics this winter, I have been studying the undulatory theory with great admiration. We are, I am sure, much indebted to you for putting Herschel's Treatise on Light into a more convenient form than we can find it in England.'

To PROFESSOR AIRY, *Cambridge.*

'EDINBURGH, *December* 11*th*, 1834.

'I have at length found leisure to read with great attention, and consequently with very great pleasure, your undulatory tract, which quite fulfils my expectation as to the nature and extent of the evidence on this marvellous subject. I have been getting sundry pieces of apparatus made, and can now profit by your valuable practical lessons, as well as by the papers with which you have from time to time favoured me, and which I am now better prepared to appreciate. Allow me to ask you a few practical questions. . . . I hope you will not be overwhelmed by my questions. I have been working lately with Melloni's Thermo-multiplier, and have verified some of his most curious results regarding radiant heat, which in their connection with light are extremely remarkable. It is unfortunate that that point to which his researches chiefly tend is the most obscure in the theory of light, namely, absorption. I have lately succeeded in establishing, as I think for the first time demonstratively and quantitatively, the polarization of non-luminous heat. I abandoned the method of reflection, which is the only one hitherto employed, and adopted that of transmission through piles of thin mica plates, for which the Thermo-multiplier is well adapted; and with entire success. I have also been endeavouring to determine numerically the refrangibility of non-luminous heat. I discovered, what I now find that Melloni had previously done, that the tourmaline transmits almost as much heat when two pieces are placed with their axes crossed as when parallel. Melloni saw quite as much as I also at first found, but I afterwards detected a slight difference.'

To the REV. DR. WHEWELL.

'EDINBURGH, *January* 1*st*, 1835.

'. . . I am quite full of polarization. I do not exactly understand what you mean by double refraction existing without polarization, but at all events since I wrote to Airy I have made great progress. My original experiments related to polarization unconnected with double refraction, but I have since extended the proof to every recognized species of polarization; and non-luminous heat must be doubly refracted, because it can be polarized by tourmaline and depolarized, or di-polarized if you will, by other crystals. Hence we are entitled to conclude that we might feel brushes and rays of dark heat if our hands were delicate enough thermometers. I can also make some approximation to the length of a wave. I have as yet communicated this to no one. I shall read it on Monday to the Royal Society, when it will be immediately printed. I have proposed Airy as our honorary F.R.S.E., and hope soon to communicate his election. Pray tell him I shall be happy to have the polarizing grinder any time before the middle of March. . . . Have you anything to say about the Rumford Medal? I think Melloni ought certainly to get it, for his two masterly papers in the *Annales de Chimie.* There have been few of the adjudications for researches so accurately fulfilling the founder's intentions. . . .'

To SIR J. F. W. HERSCHEL, *at the Cape of Good Hope.*

'EDINBURGH, *February 5th,* 1835.

'I had a letter from Whewell the other day, communicating your obliging message to me about your very interesting meteorological results. Still I am a little at a loss what to say about them. The oscillation appears very small. My formula: $\cdot 1193 \cos^{\frac{5}{2}} \theta - \cdot 0150$ gives ·060 Eng. inches. Mr. Whewell mentioned to me only ·025 of variation from 9 to 3. The permanent low pressure at Cape Horn observed by Captain King is confirmed by Foster's voyage. The annual variation of mean pressure and also of hourly oscillation you mention is noticed by Humbolt in equatorial climates. Is the barometer highest in summer or winter? I fear we are likely to find little analogous to your observations at the Cape in the Mediterranean. The oscillation is undoubtedly greater: and I do not think the barometer is highest in bad weather. The variable pressure in different latitudes is a very important and to me, till lately, an unexpected fact. I hope that you will be able to bring your barometer safely home again, and so determine the height of your observatory. I hope you have your actinometer with you; here it has a sinecure, there being no sun worth measuring.

'I am very much to blame for not having written to you sooner. In truth, I have been in an almost constant state of exertion since we parted at Cambridge. The result of my last labours I enclose, and should be anxious at your leisure to have your opinion upon it. Setting on foot a six months' course of experimental lectures has been, you will believe, a laborious task. It has occupied most of my time since I saw you last, but is now nearly over. I made a short tour in England last summer, but was kept in a state of excitement by preparing for the Association meeting here, which went off well, and which you may believe interested me much. I suppose your Cambridge friends supply you with news from head-quarters, which would make it presumptuous in a hyperborean like me to offer you any. . . .'

To the REV. DR. WHEWELL.

'EDINBURGH, *February* 22*nd*, 1835.

'. . . I feel quite gladdened at the interest you are disposed to take in the subject—polarization of heat—for as there is not an individual here capable of fully appreciating it, it is naturally to England, and especially to Cambridge, that I look for that sympathy which is a superadded enjoyment to that of the mere perception of truth. . . . I was sorry to see that Trinity lost the Senior Wrangler, as in May last Goulburn was very confidently pointed out to me. . . . I was so engrossed during the earlier part of our session with my experiments on heat, that I am obliged to work hard with the business of my course, and am soon to begin optics, which I have not yet lectured upon. I shall imitate Airy in polarizing and tormenting light on the large scale. I continue my practice of lecturing on the higher branches to those who choose to attend,—and though often not to more than ten or twelve, I feel myself well repaid. I shall thus be able to introduce the undulatory theory for the first time in Scotland. Airy's Polarizing Grinder, as he calls it, delights me.

'Your plan of a fusible surface for detecting rings had occurred to me, but, you will see by reading my paper, would be hopeless from the extreme minuteness of the quantities with which we have to do.'

The summer of 1835 was spent abroad, and therefore falls to be noticed in another chapter. On his return to England after a visit to Fettercairn, he thus writes from Edinburgh to Sir John Herschel, who was then engaged in scientific observations at the Cape of Good Hope :—

'EDINBURGH, *October 25th*, 1835.

'The results of your table appear to me very satisfactory and interesting. One thing strikes me as requiring new investigation, viz., what are really the hours of maxima and minima, which in your fine climate might be easily fixed; for I have no doubt that the diurnal curve is very different from ours. I argue this rom the circumstance that the barometric pressure at noon, instead of being the mean of the day, coincides very nearly in your table with that of 9 A.M., which therefore in all probability is not the hour of maximum.

'I assumed the hourly variation to be a function of the latitude, simply because the tables of observation seem to indicate it; and, whatever may be the true theory, it does not seem to me that the mere distribution of sea and land can be regarded as the main cause of the variation. In conformity with your wish, conveyed by Mr. Whewell, I endeavoured to get observations established at Malta; and wrote to Dr. Davy for the purpose. . . . He has returned, however, to England, but I will try to get new ones set on foot here at the College. The formula given by me was empirical. . . . The Association is itself the best Meteorological Society that ever was formed, and capable of doing almost anything. . . . I have been led to study the undulatory theory, particularly lately in reference to my own inquiries and lectures, and have received from it the same pleasure which everyone must do who approaches the subject with perseverance and candour. . . . I have exhibited all the chief phenomena of polarization upon screens on the large scale to my class, for the first time in Scotland. You must have heard so much about the comet, that I need not add my account of it. It has really been a fine object.'

To the REV. DR. WHEWELL.

'THE DEAN HOUSE, EDINBURGH, *November* 12*th*, 1835.

'. . . I have got a fund of new experiments on hand, and a famous supply of rock salt from Cheshire.

'I blame myself for not having sooner taken up your suggestions about tides, whilst you have been so attentive to my small matters. I feel confident that nothing of the kind exists here, but I will endeavour to get it established.

'Is Professor Airy at Greenwich? He did me the honour of requesting from me information and advice. I gave him very little of the former, though I adventured some of the latter. . . . I spent a day with Mr. Harcourt on my way down. I pressed upon him the necessity of making great exertions to secure a good attendance at Bristol. I think the place ill chosen, but yet that it may be one of the best in point of science. . . . I fully propose being there, and also being in Scotland in the early summer. Do arrange to come down and make your visit to Orkney: my sisters earnestly second my cordial invitation to pay us another visit. You will find us in an old chateau to the north west of the town, with gloomy walls, winding stairs, and painted ceilings, but a hearty welcome. By a curious accident we lived here thirty years ago, and my sisters were born here. . . . I do not know whether I have anything to recommend specifically about your Mechanics; at least it will be some time before I shall have leisure to examine it. Only pray don't enlarge it. I have no doubt you will improve it. My class is enlarging. I have issued a programme for a prize essay on the undulatory theory, which I will send you. I like my class and my work very much, and already perceive an improvement and a desire to improve. I still propose publishing on the Pyrenean hot waters. . . .'

To the Same.

'EDINBURGH, *January* 7*th*, 1836.

'. . . My special thanks for Hopkins' paper, which arrived at an admirable moment. I was reading a paper to our Royal Society about Auvergne, and particularly upon elevation craters, which was quite in point. I am writing a paper just now which I intend for the R. S., London, on the Pyrenean springs, their temperature, geological relations, &c.; and on the former point, temp., I am vain enough to hope that it may prove a sort of model to future observers: at least no one has hitherto so observed, I believe. I have also in hand a little Gothico-mathematical speculation which perhaps you may laugh at; but I give you leave beforehand.

'But these are only secondary occupations, which, with my lecturing labours, only revolve round my primary, the polarized heat. I have managed to magnify the effects so as to be, I hope, beyond cavil. . . . I think that experiment is a quietus for Biot. . . . Excuse this very long story, and pray come down and see my experiments. When I have finished what I am doing I should be most happy to look into your conductivity. . . . If Mr. Airy is with you, pray give him the above numerical results.'

On February 2nd, 1836, we find him writing thus to Dr. Whewell on the first blush of a new discovery he had just made:—

'EDINBURGH, *Feb. 2nd*, 1836.

'I cannot help writing two lines in a hurry to tell you that I succeeded yesterday in making the most curious discovery respecting heat, it seems to me, that I have yet arrived at, and one quite decisive of the identity of its character with that of light. I found that dark heat is copiously reflected within rock salt at an angle too great for its emergence. This I had foreseen last summer before I was aware that Melloni had actually tried it, and at the same time I conceived the possibility of trying whether two total reflections would produce the same effect in the case of heat as in that of light. I have had a Fresnel's rhomb made of rock salt with angles of 45°—one of the critical ones, nearly, calculated by this formula, giving μ its proper value for light. I placed it between polarizing and analysing plates of mica, as described in my last. When the plane of total reflection coincided with that of primitive polarization, or rather was perpendicular to it, the heat was as much polarized as before the rhomb was interposed; when it was inclined 45° it was wholly unpolarized, apparently, or even the longer axis of the ellipse turned a little the other way, corresponding to μ for heat. This I made out even with a very imperfect rhomb, and with heat wholly unaccompanied by light. . . . I congratulate Scotland upon Smith's distinction. . . .'

'EDINBURGH, *March* 10*th*, 1836.

'. . . Pray be the depository of these facts in case Biot and Melloni, with the odds of two men at leisure against one man with his hands full, take them from me. . . . . Since writing the above I have solved, partly, a doubt which has much puzzled me. The action of metal gives a maximum polarizing angle for heat greater than for light, whence I concluded that the index of refraction must be greater for the former than the latter, contrary to my general views. I have just found in Brewster's paper that a precisely similar fact occurs in the action of metals on light; the red ray is polarized at a greater incidence than the blue. It is most satisfactory thus to find the truth of experiments confirmed by an apparent anomaly, as has several times occurred to me lately.'

The following letter to Mr. Leslie Ellis, before he entered on his Cambridge career, will interest many who loved and admired that remarkable man:—

'EDINBURGH, *February* 14*th*, 1836.

'. . . . I assure you also that I had been long looking forward to the fulfilment of your promise to write to me, and began to feel some anxiety as to your state of health. . . . Let no stimulus of fame or advantage induce you to make a sacrifice of the first of earthly blessings. I hope you will not go to Cambridge, unless you are equal to the fatigue of such a career as your tastes and talents would enable you to pursue.

'I hope that under any circumstances you will not lose sight of your physical pursuits in purely mathematical ones, which are of a comparatively narrow character. It is in the field of contingent truth that the triumphs most congenial to the human mind in a healthful condition are to be gained. The disposition at Cambridge strongly aims in this direction, and I am convinced that you will reap as much credit and more advantage by studying mathematical physics as pure mathematics. I cannot conceive a better exercise than Airy's tract on Light, which contains some hard mathematics, but the acquisition of the clear physical views it presents is much harder. I am glad you should feel any interest in so unpopular a subject as polarized heat. I have now greatly extended my experiments, made the effects much more obvious, and made some new singular discoveries.

'The Royal Society of Edinburgh have done me the honour to award me their Keith Medal.

'My summer was chiefly in the Pyrenees and Auvergne. In the former I studied hot springs, in the latter volcanos. I do not wonder that you were appalled by the difficulties of the measure of absorption of the atmosphere, which in fact involves the same difficulties with the theory of refraction, which has been a celebrated problem amongst mathematicians. . . .'

When his third professorial session was over his correspondence was resumed.

To M. QUETELET, *Observatory, Brussels.*

'*May* 17*th*, 1836.

'. . . I have been so much occupied with my experiments upon polarized heat, and with several papers which I have communicated to our Royal Societies, as well as with my annual lectures, that I have had but little time for correspondence or for reducing old observations. I have, however, got a pupil to calculate my magnetic intensity observations. I have weighed and measured 800 individuals myself. I have had the mortification to discover that the construction of the dynamometer was extremely insufficient, but this can scarcely affect the relative results. I have separated the English, Scotch, and Irish. The ages are chiefly from fourteen to twenty-four.

'On the 15th the solar eclipse was most admirably seen here. . . . I observed with a 7-feet reflector the immersion and emersion of the spots, of which there were several, but I could not observe the slightest distortion produced by refraction upon those delicate objects. My attention was chiefly directed to this object: to examine the light from the sun's edges, at and near the annular period, in order to ascertain whether the dark lines in the spectrum were more numerous or stronger in the light which must have traversed the greatest thickness of the sun's atmosphere, and which have been supposed by Sir D. Brewster and others to be due to the absorptive action of that atmosphere. An attentive examination assures me that no material difference could exist; indeed, I did not perceive the slightest. I therefore conclude that the sun's light is originally deficient in those rays.'

To the Rev. Dr. Whewell.

'Edinburgh, *May 20th*, 1836.

'. . . One interesting point which seems to me to admit of development is this—the variation in the kind and quantity of proof required as demonstration by the human mind in different ages. That the educated part of mankind are more acute than they were a century ago seems probable. Paradoxes cannot be so successfully maintained now as they were then; and people are not now allowed to take up the querulous and futile objections which passed current 150 years ago, and which great men condescended to answer. Some parts of Newton's

Optics contain surfeit of demonstration which would not now be tolerated; and as the habit of estimating both mathematical and experimental evidence advances, the chances of error by abridged processes of reasoning are diminished, higher powers of mental abstraction are produced, and a degree of sagacity brought to bear upon all kinds of questions which perhaps more than compensates for what the sciences may have lost in the ambitious display of continuity of reasoning. . . .'

To MISS FORBES.

'EDINBURGH, *May* 21*st*, 1836.

'. . . The eclipse was admirably seen here, and seemed to strike every sort of person much more than they expected. I was making optical experiments in a dark room most of the time, but ran out for half a minute to see the ring, which was a wonderful sight. I sent you an account in the Advertiser. Dr. Chalmers preached, and I managed to hear him, too. Evening service was postponed in the churches and chapels, except Mr. Bagot's, and the smoking of glass and the burning of fingers and blacking of faces was wonderful. . . .'

His travels during the summer of 1836 were confined to his own country, followed by a short visit to England. In the same month, May, as that in which he watched the annular eclipse of the sun at Edinburgh, he explored the Lead Hills, returning by Moffat and St. Mary's Loch.

The following letter refers to this tour:—

To PROFESSOR JAMESON.

'*June* 1*st*, 1836.

'. . . On occasion of a late visit to the district of Lead Hills I suggested to my friend and former pupil, Mr. Irving, of Newton, the importance of determining the temperature of the springs in the bottom of the Lead Hill Mines at this particular epoch. The working having been discontinued since the end of March, any supposed influence of animal heat and light is avoided, and yet the pumping of the water has been regularly carried on. Mr. Irving immediately and zealously undertook the inquiry; and descended to the deepest part of the mine on the 16th of May and found the temperature of the water in the bottom to be 49°. This was at a depth of 95 fathoms below the entrance to the Susanna Vein. . . .'

In June he traversed on foot the greater part of the Highlands. Beginning with Braemar, he passed by Glen Feschie to Aviemore and Inverness, thence up the Beauly to Erchless, where he parted with his companion, the Rev. H. W. Sheppard. Alone, he seems to have turned back to Ben Mac Dhui and Loch Aun, whence he crossed Scotland by Loch Laggon to Glenfinnan, Knoydart, and Skye, ascended the Coolins, thence to Loch Maree, passed through Inverness, and on to Glencoe. The rest of the tour shall be given in his own words:—

To Miss Forbes.

'Dalmally, *July* 23*rd*, 1836.

'. . . I have been in full action for a very considerable time, and the repose of Saturday and Sunday last which I spent at Inverness was much trespassed upon by the misfortune of a tremendous wool market, which gathered all the North of Scotland. I came along the course of the Caledonian Canal, not by steam, for the most part, but walking, in order to visit the adjoining glens. I took Glen Moriston one day, Glengarry another, and Loch Arkaig a third. On the whole I was as much pleased as I expected. From Fort William I went to Ballachudilish, which is most charming indeed; and yesterday I devoted to a leisure survey of Glencoe, which extremely delighted me, and is really the best scenery on the whole I have seen in Scotland, and worthy of comparison with any I know. I walked over to-day from Kingshouse down Loch Etive to Bunawe. Ballachudilish is a glorious contrast to Glencoe, and I experienced much hospitality from the Stewart family, who are very primitive, patriarchal people. . . .'

In these tours he was not only gathering strength and health for further mental work, but rapidly as he passed along he was making geological and other observations with a rapidity of eye for which he was unsurpassed.

After visiting London in the dead month of August, and finding Oxford still more deserted, he passed on to Bristol to attend the meeting of the British Association there.

On his return to Edinburgh he thus writes at the end of September:—

To M. AUGUSTE DE LA RIVE.

'EDINBURGH, *September 26th*, 1836.

'. . . Ever since 1832 I have been either travelling or in a state of the most active mental occupation—the former happily providing the health and spirits necessary for the latter. . . . The subject of Mr. Cross[1] is, I confess, rather a disagreeable one to me. You will readily enough conceive how much people more conversant with geology than electricity must have been struck by hearing most eloquently expounded a series of experimental discoveries, for they were perfectly original to Mr. Cross, silently prosecuted for many years by a retired country gentleman in Somersetshire, and only elicited by chance in the course of discussion. From the first moment that the matter was mentioned to me, and on every succeeding occasion, I really believe not less than fifty times, I have patiently vindicated the claims of Becquerel, which only require to be mentioned to be acknowledged. I own I felt somewhat indignant on the subject, because, having seen Becquerel's magnificent preparations and conversed at great length with him on the subject, I had been led at various times, publicly and privately, for several years, to draw the attention of geologists to one of the very best things ever done for their science. . . . I have, however, great pleasure in assuring you that everything that could be done was eventually done to put matters straight. The Bristol meeting went off ex-

[1] Relating to the production of artificial crystals by electricity.

tremely well, and promises admirably for the permanency of the Association. £2,700 have been voted for objects of a truly scientific character during next year. . . . My discoveries in heat have been warmly contested in Paris, but, as you are aware, MM. Biot and Melloni have fairly given in. . . .'

Of Session 1836-37 there remains the following short notice, jotted down by Forbes at a late date :—'Had a very pleasant group of students—Batten, Cleghorn, J. Anderson, J. Rankine, Harrison—the pleasantest I ever had, much occupied with experiments on radiant heat.' Some of the students above named must have been attending his advanced class, or his regular class for the second time.

The few letters that he wrote during this session were, as usual, almost entirely confined to those scientific subjects that were filling his thoughts, either in his lecture-room or in his investigations.

To M. ARAGO.

'*January* 3*d*, 1837.

'I write to mention some results respecting terrestrial magnetism at which I have lately arrived. In 1832 I made an extensive series of experiments with Hansteen's Intensity Apparatus in the Alps, and in 1835 in the Pyrenees. One principal object was to ascertain the influence of heights. I doubt extremely whether any decided result can be drawn from preceding observations. . . . Those of M. Kuppfer seem to be of little value. They were not made at the summit of the Caucasus. . . . I have referred the positions of my stations in the Alps and Pyrenees to the three co-ordinates of latitude, longitude, and height, and deduced the influence of each by the method of . . . . I have in the first instance confined my calculation to horizontal intensities. From three different series of observations, made with two needles, I find always a negative coefficient of the height, indicating, at a mean, a diminution of ·001 of horizontal intensity for 3,000 feet of vertical ascent. If, as Humboldt states, the dip diminishes in ascending, the diminution of total intensity will be somewhat greater. You will judge of the extent of the inductions upon which this is founded when I mention that the sum of the heights to which I have carried Hansteen's apparatus exceeds 160,000 feet, or thirty vertical miles, twelve *lieues*.

'I have lately been engaged in procuring thermometers similar to those at the Observatory at Paris, to be sunk to different depths in various soils. I have three sets from three to twenty-six feet long; one set to be sunk in trap-tufa, a second in sandstone, a third in pure loose sand. . . The observations in the Lead Hills are being continued.'

To the REV. DR. WHEWELL.

'*Jan.* 31*st*, 1837.

'. . . I feel gratified by the prominent place you have given to my experiments as bearing upon the theory of Heat, in which you have done me full justice. . . . But I must mention for yourself, if not for your book, that the discovery of the polarization of heat was not the necessary consequence of applying the thermo-multiplier to the investigation, which would have been a poor achievement, seeing it was another man's invention; but that Melloni had first applied the instrument to the tourmaline question, and answered in the negative (*Ann. de Chimie*, vol. 55); then Nobili, the inventor, attempted to repeat Berard's experiment with the most improved piles, and with results quite null (*Bib. Universelle*). So that I conclude that, when I published my experiments, the question of polarization was negatively answered by persons operating with every advantage which I possessed, and indeed seemed to be set at rest. My discovery was the application of mica as a polarizing substance, first by transmission, then by reflection; and I have shown that repeating Nobili's experiment—the same as Berard's and Powell's—the quantity of heat reflected from glass is so excessively minute that the errors might well equal the total effect. I think you have not mentioned total reflection and circular polarization.

'As to simple reflection, Melloni should be mentioned alone, but I claim double refraction.'

The following extract from a letter to Professor Airy shows what his occupations were, and how closely his lectures and researches were combined:—

To G. B. AIRY, ESQ.

'THE DEAN HOUSE, *March* 15, 1837.

'I have been exceedingly busy, and not very well, which have been the causes of my silence. Amongst other occupations I have had to read five essays, which I have received in competition for a medal I proposed, on the Undulatory Theory of Light, a new subject in Scotland, which I am delighted to find has stirred up our youth, and I have got some really respectable compositions. This is a proof to me that things are mending, and that exertion, private and personal, is not thrown away, even where public sympathy or support is not to be looked for.

'I have not abandoned my polarized heat, but have been much driven about this winter. I have got twelve thermometers sunk in different soils from three to twenty-six feet deep, to measure conduction. Shall you be certainly at Greenwich the last days of April?'

The following letter, written about this time, proves that the intercourse with Sir David Brewster, for a time suspended, was now cordially renewed:—

'MY DEAR SIR DAVID, 'EDINBURGH, *April 28th*, 1837.

'. . . . Your experiments on absorption must be most interesting. I think Wrede, the first pages of whose paper Taylor has lately translated, has done something of the kind you allude to, if I understand it correctly. If I recollect well, he imitates the phenomena of absorption by combinations of thin mica plates, that is, by the colours of thin plates.

'I will do my best to capture a Wolf's lens for you, on condition that you will not require an affidavit that I saw the wolf make use of it. To stare a wolf in the face in the Black Forest would be enough to throw any optical philosopher into a fit of reflection.

'With best regards to Lady and Miss Brewster, believe me, my dear Sir David, yours most sincerely.'

Forbes was not a man who, while he belonged to a corporation, could confine all his energies to his own peculiar subjects, without regard to the working of other departments and to the general well-being and well-working of the whole body. He had not been long established in his chair before his action began to make itself felt, beyond the bounds of his own class, on all the machinery of the University. The Scottish Universities, like those of the sister kingdom, had suffered from a long torpor, which lingered somewhat later in our northern seats of learning, after it had passed from Oxford and Cambridge. No doubt, during the last century and the first three decades of this, there had arisen, here and there in the four northern Universities, men of real genius, lights of their times, not likely to be surpassed by those who now fill or may in future fill their places. Principal Robertson, Dr. Thomas Reid, Dugald Stewart, Sir John Leslie, Dr. John Hunter, Professor Wilson, and many more, were men of whom any university or country might be proud. They not only communicated life and energy to the students who heard them, but they enriched the literature, the philosophy, the science, and the scholarship of their country. But their action was wholly isolated and individual. It began and ended in their own classes, stimulating, or, it may be, only delighting their hearers, but not necessarily producing in their students any solid or certified attainments. Of the several functions of a university life, the professoriate was the only one which was really alive, and it had swallowed up all the rest. Graduation was as good as dead, a mere form, little valued, and seldom sought for. But the teaching of able professors, excellent as a mental stimulant, requires, if it is to produce solid fruits, to be supported by three additional conditions. It must be received into minds previously prepared by adequate school training; it must be supplemented and solidified by careful and methodic getting up of books during the college course; and lastly, the joint result of professors' lectures and private reading requires to be tested by thorough examination. In Scotland, when Forbes became a professor, these three necessary buttresses to the professors' lectures were wholly wanting. The professoriate reigned solitary and unsupported. Consequently, of the large amount of mental force annually let loose from those Scottish Chairs, who shall say how large a proportion lost itself in air?

In Oxford the system of systematic examination for degrees had been revived as early as the first decade of this century, when Sir Robert Peel was an undergraduate. Slowly the sense of the need of a like revival crept northwards; and by the fourth decade, when Forbes entered the Natural Philosophy Chair, he took it up and pressed it on his colleagues with characteristic energy and perseverance. For the following sketch of his exertions in this direction I am indebted to the kindness of Professor Kelland, who has for more than thirty years filled the Mathematical Chair in Edinburgh University, and done so much to promote thorough and accurate teaching, not only in his own class, but throughout Scotland. If the general reader finds in it some details which may not interest him, I must hope he will bear with them, in consideration of the value which these possess to the many who, from their interest in the further improvement of Scottish education, will welcome any light which may be thrown on the history of the steps by which it has reached its present condition.

After observing that in the early years of this century degrees in Arts had been so little in demand, that it was almost a favour conferred on the University when an able student proposed to graduate, Professor Kelland goes on to state that as early as 1814 the University of Edinburgh made an attempt to frame some very rudimentary rules for graduation, and again in 1824 there was an endeavour still more to shape these rules into a system. Both of these attempts, however, proved abortive, and a third made in 1831 had no better result.

'Up to the date of Forbes' appointment to the chair of Natural Philosophy, things continued in the old loose and unsatisfactory condition.

'It was reserved for Forbes to institute that complete working system of examining by means of printed papers, and of judging the results by marks, which is in force at the present time. To Forbes belongs the merit of having grouped the subjects of examination under three heads—the Classical, the Mathematical, and the Philosophical. The effect of this grouping is, though perhaps this result hardly came into Forbes' calculation, that a moderate amount of knowledge at any rate is exacted of one, at least, of two or three kindred subjects. The system was a well-devised and admirable one. There was, however, one error of detail, arising no doubt from the fact that seven separate interests were involved. The requirements were too high, and the amount of Greek demanded was more than an average student could bring up; and in other departments, Natural Philosophy among the rest, there was an indefiniteness of programme which must have operated to alarm the conscientious candidate.

'The first trial of the new scheme took place in April 1836, when six candidates presented themselves for examination, and all passed. Next year, 1837, Forbes was appointed Dean of the Faculty of Arts, as not only well qualified in all respects for discharging the duties of the office, but more particularly as having taken so prominent a part in maturing and establishing the new system. It was thought fitting that he should have the chief care of watching over the success of what might be called his own experiment.

'Immediately on his appointment he introduced some important modifications into the scheme for estimating a candidate's proficiency. These need not here be detailed. Spite, however, of the excellence of these regulations, candidates for the degree came forward but in small numbers. In 1837 there were seven candidates; in 1838, six; in 1839, one only. The Faculty agreed to recommend this gentleman for a degree without examination. In the two following years there were five and two candidates respectively.

'These numbers show that the first effect of the new system was rather preventive than encouraging. But however great the difficulties, Forbes did not lose heart. Various modifications were from time to time introduced into his scheme, chiefly by himself, such as the allowing the different branches to be taken in separate years, and awarding honours in the several departments. To Forbes belongs the credit of having devised and brought into working order this well-appointed scheme, and for this the University owes him a lasting debt of gratitude.

'It will be seen from the above that one great feature in his character was order or method. He was on system a thorough disciplinarian, as well in his own class as in University matters. He was orderly in the extreme. His class examinations were fixed year after year for corresponding days, and his colleagues were compelled to accommodate themselves to his unbending requirements. It must not, however, be inferred from this that there was anything harsh or unkindly in Forbes' dealings with his colleagues. On the contrary, he was thoroughly kindly, though somewhat cold in manner. His principle of action was 'straight forward'—a principle admirable in itself, but apt to carry its bearer rather sharply against an opponent, who, even if adopting the same principle, may have thought fit to travel to it by a different route. On every great question he had thoroughly made up his mind—had so carefully considered the subject and so completely mastered it, that he could not appreciate the determined opposition which his views were apt to meet with from some of his colleagues, who equally with himself had the interest of the University at heart. When two such unbending natures as those of Forbes and Sir William Hamilton came into contact, as they did more than once, the shock was a rough one, and the result not generally beneficial. A middle course, which would probably have been the right course, was, from the nature of the opponents, an impossibility. In looking back on these conflicts, each of the combatants referred to stands out as a man of immense power, of high principle, of honest purpose—with only an excess of one virtue so great as to make it cease to be a virtue—self-reliance. With their wide differences, each has left a permanent mark for good on the University.

'In 1841 came up the subject of the disposal of the magnificent legacy left by General Reid for the endowment of a chair of the Theory of Music, and for the library and general purposes of the University. In the controversies to which the disposal of this fund gave rise, and which for years kept the Senatus in the law-courts, Forbes took a prominent part. He honestly considered a chair of Music to be a mere ornament, and therefore resolutely opposed all votes of money for supporting the chair. He believed that there were far more urgent needs to be supplied than this. It was notorious that certain professors from age and long service had great difficulty in conducting their classes. Forbes thought that the Reid Fund could not be better employed than in providing retiring allowances for these aged professors, and supported this view with all his energy. Of his colleagues, some pleaded for the museums, some for the library, while only a very few stood by the Music Chair. Hence came the rupture with the holder of that chair, and ere long with the Patrons as the governing body of the University. It is needless to dwell on the litigations that ensued between the Senatus and the Town Council and the Professor of Music. The result was that, after years of contention, the provision for superannuated professors, and other most desirable objects which Forbes had at heart, were stopped, and the

Reid Fund was by the law-courts restored to the support of the Music Chair.

'In 1842 another large fund, the Straton Fund, came to be applied by the Senatus. Here again Forbes took the initiative, and was mainly instrumental in inducing the Senatus to establish Fellowships out of the combined Reid and Straton Funds, to be held by distinguished students, after graduation. Ultimately these Fellowships were founded out of the Reid Fund, but, owing to the decision of the law-courts, they had but a short period of existence.

'It only remains to be added, that whilst Forbes was a stern and unflinching opponent, he did not allow his opposition to interfere with his friendships. The colleague who most systematically and consistently opposed his views regarding the Music Chair, enjoyed his warmest friendship and had the kindliest intercourse with him up to his dying day.'

When a man of vigorous mind and resolute will throws himself into public action to carry out his views of what is right, it must needs be that collisions come: and if he meets with men of views as decided and wills as strong as his own, the collision is sure to be a severe one. Forbes never shrank from such collisions, however painful, if they met him in the way of what he conceived to be duty. But he was by nature no polemic. He did not love the battle for its own sake; indeed, it cost him more than most men to enter into personal conflicts. Whatever ambition there was in him found scope in wrestling with the difficulties of Nature and extorting her secrets from Science. Though he never flinched from opposing men when he thought he ought to do so, he felt very keenly the hard words and severe blows which such encounters call forth. Therefore, however manfully he may have stood to his guns in the Academic combats to which Professor Kelland alludes, it was in the intercourse of the class-room and of private life with like-minded students that he found the field most congenial to him.

Professor Kelland alludes to the encounters which sometimes took place in the Senate when Forbes and Hamilton took opposite sides. Physicists and metaphysicians do not generally think very highly of each other's pursuits. This natural want of sympathy would not be diminished by the disparagement and scorn with which Hamilton spoke and wrote of some of the processes most in favour with mathematicians. But it is needless to revive these buried controversies. The mention, however, of these two colleagues side by side cannot but recall the striking but contrasted appearance of the two men as during these years they passed along the Bridges to and from College. Forbes, with tall, thin, but lathy frame, in the invariable suit of black, and broad white neck-tie, his head stretched forward, his long arm swinging resolutely by his side, strode rapidly along like one bent on some determined purpose from which no man could turn him. An hour or two later, coming the same way, might be seen Sir W. Hamilton; under one arm his lecture portfolio, the other hand thrust deep into his pocket, hat pushed well back, and exposing a noble breadth of forehead, prone in meditation, from beneath which looked out those large brown eyes so loaded with intellect that the youth on whom they were turned almost shrank beneath them, oppressive as they were with their weight of thought. A third there was, an elder colleague of these two, tall, and still erect, broad-breasted, with an herculean frame, now become massive, almost portly, moving no longer with the restless step of youth, but with the firm, measured tread of ripe middle age. In that well-known countenance, so vivid, yet so benign; the eagle eye, with its fire mellowed, but not abated; the long golden hair, just faintly silvered, streaming back over his shoulders, as if the wind and no hand of art had shed it there, all who turned to look on him—and who did not?—saw for once before them the embodied ideal of an inspired poet. Such was the look these three colleagues wore to the eyes of their admiring students, all of whom could appreciate their striking outward appearance, little as they might sympathise with their inner man. It is not often that one University at the same time possesses three such representatives.

Before Forbes had completed his third professorial session, his eminence as a lecturer and his growing reputation as a discoverer had so awakened the enthusiasm of his students, that they publicly expressed their sympathy with him in his researches, and their pride in his achievements, by presenting to him the following address:—

'UNIVERSITY OF EDINBURGH, *February* 16*th*, 1836.

'The students in the Natural Philosophy Class in this University beg to convey their congratulations to Professor Forbes on the well-merited honour which he received last night as the successful demonstrator of the polarization of heat. The students would express their attachment to Professor Forbes as to an excellent and amiable teacher, their personal interest in his scientific pursuits, and their personal pride in the successful results of these. That he has not only succeeded in displaying the polarization and depolarization of heat, but that he has also detected signs of its circular polarization, strongly excite their expectation that the evidences of interference will not long remain undetected by the delicate and judicious experimentalizing for which Professor Forbes is so justly celebrated. They hope to see the undulatory theory of heat as fully demonstrated as the undulatory theory of light, and Professor Forbes' name raised to the highest eminence among the philosophers of the day.'

However cordial may be a professor's relations with his whole class, there will always be some with whom, either from circumstances or from congeniality of pursuits, the intercourse of the class-room will lead on to closer friendship. Of such intimate student friends Forbes had a goodly number; and I can remember the peculiar tenderness with which, either in public addresses or in private conversation, he alluded to these in later life.

A few of his letters about this time will illustrate the nature of his intercourse with these young friends:—

To THOMAS CLEGHORN, ESQ.

'THE DEAN HOUSE, *Oct.* 22*nd,* 1836.

'. . . When I had the pleasure of seeing you here in May last, you suggested the possibility of establishing a society of an academical character for the special object of encouraging a taste for physical science. You may perhaps not have thought much on the subject since, but I assure you that the more that I reflect upon the chances of success of such an undertaking, and the importance of which it might ultimately prove, the more I feel disposed to encourage any well-devised scheme for attaining the object in view. I have often and maturely thought of it since, and I write to say that if you feel disposed to prosecute your own suggestion, you may reckon upon my cordial co-operation.

'So far as I recollect, the desideratum which you and some of your companions have felt was of a purely academical character. It had no similarity to those societies of which one or two already exist, and whose object is to imitate exactly the great scientific associations of the country. Your object seemed to be to have a society which should encourage the taste and give facilities for physical studies by the union of kindred minds, rather than to pretend to the more arduous task of extending the boundaries of human knowledge. . . .

'I have thought of no means so fit as your suggestion of a society which might act energetically in two ways: (1), by inducing a careful study of experimental essays—Newton's Optics, for example—in order to take a share in a discussion which might be raised upon any disputable point in experimental investigation, of which there are thousands; and (2), by providing from the funds of the society instruments of a simpler kind, but which are yet too costly for the easy purchase of an individual; for instance, a barometer and an electrical machine. Also, the union of several individuals may accomplish what one might want zeal or leisure to perform. For example, Sir John Herschel has desired the co-operation of observers to note the barometer, thermometer, &c., for thirty-six successive hours occasionally. My own occupations prevent me from undertaking this labour, and I believe that it is not done in any part of Scotland,—when I was your age, or younger, I used frequently to make hourly observations for twenty-four successive hours without assistance.—but I would willingly furnish a committee of such a society with the necessary instruments, and assist in the use of them. . . .

'As the suggestion was your own, so I wish the execution to be; but you may make me of use in any subsidiary manner that you please, and in conversation or by writing I will endeavour to further your objects.'

To J. T. HARRISON, ESQ.

'EDINBURGH, *September* 17*th*, 1838.

'. . . There has been a considerable break-up, of course, amongst your associates in the Nat. Phil. Class. Still, however, I have kept my eye pretty well upon those with whom you were more particularly associated, and the Physico-Mathematical Society prospered last winter remarkably well. Batten is now at home in Somersetshire, but comes to London in winter to study law, having entered at the Middle Temple. I hear from him constantly, and I hope that, if you ever reside in London now, you will find him out. Cleghorn and Anderson are both occupied, one with law, the other with medicine; they have, however, by no means forgotten their physical studies, and the hourly meteorological obs. go on chiefly under their auspices. Geo. Irving has reached the dignity of an advocate's gown. John Rankine is more zealous than ever, and has completed an admirable set of anemometer observations. Lewis Gordon is now studying at Freiberg, in Saxony: I heard from him not long ago.

'I shall be glad to hear, though I scarcely expect it,

that you have not in the midst of your professional pursuits entirely lost sight of the general scientific principles which form its surest foundation. I do not doubt your good-will or the clearness of your views of what befits a liberal and enlightened prosecution of your profession. That I am sure you will never do; but I rather fear that the very success to which your talents and application so well entitle you, may have already forced you to travel upon the narrow railroad of every-day applications. . . .'

To the Same.

'*December 6th*, 1838.

'. . . At the very time you were writing I was on a visit to the North of England, by way of a little stretch, before buckling to my winter's work. I had the gratification of making the acquaintance of your family and of accepting Mr. Harrison's most cordial invitation to spend a day with him. I visited your brother's most splendid viaduct across the Wear—Victoria Bridge—and was much gratified altogether by my excursion. Indeed, Newcastle and its vicinity affords a great field for persons who have any taste for the arts whatever. I was much pleased too with the Monkwearmouth Bridge; and I visited Dr. Cowan's school, where your younger brother, and I believe you also, were educated. I cannot sufficiently wonder that I took no pains to find you out when I was last in London. I took it dreamily for granted that you could not still be there, though it seems you were. I am interested by the honest account you give of your fortunes in London. It is a place of all others to find one's level; and though, from what I heard from your father, I fancy you state the case rather too strongly against yourself, I doubt not that even in these very reverses you will find the germs of future success. I am sure that the honesty of your character and the strength of your principles must be the foundations of high professional character, and that as you seem to be satisfied that Mr. Brunel means to act justly and favourably towards you, your conscientious perseverance in what you believe to be right will ultimately be rewarded. I am not surprised at what you say as to the effects which the non-observance of Sunday have had on you. An engineer is placed perhaps in as great temptations of this kind as any man, and it must be conceded that a mere engineer is not brought in contact with subjects calculated to mould his moral character and prevent his feeling from being hardened by the trying ordeal of such a place as London, of which selfishness and mechanical utility are the predominant springs and ends of action. . . . Believe me that it will ever and at all times give me pleasure to hear from you and of your success in life. I should like to know whether you expect to leave London, and whether Mr. Brunel is doing all for you that you expect.

'Never mind filling a sheet, but write me a few lines when you can, as I do.'

To THOMAS CLEGHORN, ESQ.

'THE DEAN HOUSE, *September* 23*rd*, 1839.

'I have just received your very kind letter of the 15th June, and the enclosure which it accompanied. I state the bare truth when I say that few things have ever gratified me more than this mark of your affectionate regard. I find myself reaching that time of life when mere compliment and external demonstrations of the good opinion of one's fellow-men lose much of their first attractions, and I continually attach more and more weight to the personal attachments of those whose moral and intellectual qualities I can admire, and whose amiable dispositions I can love. The terms of your dedication, and of your letter, alike prove to me that you have justly apprehended my feelings on this subject, and are aware that in truth I attach more importance to the friendships which my position as a teacher and cultivator of science may enable me to make, than to any celebrity which I may acquire in that department. As a worldly man, I am gratified by your having dedicated your 'Thesis' to me, rather than to many other persons whose countenance might have been professionally useful to you. As a man of feeling I am infinitely more rejoiced at the expressions of personal attachment which it contains.

'Believe me, my dear Cleghorn, ever most sincerely your friend,

'JAMES D. FORBES.'

To H. G. CUMMING, ESQ.

'EDINBURGH, *November 22nd*, 1840.

'I was very much gratified by your letter of the 27th October, and now that a month has nearly elapsed I should like much to hear more particularly what you are doing. I have not very precise notions of what a broker's business is, and I will thank you to enlighten me. I am given to understand, however, that it does not very materially differ from a merchant's as far as learning the profession goes, and therefore I hope that even if you should be unable to make up your mind entirely to like it you will yet be able to bear with it. There are two very good grounds on which you may be encouraged to do so: first, that things the dullest and most repulsive, steadily pursued, gradually and insensibly, and in spite of one's self, become interesting in a certain way, chiefly from the satisfaction which always attaches to a sense of steady effort to do right; and secondly, from the consideration of alternatives, namely, that in this world a majority of people are compelled to find happiness as they best may, in doing what they cannot altogether help, and in choosing what appears in prospect the lesser of the two evils, though it may ultimately turn out a very real good; whilst on the other hand the few who really get their own way and have all externals and full luxury of choice for procuring happiness, may much oftener fail in doing so than those for whom some inevitable destiny or strong motive has chalked out their course of life. I am not writing from theory, but from very real observation, which I could substantiate by instances of A, B, and C. . . .

'I wish I could send you any news worth having; but you know my winter habits are not such as to furnish me with much chit-chat.

'Young Mackintosh, of Geddes, is attending my lectures. He brought me a letter from Albyne. I like him extremely. He tells me that he is acquainted with you, and he had heard somehow or other that you did not like your present situation, and, like me, he was sorry for it. We had a very pleasant walk together yesterday.'

The summer of 1837, from May to October, was spent in an extended tour through North Germany and Austria, and in the summer of 1838 he returned to the same country for a less extensive tour and for a shorter time.

The summer of May 1839, from May till August, was spent in the South of France and more or less among the Alps. In August he was recalled by the meeting of the British Association, which took place at Birmingham. These foreign summers fall to be noticed not in this, but in another chapter.

The winter of 1837-8 was busy with experiments on radiant heat, till these as well as his lectures were interrupted for a time by an attack of scarlet fever, in March 1838. These experiments were, however, again recommenced, and continued to engage him during the winters of 1838-39 and of 1839-40. At the close of the latter winter session they were dropped for a considerable length of time, and though afterwards taken up, never again received a long and undivided attention.

The following letters belong to this period, and bear on this subject:—

To M. Cauchy.

'Edinburgh, *December* 11*th*, 1839.

'. . . I receive with the greatest interest your suggestions on the subject of radiant heat, and your promise to pursue the theory. In my third paper I have proved to demonstration, I think, that all kinds of radiant heat are not equally polarized—contrary to the opinion of Mr. Melloni—and consequently it is possible, as I have suggested at page 9 of that paper, that there are normal vibrations combined with transverse, and more abundantly, as the temperature of the source is lower. . . . I see no difficulty in constructing graphically the ellipse in the case of heat. Would not this be the most satisfactory way? Can you suggest any mode of determining the length of the wave in the case of heat? In depolarization we only get the ratio of $\frac{o-e}{D}$. I have sometimes thought of applying the ellipticity of the ray in the case of Fresnel's rhomb, when the angles do not give circular polarization, to this purpose. . . .'

To PROF. PHILLIPS, *York.*

'EDINBURGH, *December* 11*th*, 1839.

'. . . I believe in my conscience that I am better employed now than in writing reports. I have fields, not of untrodden snow, but of untrodden heat, before me, all promising a rich harvest. Cauchy has just set me a task to which I must buckle myself. . . .'

To PROFESSOR POWELL.

'EDINBURGH, *December* 27*th*, 1839.

'. . . I own that I should be very glad to see a good synopsis of what has been done during the last eight years, well aware that, so far as my own contributions are concerned, they were published in a form repulsive to the generality of readers, and even to many men of science who would willingly glean from your critical pages how much has been established, and on what kind of evidence, without troubling themselves about details. . . .

'I hope that you will not treat Melloni's papers so slightly as you propose. His earlier ones are full of interesting and original experiments. So far as I am concerned, I have nothing to wish but that everything Melloni has done should be fully and accurately known, for by a rare good fortune there is not a single debate in point of priority between us, and every experiment we have in common, with a single exception—the variable polarizability—confirms each other's results. It is for you as a critic to judge how far, if in any case, my right conclusions were founded on false grounds, for this is the sum of Melloni's captious criticisms.'

To the ASTRONOMER ROYAL.

'EDINBURGH, *January* 11*th*, 1840.

'Enclosed is a memorandum about some experiments on heat, which are new and unfinished, and I hope you will think important. I cannot help thinking this affection of heat by mechanical surfaces and textures one of the most singular and important yet noticed. For instance, is there any way in which we can trace the action of striæ on transparent surfaces upon heat or light different from physical interruptions or spare spaces like fine wires? Has anyone thought of proposing this question in the case of light, viz.: "Is the illumination of a screen by parallel rays of light passing through a grating, always determinable by the area of the interstices compared to the whole area of the grating?" Can you refer me to any investigation of this? Is there any *primâ facie* absurdity in supposing it should not be so? If striæ or scratches act really as opaque lines would do, my experiments lead me to think that in the case of heat the interrupting power of the screen depends on λ. But I have not succeeded in getting any wire gauze fine enough to test it in that case.'

To PROFESSOR WHEWELL.

'EDINBURGH, *February* 8*th*, 1840.

'. . . I have lately been making a few preliminary experiments on the form of the elliptic vibrations of heat, in verification of some formulæ of Cauchy, which he wrote to me to endeavour to test, and I find no difficulty in doing so. The results seem to come out well, and by a graphical process I can readily project the ellipse, find the direction of the greater axis, the excentricity, &c. You will believe that I truly rejoiced in my friend Ellis's success, as well as in the honour of Trinity. I fear I shall scarcely know Cambridge soon, if my old friends take wing as they have lately been doing. Gregory gives me the hope that if I go to Cambridge in May, which I have some thoughts of doing, I shall find you lecturing on Moral Philosophy.'

To the Same.

'EDINBURGH, *February 22nd*, 1840.

'You cannot gratify me more than by leading me to imagine, what I wish I could persuade myself of, that I can ever render you any service worth giving. . . . I look forward with much interest to the appearance of your book, and on several accounts. Some views you once stated to me as we walked down the north bank of the Cam, some two years ago, have been sticking by me since, and I expect to see them developed. Will it be one volume? If so, it must be nearly complete.

'I have not quite, but nearly done tormenting heat with gratings and dusty diaphragms. All I can do, all kinds of heat get through the finest gratings (metallic) in precisely equal proportions, and that equal to the area of interstices; and yet grooved surfaces exert a powerful specific action. I am reluctantly forced to the conclusion too, that pure metals may be reduced to powder so fine as to affect the quality of the heat transmitted, though the thinnest gold leaf permits no appreciable portion of heat to pass.'

To SIR J. F. W. HERSCHEL, BART.

'THE DEAN HOUSE, EDINBURGH, *March 9th*, 1840.

'. . . I should like very much to know the degree of sensibility of your paper to heat; whether, for instance, the heat of the hand affects it. It has long been an object with me to get a surface capable of detecting heat pictures, such as those which polarization and diffraction would indicate. Mr. Talbot gave me hopes at Birmingham of providing me with such, and he actually sent me a primrose-coloured paper which was transiently affected by a pretty violent heat; but it was far too insensible to be of the slightest use. It was, I think, iodide of silver. Is the impression on yours permanent, and would you allow me to try a small piece of it? . . . Have you ever applied photography to the beautiful figures of polarization and diffraction? The only experiment I ever made on the subject—which I have purposely avoided spending my time upon, seeing that it is in much better hands—was to fix the splendid image of calc spar rings and brushes by solar light, and to my surprise I wholly failed. Perhaps the calc spars had absorbed the chemical rays; but at all events my experiment, being but once made and by inexperienced hands, was not worth much.'

But much as his experiments and discoveries filled his time and attention, they still left room for graver thoughts.

The following extract from his journal was written just before he entered on the work of the session 1839–40:—

'THE DEAN HOUSE, *Oct. 20th*, 1839.

'Five years and a half have elapsed since the last pages were written! What a chasm in anticipation! What a span in retrospect! I thank Almighty God that if any one change more than another has characterized my mind, it is an approximation to the fulfilment of the wish expressed in the last sentence. I yearly feel a greater readiness to die, and become more familiarized with the idea. I wish that I could impute this to an increasing fitness for such a change; it arises, I am perfectly aware, fully more from the experience of satiety even in the more intellectual earthly gratifications. . . . Yet my grounds of calm anticipation of the future have undoubtedly in some respects greatly improved in these five years. I have a far more settled and reasonable dependence on the atonement of Jesus Christ, a doctrine of which God's grace has within these few years revealed to me the full importance, as a ground of hope and guide of action. I am now about to celebrate the tenth anniversary of my father's death: since that event I have advanced one vast irretrievable step towards the grave. I have great doubts whether I shall ever form earthly attachments closer than those I at present possess.'

It may be worth while here to note a piece of practical counsel as to the observing of the Sunday rest which he offered to his students in the lecture with which he opened the session of 1839-40:—

'By earnestness in your studies during the week, I advise you to reap the enjoyment of that beneficent provision of the Almighty, and by a sedulous abstinence in thought, as well as in act, from your ordinary occupations, to restore the tone of your minds and the capacity for vigorous exertion. None who have not made a strong effort are aware of the admirably tranquillizing influence of twenty-four hours studiously separated from the ordinary current of thought. Monday morning is the epoch of a periodic renovation.'

Sir Andrew Agnew, who at that time was prominent as a defender of the religious observance of Sunday, was so much pleased with these words that he applied to Professor Forbes for leave to publish them. Forbes replied that as they had been publicly spoken they were public property, but that if they were printed he should prefer their being given as a report of part of his lecture, rather than as a communication made directly by himself. For this would argue a love of notoriety from which he rather shrank.

Just before the close of the session, in April 1840, his eldest sister, Eliza, died at Dean House, and left him with only one home companion, his sister Jane.

The summer of 1840 was spent at home, and then for the first time we meet with allusions to his own health. In the retrospective journal written in 1850 he writes that he was very ill all the summer of 1840, and we hear of his consulting Dr. Chambers when he visited London and Cambridge in the June of that year.

In August he made with his sister Jane a tour through

Dumfriesshire and Kirkcudbright, of which there remains a diary written in a lighter and more frolicsome vein than was usual with him. Having no glaciers or other great object to study, he amused himself with noting the peculiarities or follies of the natives he met.

On going to visit the mausoleum of Burns in the largest kirkyard of Dumfries, he notes:—

'If there is anything characteristic in Dumfries, it is the sepulchral magnificence with which the churchyard abounds. Scarce a tailor can die without leaving his measure for a stately monument. The only gradation of rank acknowledged in the cemetery is the geological distinction of the primitive granite, which rises over the grave of the border chief, and the modern spongy red sandstone, which not less strikingly and more gaudily covers a clockmaker, Mr. ——, who was one of the chief magistrates in the year 18—.'

Again, in journeying up Nithsdale, when they slept at Dunscore, he notes: 'We had tea, but mine host had partaken of something stronger, for he slapped me on the back, and was sure that if we came from Edinburgh I must be travelling in the grocery line, and be perfectly acquainted with Mr. Brown, the Leith tobacconist. . . . . After spending his best care on us and on our steed, whose mouth he affectionately wiped with his pocket handkerchief, he reluctantly let us go.'

In Nithsdale he amused himself besides with watching the meeting of the graywacke and red sandstone.

The journal throughout is written in a more jocular style than was usual with him, here and there forsaking prose narrative for comical verse, like the following:—

'The banks of the Nith are both fertile and green,
And haymakers merry on all sides were seen.
More beautiful cottages nowhere one can see
Than those which the Duke has built for his fancy,
And lodged all his people in excellent style;
So, passing the Castle, we drove for a mile,
To call upon good Mr. Gamekeeper Shaw,
Who at last had succeeded in finding the law

Of the growth of the salmon, the par, and the smolt,
And showed how each nat'ralist has been a dolt.
This gained him this year the Keith Medal and prize,
Whilst Knox, Wilson, and Jamieson opened their eyes
At finding the gamekeeper's salmon so far
Advanced to a premium, and theirs below *par!*'

As they traversed the Galloway coast to the west of Nith, and explored the Piper's Cave and Needle's Eye, while thinking of Dirk Hatteraick and Meg Merrilies, he not the less had an eye on the syenite or felspar rocks interfused with the stratified slate.

In ascending Criffel he found first Little Fell: somewhat further on the Great Fell, and then Criffel above both. The undoubtedly authentic origin of these names as handed down by tradition he thus records:—'The devil, having collected the scrapings of the earth, came to lay them down in Galloway. The first pickle made the Great Fell, the next the Little Fell. Then he couped the basket over and left it, to form the Creel Fell, or Criffel!'

Further west in Galloway he met a geologist with redundant locks, beard, and moustache—a thing then very unusual. On him he made the following epigram:—

'Medusa's head of old turned all to stone,
Her snake-encircled eyes did light upon.
But times are changed: now C——'s keen glance
Would stare the mountains out of countenance;
The rocks resent the unaccustomed stare,
And Gorgonize his golden head of hair.'

The only other record of this summer which remains is that contained in the retrospective journal so often already alluded to.

'*September.*—Short visit to Arran—alone. British Association at Glasgow. Presided at Physical Section. Thence to Pitsligo, and along Banff coast to Dumphail and Altyre. Met Agassiz and Buckland there. Agreed to visit the glaciers with him in 1841. In October went to London to consult Dr. Chambers.'

In November he was elected Secretary of the Royal Society of Edinburgh, a post which he continued to fill with characteristic energy and business-like exactness, till the failure of his health obliged him to resign it.

The following letter of this date will interest many, were it only for the sake of him to whom it is addressed:—

To R. LESLIE ELLIS, ESQ.

'EDINBURGH, *December* 20*th*, 1840.

'. . . I rejoice in no slight degree that you propose to yourself so manful a course of reading in mixed physics; and allow me first to say, that I hope that in doing so you will continually bear in view that your tastes, talents, and position alike give us reason to hope that you are to be an extender as well as an occupier of the domains of science. In selecting amongst the subjects to which you have referred, and your enumeration of which almost supersedes my offering advice, I would suggest that you should pursue that department to which your taste most naturally leads you, and which you think you would choose to make your own by substantive additions to our knowledge.

'Your keen interest in physical reasoning and your clear notions about experimental evidence, happily preceded as a spontaneous act of your mind the acquisition of the admirable analytical skill which your brilliant Cambridge course has rendered available to you; so that it is speaking very much within bounds to say that very few men are so well placed as you for entering on the mixed physical and mathematical investigations which so rarely are successfully cultivated together. So far as I can form a judgment, amongst the subjects you mention, the subject of electricity is less adapted for your study than the others. Magnetism is very good. But though much is to be done by following out Gauss' method both experimental and practical, the experiments are not such as you would think of undertaking, nor such as in the hands of any one man can lead to a great discovery. The allied subjects of light and heat I consider the most hopeful and the most brilliant for a person furnished as you are with all the necessary weapons,—one whose clearness of mathematical apprehension may disembarrass them of some of the cumbrous learning and merely plausible speculations by which they have been overlaid, and whose love of experimental laws may enable him to trace analogies which will ultimately point to the identification of classes of phenomena still distinct.

'You are, I apprehend, already well read in the undulatory theory of light, though perhaps the original memoirs of Fresnel and Airy are not all known to you, and well deserve being read, as the kind of investigation which they point to and recommend is, I think, far more satisfactory than the endless memoir of Cauchy and of some others, even in this country, whom I could mention. That there are real outstanding difficulties to the theory of the transverse vibrations, none, I apprehend, who understand the present state both of the experimental and mathematical evidence, can deny.

'From this subject, which you should prosecute with pieces of mica and Iceland spar and Nicol's prism in your hand, you should turn to heat. Fourier you have probably read already. I doubt whether Poisson is worth the labour, even to you, who would read it comparatively easily. The researches of Melloni and my own on radiant heat, of which you will find an abstract in Powell's forthcoming Report, contain no doubt, as far as they are correct, the elements of a just theory of heat, parallel with that of light; and here a great deal is to be done and will soon be done; and I should be much more glad that you did it than that it should be imported from abroad. I assure you it is with great diffidence that I presume to offer anything like advice to one so competent to judge for himself and with so many able advisers besides. Should you think well of the suggestions I have made, I will endeavour, so far as I can, to aid in carrying them into effect.

'With kind regards to Mr. and Mrs. Ellis, believe me, my dear friend, most sincerely yours.'

The letter of Forbes to Dr. Whewell on his appointment to the Mastership of Trinity has a certain historical as well as personal interest:—

'EDINBURGH, *October* 31*st*, 1841.

'I do not know whether your appointment to the Mastership of Trinity has actually taken place; if so, excuse the deficiency in my address. I hope you will have little difficulty in believing that amongst the many who rejoice on this occasion, not one can do so with more disinterested sincerity than I do.

'Having learned during the ten years of friendship with which, notwithstanding our disparity in years, you have honoured me, to value and admire your personal as well as your public character, I cannot but rejoice that you should fill a station for which the previous course of your life and studies so eminently fits you. The great influence connected with that station will give additional currency to the views of education and other subjects which I am happy to think I hold in common with you, and I certainly augur the happiest results from the permanent connection thus formed between you and the University. Beside all this, I feel the liveliest pleasure in any circumstance which, like this, bears public testimony to your well and hardly earned reputation, and which must place you in a position as really desirable as any I suppose which this world affords.'

During the session of 1840–41, among the students who attended his class was one whose beautiful and engaging character soon attracted Professor Forbes. John Mackintosh had just completed his course at Glasgow College; and, having to spend in Edinburgh a year before he passed to Cambridge, he attended the Natural Philosophy Class. The Professor and the student, if in some things unlike, resembled each other in the rare elevation and purity of their natures. The intimacy which arose between them quickly ripened into a deep and affectionate friendship. At the close of the session, in May 1841, they made a short tour together in Arran, then together travelled to London, and thence to the Continent. In June and the early part of July they were travelling companions among the Pyrenees. After traversing these, they parted at Grenoble—John Mackintosh to return to Scotland, Forbes to meet Agassiz on the Aar glacier and to ascend the Jungfrau. For the next ten years they two kept up by letter an interchange of true and tender affection, to be closed only by the last letters which shall be given at the proper time.

Returned from this, to him, eventful Alpine summer, Forbes employed what remained of the autumn before his winter work began in throwing his observations and reflections on what he had seen, into the shape of an article on Glaciers for the *Edinburgh Review*. This article appeared in April 1842, and was well received, and almost at once translated into French. He was at the same time invited by Murray and Lockhart to write for the *Quarterly* on the same or on kindred subjects; but this he appears not to have done, though, in his general views of things unscientific, his sympathies went more with the *Quarterly* than with the *Edinburgh*.

In May 1841 he and his surviving sister left their old home at Dean House, in which they had lived since 1835, as it was to be pulled down to make way for some large changes in that neighbourhood. They took up their abode for a time in a house in Ainslie Place, in which, however, they passed only two winters.

The winter session of 1841–42 was, if possible, a more than usually busy one. Besides his class work, his spare hours were given to the preparation of a paper on astronomical refraction, which was to be read before the Royal Society of London. He also prepared and delivered in his class-room some lectures on Glaciers, which, being open to the public, were numerously attended. He was, in addition to all this, preparing for a new and still more vigorous campaign among the Alps next summer.

As soon as the close of the session 1841–42 allowed,

he set off for London, and after a few weeks spent there he proceeded to Paris. Just about the time of his arrival there he was elected a corresponding member of the Institute of France,—an honour which is regarded, I believe, by all European men of science, as one of the highest which they can receive.

Of the eventful summer of 1842, as of all his other foreign summers, the full record will be found in another chapter. I shall only here give the entry in his journal when looking back to it in 1860, at a time when he knew that for him Alpine adventure was over: 'This was the most active and, except the summer of 1833, when writing my lectures, the busiest summer I ever spent. It was also, I believe, the happiest. It still thrills me with delight to look back to it.'

When this energetic summer was ended, he left Chamouni on the 29th September, but did not return to Edinburgh till October, a short time before the session opened. All the next winter, 1842–43, he was intensely occupied in writing his book of Alpine travels and in reducing his observations. Murray, the well-known publisher, had undertaken to bring out the book, but some misunderstanding arose between the author and the publisher. On New Year's Day 1843, we find him in London conferring with Murray and trying to settle differences that had arisen. But this attempt did not succeed; for in the end the book was published, not by Murray, but by the Edinburgh firm of Messrs. Black and Co.

On returning from London, all the spare hours of the following spring were devoted to the book. It is curious to remember now, that this was the most agitated winter which Scotland had seen for many a year. Edinburgh was absorbed with the turmoil and excitement which preceded the disruption of the National Church. Undisturbed by the din of ecclesiastical collisions, Forbes went silently on with the work he had set himself. By the close of the session the book was nearly completed; and in the end of April he refreshed himself from his winter's labours by a visit to Makerstoun and by a journey thence on foot to Melrose and Peebles.

On May 18th we find this entry in his diary: 'Witnessed disruption of the Church of Scotland at Assembly in St. Andrew's Church. Same day asked Alicia Wauchope to be my wife.'

On July 4th his marriage took place with the lady here mentioned—the eldest daughter of the late George Wauchope, Esq.

The last lines of his Travels in the Alps had been written on the immediate eve of his marriage. The dedication to his travelling companion, M. Bernard Studer, is dated 1st July, 1843, and his marriage took place on the 4th. Having therefore completed his work and committed it to the publishers, the critics, and the world, he set off with Mrs. Forbes on his marriage tour to the Continent.

The following letter, addressed to Mrs. Forbes, from the Astronomer Royal and President of the Royal Society, which, though bearing a recent date, is mainly a retrospect of intercourse with Forbes during these early years, will form a fitting close to this chapter:—

'ROYAL OBSERVATORY, GREENWICH, LONDON,
April 11*th*, 1872.

'DEAR MADAM,

'When I wrote last to you, I hoped to have a little time at command for bringing together some notes regarding my late friend, as soon as the short refreshment which I required should be terminated. But I have never had my thoughts free, and now I fear that the very few ideas that I can put together may be too late.

'I cannot certainly say when or how my acquaintance with Mr. Forbes began. I have some obscure idea of having seen him before my marriage, which took place in March 1830. But I certainly saw him at the meeting of the British Association in Oxford in 1832, on terms which implied established acquaintance. In the latter part of that year he was candidate for a professorship—I believe the same which he subsequently obtained—and I had the pleasure of offering to him my certificate of competency. At the beginning of 1834 he was candidate for a different office; and I think that I then attempted to take a step in his favour as before.

'In the years from 1830 to 1835 I was much employed on physical optics, in its two divisions of diffraction and polarization.

'Professor Forbes took great interest in these subjects, and I had much correspondence with him. I had arranged for my lectures at Cambridge a rather rude machine for showing the relation between plane-polarized light and circularly or elliptically-polarized light, and Professor Forbes entered heartily into this matter, and requested me to furnish him with a duplicate of my machine.

'The undulatory theory of light was then struggling into existence, or at least into reception; and the clear understanding and hearty support of it by Professor Forbes must have contributed materially to its successful establishment in the locality with which he was more immediately connected.

'In 1835 the question was raised by me of the establishment at this Observatory of a system of continuous magnetic observations. I had from the first the advantage of frequent correspondence with Professor Forbes on the various points entering into consideration.

'In 1836 Professor Forbes communicated to me, as one of the first persons, I believe, his splendid discovery of the polarization of radiant heat.

'From this time almost every important step made by myself, or in this Observatory, was made known to Professor Forbes, and almost every scientific enterprise undertaken by him, and his general proceedings at Edinburgh, were soon communicated to me. I forget in what year it was that I testified officially to the importance of the magnetic determinations which he had made with so much labour at different elevations above the sea.

'The record of his experiments on the conduction of heat through metals was lodged, at his request, in this Observatory.

'An early copy, I believe the first, of Professor Forbes's Travels in the Alps was presented by him to me; and I read with great interest the observations on glaciers, especially the part beginning with the chapter which bears the appropriate motto,

> "The glacier's cold and restless mass
> Moves onward, day by day."

'And I have often contrasted, in my mind, the well-directed and careful observations there detailed, and the cautious deductions from them, with the paltry and doubtful objections that have been made to subordinate points of the theory. As a whole, this essay must hold a very high place.

'No person, perhaps, could testify better than myself to Professor Forbes's scientific character. In every investigation of his which I saw, he was careful, accurate, and truthful. He would not change an idea or a term for any consideration but his own conviction.

'This uprightness and manliness of character was impressed on all his transactions, and in time insured to him the respect of all who had sufficient opportunity of witnessing it. It was matter of great pleasure to Mrs. Airy, as well as to myself, to enjoy the hospitality of your house, or to offer our own, to take part with him in an expedition for sight of a solar eclipse (1842), or in an excursion to our northernmost islands, or to put ourselves under his guidance in Scotland or in Savoy.

'In the systematic pursuit of science, and in preparations for accurate geography and topography, Professor Forbes prepared himself with powers which sometimes almost assumed the character of the lighter accomplishments. His mapping of the Mer de Glace of Chamouni is an excellent specimen of surveyor's work. In contemplation of an excursion, I found that he was taking regular lessons in drawing, and the fruit of these appeared in those most admirable depictions of Alpine scenery which adorn his book of Travels in the Alps. I suppose that it may be asserted that the present popularity of

Zermatt, a place which was before scarcely known, is almost entirely due to Professor Forbes's picture of the Matterhorn. The power thus obtained was of course used also for other purposes, and I know of at least one excellent sketch of scenery and antiquarian remains which still adorns a lady's portfolio.

'It is needless for me to say after this how much the loss of my valued friend was felt by Mrs. Airy and myself.

'I am, dear Madam, very faithfully yours,

'G. B. AIRY.'

CHAPTER VII.

MARRIED LIFE.

FORBES and his newly-married wife had not advanced far on their wedding tour when they met with an unlooked-for interruption. Scarcely had they reached Bonn on their way up the Rhine, towards Switzerland, when he was seized on July 20th with a severe fever, which brought him very low, and detained him there till the end of August.

During this detention at Bonn, the first tidings of the reception his book on Glaciers had met with at home reached him in the shape of an article on it in the *Edinburgh Review*, from the pen of Sir David Brewster. That this article must have been a tonic to his spirit, the following letter acknowledging it seems to show:—

To SIR DAVID BREWSTER.

'GENEVA, *August 3rd*, 1844.

'Having now first had the opportunity of reading your article upon my volume of Travels in the *Edinburgh Review*, you will understand that it is impossible for me to rest until I have expressed to you some small share of the glowing feelings of pleasure with which I read it.

'During my short scientific career I have had yet ample occasion to know the rarity of well-weighed praise, the still greater rarity of well-timed support. Even the best of us are apt to commence from friendship rather than from a studied conviction of merit, and to spend that commendation in generalities and circumlocutions, until a decided and general opinion in favour of a new view or discovery shall have placed the wary advocate on the sure ground of participating in a victory, and not, even by remote possibility, of being plunged in the consequences of a defeat.

'You have done three things for me, any one of which would deserve my sincere thanks. 1. You have read my book carefully, and really studied the theory you undertook to judge. 2. You have allowed yourself to be convinced by arguments, unswayed by the force of opinion against them. 3. You had the moral courage to stand forth, whilst most if not all hesitate, to express your convictions with a force and clearness which alone must carry much weight in public opinion, and exactly at a time when such assistance and public acknowledgment are most valuable and most hard to be had.

'The flattering terms in which you have expressed yourself as to the difficulty and importance of this and such like generalizations, exaggerated as they will no doubt appear to many persons, are such as find an echo in the heart of the person most immediately interested, who feels his own travail of mind portrayed with a force and reality which seems more like a dream of egotism than the public sympathy of a kindred spirit. It was this which so deeply touched me, and you know by experience the value of so rare and entire a sympathy. It is impossible for me to express all I have felt on the subject, and I will not attempt it, but conclude by assuring you that you have added a very deep debt to others which I already owe you. I was quite struck by the apparent facility with which you had seized the bearings of this intricate subject, but which must, I know, have cost you much time and labour. The tabular views add much to the clearness.

'I have had a pretty extensive correspondence on the subject with different continental philosophers, who have proposed to me their difficulties with much frankness;

and though the progress of the theory is slow, it seems to be solid, and the answers I have given have been generally well received. Several of these I have embodied in supplementary letters published, or to be published, in the *Edin. Phil. Journal*, which I hope you may have leisure to look at. The most interesting circumstance occurred to me, however, two days ago. I called at Bex on M. de Charpentier, who, you are aware, is old, and is usually supposed to be sufficiently wedded to his own opinions, and last year I found him quite indisposed to admit the smallest scepticism as to the truth of his dilatation. Now, however, he received me with great warmth, and told me that he had read my book and visited the Mer de Glace, and he believed that every fact admitted of being as well explained by my theory as by his; that, moreover, he expected soon to be able to announce his entire conversion. He even presented me with a book inscribed from "son bientôt converti serviteur et ami S. de Charpentier."'

The substantial merits of Forbes' book of 'Travels through the Alps of Savoy' will be discussed in another part of this work. Only one word concerning it which occurs in Canon Kingsley's Miscellanies may here be given :—

'We have heard Professor Forbes' book on Glaciers called an Epic Poem, and not without reason. But what gives that noble book its epic character is neither the glaciers, nor the laws of them, but the discovery of those laws; the methodic, truthful, valiant, patient battle between man and nature, his final victory, his wresting from her the secret which had been locked for ages in the ice-caves of the Alps, guarded by cold and fatigue, danger and superstitious dread.'

By the end of August he was so far restored as to be able to go on by easy stages towards his beloved Alps. A month was spent among them partly in showing to Mrs. Forbes the more accessible of his favourite haunts, partly in carrying forward some of his old work at

Chamounix, and on the glacier of the Grindelwald. On returning to Bonn to consult Dr. Nassé, the physician who had attended him with much care and skill during his late illness, the advice he received was that to undertake his winter's work was inexpedient, and might prove dangerous, and that to ensure recovery he ought to winter in Italy. On receiving this advice, he acted, as he did so often afterwards, and at once surrendered to the judgment of his physician. He wrote to the then Lord Provost of Edinburgh, as the representative of the Town Council, the patrons of the University, telling the state of his health, and the advice of his physician, and applying for leave to appoint a substitute to teach his class during the ensuing winter. The request was at once granted, and Forbes set out with his wife for Italy. Naples was reached about the middle of November, and Rome by the beginning of January, which places were their head-quarters till the end of May. September 1844 was far advanced before they reached the first home of their married life, at 15 Ainslie Place, Edinburgh.

Soon after his return he received the following warm welcome from a former pupil, whose words no doubt expressed the feelings of many besides himself:—

From MR. RANKIN.

'*November* 14*th*, 1844.—My heart leaped when I saw your well-known handwriting again; often while you were in the South I thought of you, and longed and prayed that you might soon be restored to health and professional duty. . . . However, I was gratified by seeing in the public prints that you had returned much benefited by your *séjour* in the South. . . . Allow me to thank you for your kind interest in my well-being. I feel indebted to your kindness more than can be expressed, for it was this kindness which first awoke my faculties into energy, and enabled me to taste the pleasures flowing from their lively exercise. Indeed, often since my settlement here, when some extra work was accomplished by dint of extra application, I have thanked you in my heart for teaching me what difficulties steady mental energy can overcome.'

He had, however, scarcely returned, and before he had time to settle down, than he was called away to attend the gathering of the British Association, which that year, towards the end of September, took place once again at its earliest meeting-place, York. Some of the incidents of that meeting as they occurred to Forbes are given in the following letters written at the time to his wife:—

'BISHOPTHORPE PALACE, *Wednesday evening.*

'. . . We arrived here to-day at twelve—I mean at York, in company with Sir D. Brewster, Lord Enniskillen, Dr. and Mrs. Alison, Sir A. Agnew, and many other stars. It is very cold. My cold is not better. The people staying here are all most pleasant. We have Brewster, Murchison and Mrs., two Archdeacons Wilberforce, Liebig, Peacock (Dean of Ely), Lord Northampton, &c. &c. Whewell and Mrs. were at dinner. The Archbishop is wonderfully well. The William Harcourts are a very charming family, Mrs. very, and one of the sons a most amiable youth. How I wish you were here. I am so sorry I can't send this to-night.

'*Thursday morning.*—I am better to-day. Tell me all about yourself. I have had some talk with Whewell about glaciers; he said my note expressed exactly his opinion about Hopkins and his papers. I have not settled whether to make a communication or not: I want to avoid a collision with Hopkins, who is here.'

'BISHOPTHORPE PALACE, *Sept.* 27, 1844.

'I have received your two kind notes. I wish you were with me. I am better to-day. The first evening meeting and opening of the Association was last night; I did not go, but stayed cosily with the Archbishop. We stayed cracking for three hours, and then he went and fetched me lozenges for my cold.

'5 P.M.—I am much better to-day. We have had such a lively discussion in the Section on Glaciers. Lord

Fitzwilliam gives a grand dinner to 200 people; I am just going to it.

'*Saturday, Sept.* 28, 1844.—I wrote you very hurriedly yesterday. Thanks to the previous evening's repose and the Archbishop's lozenges, my cold was much better, and I was able to get well through the day. My notice on Glaciers was, I do think, very well received by a "select and fashionable" audience, not to say a learned one, which it was; and the impression seemed general that viscosity was established. Whewell explained my theory of the Veined Structure and the Frontal Dip very clearly.

'*Monday.*—My cold is much better; but we had a postponed discussion on Glaciers on Saturday morning, when Hopkins and I did battle, and I am sorry to say I felt it exceedingly; it discomposed my nerves, and made me very uncomfortable indeed, until I was soothed by the Minster service yesterday, which was beautiful. But I see it will be quite necessary for a long time to avoid excitement as much as possible. I need hardly say that the discussion was not prolonged by me, but by the desire of the Section. I am sure, darling, you would be pleased by the kind way people, not scientific, speak of my book. The two Archdeacon Wilberforces and the Bishop of Ripon are much interested.'

He had not long returned to Edinburgh when his eldest child, a daughter, was born. It is thus he communicates the event to one of his dearest friends:—

'*October* 20-21, 1844.—. . . I lose not a moment in telling you the joyful news. I pray God that our children may partake of the affectionate bond which unites their fathers—a bond which, let me say it at this solemn moment, will endure, I trust, even beyond the grave. In the morning I will give you further news.

'*Quarter to one.*—I have just seen my little babe, such a nice lively little animal.'

To the same friend, writing a month later, he thus expresses his sentiments on first reading the Life of Dr.

Arnold, by the Rev. A. P. Stanley, now Dean of Westminster. That book has long since been stamped by the world's approval, as—Lockhart's 'Life of Scott' alone excepted—the best biography of modern times. It is curious, however, to see the estimate formed of that biography, before the world had given its verdict, by one the bent of whose mind was so different from that of Arnold's, so averse to that busy, even restless, discussion of moral and religious subjects which to Arnold was the breath of life. To those who now pass for the 'advanced thinkers' of the time, Arnold's views on fundamental principles appear even unduly conservative. It must, however, be remembered that thirty years ago this biography sounded almost the earliest note heard in this country of that so-called 'liberal religious movement,' which has since become rather a weariness. If some devout minds were then attracted by it, many more were startled, not to say repelled.

'*Nov.* 1844.—. . . I am much more ready to do you battle on the subject of Arnold than of the British Association. It is a book well calculated to bring out the deep gushing well of hidden virtue in a man whose overt acts must evidently have made him unpopular and misunderstood. In the first place, then, it is a work of biographical justice to his memory. In the second place, it may commend fervent piety to some men of Arnold's intellectual force. In the third place, it may show how much violently original opinions like his inevitably tend to be sobered down by the experience of life; and how much "disquieting oneself in vain" might be saved by believing that opinions held by men whose principles we approve of are as likely at least as our own to be true on subjects on which they have had practical experience, and we have not. The same candour which acknowledged the generally admitted barbarism of a Latin grammar in Latin to be practically best would have found something of similar practical wisdom in some religious and political institutions against which his fervid and presumptuous judgment rebelled. You rightly say that as his age matured "his notions were getting complete and correct;" but this makes one regret the more that he was not content with holding his crude opinions himself, but must deliver them as a testimony which the progress of the age demanded of him, although he might have been led to suspect them when he found that he had not one adherent. This terrible obstinacy and intellectual pride invariably wrecks itself, leaving no trace behind it, because it utterly contemns the support or advice of any class of men. In some cases this desperate struggle seems to have worked him up into a sort of mania, which dictated such a passage as that at vol. ii., foot of p. 425, third edition. On the whole, except the uses I have mentioned above, I think those of the book are of an entirely negative kind. In truth, except his warm and affecting piety, and his fond zeal for his school and scholars, what are the fixed principles to be found in his life? what are the lessons of polity, civil or religious, which your £500 lady wishes to impress upon the governors of the country?

'For myself I have read the book with the sincerest pleasure and no little admiration; and I have gathered, I do trust, some useful and enduring lessons. There are some very charming passages in his journals, and his biographer appears to have executed his task with extraordinary fidelity and impartiality. In very many of Arnold's opinions on disputed subjects I find an echo of my own, and his opinions were so heterogeneous that almost everyone probably does the same, and hence perhaps a clue to the popularity of the book. But on this very account I do not lean to the coincidence as very strengthening to my own mind.'

The session of 1844–45 found Forbes once more at his post in the Natural Philosophy class-room; but that his illness had not left him the man it found him he felt himself, and confessed to his friend Mr. Batten in the following letter:—

'EDINBURGH, *Jan.* 23, 1845.

'. . . There is no question that my powers of resisting the tear and wear of work have much diminished, and I have very carefully set bounds to my ambition and my anxieties. The great tranquillity of my life has been of the most marked benefit, and I am much improved indeed since the session began, and the mechanical act of speaking does not now fatigue me at all.

'I have lately been reading Wilkie's Life with great pleasure. He was really a fine character from first to last and his sketches of Art, of Italy, and of the Holy Land carry one on with unabated interest. I have now begun Horner's Life, which I find a very different composition indeed. The half of the first volume which I read, I find dull, pompous, and without any freshness of character or alleviation of adventure,—a boy thinking as if he were a man, and treating all men past and present as his equals at most. . . . Do you hear anything of this marvellous new invention of multiplying books, prints, deeds, bank-notes, signatures, and all indefinitely, and almost for nothing? You, who are a man about town, and frequenting the Clubs and Westminster Hall, ought to send the poor country mouse all that is worth knowing in the world.'

Again he thus writes to the same trusted friend from his final Edinburgh home:—

'3 PARK PLACE, *February* 13*th*, 1845.

'. . . No external spectator can know what it costs me to purchase health at the expense of every exciting occupation. I have sacrificed my own researches, and can write but very little on any subject which interests me without a headache and other unpleasant feelings. But I really enjoy our most tranquil life and the lighter occupations, such as copying my foreign sketches, revising my notes on Naples and Sicily, and arranging my books and minerals. We never go out in the evening anywhere, and we have no rattle of carriages nor perpetual door-bells, as in Ainslie Place. The dear little one is well, and now wears Mrs. Batten's pretty socks.'

That illness at Bonn had given the first rude shock to his strength; and though he recovered so far as to be able to return to his work as zealously as before, and twice again to revisit for a brief season his well-loved Alps, yet he never again was entirely his old self, and the inroads of illness, kept at bay for a time, gradually closed in upon him. Though his muscular frame had been lithe and vigorous beyond that of most men, there was an internal weakness which the severe strain of his life had at length brought out. Ever since he had entered college as a student, it had been with him one long unremitting effort, unbroken by few, if any, seasons of entire repose. Professors who have been long-lived have generally compensated the strain of the winter session by some periods of entire idleness in summer. But Forbes had thrown aside his labours in the class room only for still severer among the Alpine snows. If there were any weak place in a man, such a life would be too much for him. Unfortunately there was in Forbes, for all his vigour, such latent weakness, and too surely his glacier bivouacs had found it out.

In the December of 1844 he removed from his house in Ainslie Place to one nearer to the College, and 3 Park Place continued to be his permanent home till he finally left Edinburgh.

His eldest daughter was born in Ainslie Place; the rest of his children, except the youngest, were born in this new and more lasting home in Park Place. These details are necessary to throw light on the two foregoing, as well as on some of the following letters.

The summer of 1845, from the middle of April to the first week of June, was spent with Mrs. Forbes in the West Highlands, in a tour ranging from Bute to Skye.

In that latter island he explored the Coolin mountains with M. Necker, who was then living at Portree; and there, 'amidst the splendid hypersthene formation, found indisputable traces of glaciers.' These explorations were afterwards embodied in a paper on the Geology of the Cuchullin, or Coolin hills, for the latter is the name by which they should be called.

A letter written from Skye on a very different subject —the removal of University tests—may here be given.

To the Rev. Dr. Whewell.

'Portree, Isle of Skye, *April* 1845.

'. . . You ask about our University tests. As on all similar subjects, men's minds seem to be undergoing a curious revolution on this. The result of the late discussion in our own Senate surprised me excessively. From the first I had resolved to take no part whatever in it. As an Episcopalian, I felt too much interested in the removal of them to trust myself to vote on that side, my prepossessions being certainly favourable to some species of test generally; and besides, having been myself admitted by the liberality or the laxity of our usages into a position somewhat equivocal, I would not take advantage of that position to lift a hand against the Establishment which had connived at my entrance. Great was my astonishment when I found how much this delicacy seemed thrown away, for every professor present, including old Presbyterians, Free Churchmen, lawyers, several Conservatives, and all the Episcopalians, voted for the removal of the tests, except Principal Lee and myself, who did not vote; but it was plain that even the Principal did not dissent from the grounds of the vote, but kept back merely from a consideration of his position in the Church. This curious result satisfied me that the minds of moderate sensible men in Scotland attached to Presbyterianism are in favour of the removal of the tests which, in our University at least, have so long been nominal. But I am glad that no part of the weight rests on my shoulders. Personally, I shall feel more comfortable when these tests are removed, which I suppose will soon be done, since so many parties unite. At all events, I suppose there is no chance of their stringency being increased, and that either Episcopalians or Free Churchmen will be displaced. I am clearly of opinion, with the *Edinburgh Review*, that any attempt to exclude the latter must *à fortiori* act upon the former.'

About the middle of June, he set out alone for Cambridge to be present at the meeting of the British Association, and to visit Dr. Peacock, Dean of Ely, in his island deanery amid the marshes. The following letters to Mrs. Forbes, who remained in Edinburgh, preserve pleasing recollections of the persons who then flourished as the leaders in the world of science, most of whom have since then disappeared.

'CAMBRIDGE, *Thursday, June 19th*, 1845.

'. . . Here everything is so nice, and so comfortable, and so hospitable, that I wish every hour you could have come with me. It would have been such an opportunity for you to see the lions, who are all so pleasant and kind. I shook hands with Hopkins the first opportunity. Old L. von Buch is here, but I have not seen him yet. Mrs. Whewell does not come on account of her father's death, which throws a shadow over Trinity; but we dine with the Master on Saturday. The Airys arrived last evening in the greatest health and spirits, and I saw the comet through the great Northumberland telescope. I shall finish this in the afternoon. We are likely to have hard work. The Magnetic Committee, on which I am, is to meet at 8 A.M. to-day. We dine at Miller's.'

'CAMBRIDGE, *Friday*, 7 A.M.

'. . . We had a very long night of speeches at the Senate House. I could not get away, as all the gentlemen of our party slipped off and left me alone in my glory with the ladies. Herschel's address was very good, but imperfectly heard. Peacock resigned the chair, and he took it.

'*Afternoon.*—I have had a very busy and a very exciting day, but am standing it very well. We have a lecture from Airy to-night, and we shall not be back to the Observatory till 10 P.M., having left it at half-past seven this morning. We have Baron Ettingshausen on Etna, and Leopold von Buch in the Geological Section. I got a very cordial reception from the old geologist, which was no doubt partly owing to my having warmly espoused the Elevation theory, magnified Elie de Beaumont, and defended the claim of his countryman Ettingshausen. I am happy to tell you that in all quarters I have been so kindly, I may say warmly, received. We have the Archdeacons Wilberforce, the Vernon Harcourts, and many other pleasant York acquaintances. I do regret most deeply that you could not come, you would have seen so much in a short time. I am very happy that sweet baby is so merry, but I fear the little monkey has a short memory for her friends, she seems to forget them all as fast as anything—even poor nurse; what can papa expect?'

'CAMBRIDGE, *Saturday*.

'Yesterday we dined at Trinity Lodge. I sat beside Von Buch, which was the greatest fun; he spoke so kindly of my book and map, and told me a great deal of gossip so pleasantly and amusingly. To my infinite amusement, he told me his side of the story of the discussion with Charpentier at the Monte San Primo, which you recollect was so graphically told in our old friend's last letter.

'In the evening there was a most brilliant party in the Master's rooms, which are really splendid. Lady Herschel did not come, which I regret; perhaps I may meet her at the Dean of Ely's next week. My old and good friends the Vernon Harcourts were there, and Miss Harcourt, who gave me a most kind invitation to the Archbishop's place at Nuneham, near Oxford, but which I declined. Wilberforce, the new Dean of Westminster, is a very taking person; he has been most kind to me, and I had a hearty greeting from old Sir Robert Inglis. Kiss dear babe many times for me. . . .

'*Sunday, June 22nd*, 1845.—To-day, Sunday, is a real day of rest, so quiet and still. It is a mile and a half from Cambridge. We go into Trinity to hear the Dean of Westminster preach.'

'DEANERY, ELY, *Sunday, June 29th*, 1845.

'. . . And so I have retreated to my little bedroom in the top of the Deanery to have a little talk quietly with you, and I will tell you more than I have done about this place. The Isle of Ely is a comparatively high ground, which rises out of the adjacent fen country, which is almost perfectly flat, rising not more than a very few feet above the sea. It was formerly an unhealthy swamp, but is now generally drained and very fertile. From its situation, Ely Cathedral has a grand effect, and really the ground just near it is very undulating and pretty, more varied than near Cambridge. The Deanery is close to the Cathedral, a curious old-fashioned patchwork house, most comfortably fitted up by our admirable host, who is really a very uncommon person. He was, you know, tutor at Trinity College, and was generally liked. It is impossible to put Peacock out; he has time for everything, and a civil word for everyone; he is a first-rate man of business, and what he has done for Ely shows no common taste and enthusiasm. He has improved and is improving the Cathedral in every part, laid out the neighbouring grounds and gardens afresh, and he is passionately fond of architecture and music; his patience never wearies in showing off the beauties and architectural singularity of his cathedral, and in directing the chanting, and visiting the lessons of the singing boys. The choir is, therefore, one of the most practised in England, and the service the most complete —so complete as to be almost wearisome to one who has not such a musical taste, yet undeniably most beautiful. This morning we had a Credo of Mozart, most magnificent, an anthem, and all the responses chanted, including the Litany, so that the prayers, without the sermon, lasted two hours.

'I never was in a pleasanter house, nor perhaps so pleasant a one, and one seems even not to miss a Mrs. Dean, for the poor Dean never was married, nor will be, I suppose, except to his Cathedral. One cannot have everything in this world, and I suppose he would not have had his Deanery if he had left college and married. The party when I arrived was a large one, consisting of Sir John and Lady Herschel and two daughters, Mr. and Mrs. Vernon Harcourt, Colonel and Mrs. Sabine, Sir W. Hamilton of Dublin, Professor Sedgewick, Mr. Phillips, and Mr. Kuppfer from Russia. I wish, my own darling, you had been here. God bless you, and keep you and our sweet baby. Over again, Farewell!'

'DEANERY, ELY, *Monday.*

'. . . When you get this I shall be on the salt sea, on the way home to you. Everything has gone on so well. I meant to have written you a long letter, but the quiet and rest of Ely are even more seductive than the pressure of business at Cambridge. But my heart was gladdened, and cheerfully gladdened, this morning with two letters.

'. . . I told you what a remarkable person Sir W. Hamilton of Dublin is, but it is impossible to give you a correct idea of his character. He is a fair, fat, little round-faced man, as playful and simple as a child, with a good deal of Irish humour. He is the first mathematician in this country, and very metaphysical in his habits of thought. He is so full of his subject that he fastens upon anyone that will let him talk about equations, and yet this abstract turn of mind alternates with the liveliest fits of poetic feeling. He is always vibrating between an equation and a sonnet. You will not require evidence of the former, but I enclose a specimen of the latter in his own handwriting, which is, I think, a fair specimen of the sonnet style for those who value such things. The pure sonnet style is, I believe, very rare. I consider it a very great curiosity—the three persons alluded to are Sir J. Herschel, Sir W. H., and myself—so please preserve it carefully, and do not fold it.'

On his return to Scotland, August was spent quietly with Mrs. Forbes and his young child at Fettercairn, the place of his elder brother, who was then absent. In this retirement he wrote a paper on 'Glaciers' for the

Physical Atlas, and a review of Humboldt's Cosmos for the *Quarterly.*

Here is a description of Fettercairn and his stay there:—

To E. BATTEN, ESQ.

'FETTERCAIRN HOUSE, *Aug. 7th,* 1845.

'. . . The charm of this place is exclusively its rurality. There is a very nice garden, but no fine scenery. It lies in a flat below the round lumpish Grampians, which offer no interest except the mere exercise of climbing them, and change of air. There are no neighbours almost but the Gladstones, and it is a real retirement, and as such very enjoyable to me after having tossed about the world so many summers. I brought plenty of books and work to fill up my time, and though I often sigh for a snowy mountain, I look forward to the, I hope not visionary, hope of seeing the Alps next summer. If I did not, I should not be so tranquil. I hear often from the dear Auguste.'

Forbes' achievements in science had for some time been generally acknowledged by the scientific brotherhood. The Government, however, had not yet stamped them with the approval of the State. In due time, however, this was not withheld. In September of this year he received a letter from Sir Robert Peel, then Prime Minister, announcing the grant to him of a pension of £200 a year for the services he had rendered to science.

'*September* 19*th,* 1845.—This morning I received a letter from Sir Robert Peel offering me a pension of £200 a year, which I accepted. I obtained it through the joint recommendation of many friends. This will relieve me of much anxiety about my future prospects. I mean to reserve it principally as a store for old age, if I be spared to it, and at all events for the education, and professional education, of my children. By leaving my present income free to be used, I may pursue science with more liberty, and, if spared health, continue my travels. I trust I shall be delivered from the snare of a penurious and money-loving old age; but the night cometh, and professional men, especially professors, should retire ere they manifestly decline.'

The following letters will serve to trace the course of his life from October 1845 to April 1846, when another College session closed.

To M. DE CHARPENTIER.

'EDINBURGH, *October* 20*th*, 1845.

'How many changes and reverses have you undergone since you did me the favour of commencing your long and interesting letter of April and May last! First, your inflammatory rheumatism, and walking on crutches; then your annoyance through the unhappy political movements of Switzerland; then your recovery; then the attack of small-pox. Next I heard of you at the meeting at Geneva, and, most recently of all, of your being discovered by Studer, quite well and hearty, at Ivrea in Italy! Truly, truly this is quite a phantasmagoria of situations; and I cannot sufficiently congratulate you that the bad news has been chased away by so good. . . .

'Your account of your *rencontre* with M. de Buch on the way to Milan afforded me many days' entertainment, which was not diminished by the fact that very soon after receiving your letter I met M. de Buch at the meeting of the British Association at Cambridge, in June, when he voluntarily gave me his version of the same memorable journey, and his discussions with you. I had hardly expected that M. de Buch would have honoured me with any notice when we met, as I had heard of his rooted aversion to all authors on glaciers; but as I sat beside him at dinner at the Master's of Trinity College table, when he found that I carefully avoided any allusion to glaciers or my book, he spoke of both very good-naturedly and in flattering terms, and we are the very best of friends. Baron Ettingshausen, who has made a splendid survey of Etna, was there, and we had a full description of "Erhebungs-cratere."'

To the REV. DR. WHEWELL.

'EDINBURGH, *January 8th*, 1846.

'I am much your debtor for your very kind and friendly letter, which reached me on the banks of the Tweed, whither we had gone to spend a few days of my Christmas holiday. I return you my sincerest acknowledgment for your Christmas greeting to myself and wife. I was much and grievously disappointed, and still am, that last June Mrs. Forbes should have been compelled to relinquish the pleasure of visiting Cambridge at so interesting a time.

'I am very glad indeed that you approve of my review of Humboldt. Lockhart modified slightly a sentence or two, which perhaps, however, would have been scarcely observed by anyone but myself,—of course I thought, not for the better. Colonel Sabine has resumed his translation, and talks of adding notes.

'I hear faint echoes of Faraday's prospective discovery, and am on tiptoe about it.

'I am glad that your History and Philosophy are to be reprinted with the additions which the deliberate revisal will enable you to make. If it does not go very rapidly through the press, I will endeavour to offer any hints on the History, and shall be happy to look over any sheet in proof that you may send me. In the Philosophy, I suppose you will do battle with Mr. Mill. I cut open part of his Logic lately, and was interested in many views of education which it presented, and thought some of his cautions salutary, without attempting to form any judgment upon questions so difficult and at the same time so purely speculative, that with my present habits of thinking and analysing, I should find some pains necessary to fix myself down to an intelligent study of this very curious and interesting subject. Perhaps some day it may be more congenial.

'I am glad you still hope to do something for the Cambridge system of education, but I suppose it will be a struggle.

'You have read of course the Sequel to the Vestiges. . . . The author of the Vestiges, who is generally believed to be a denizen of modern Athens, has shown himself a very apt scholar, and has improved his knowledge and his arguments so much since his first edition that his deformities no longer appear so disgusting. It was well that he began to write in the fulness of his ignorance and presumption, for had he begun now he would have been more dangerous.

'You will see from my tenth letter on Glaciers that Agassiz has found his way into my mortar-tub at last. . . .

'I have lately written a paper on the Geology of the Cuchullin Hills in Skye, in which I think I have all but demonstrated the existence of ancient glaciers there, having been myself convinced most entirely against my will.'

To RODERICK MURCHISON, ESQ.

'EDINBURGH, *February 3rd*, 1846.

'. . . I have the pleasure of sending, on the other side, an official acknowledgment of the honour you have done us in sending your very splendid work to the Royal Society. What I have read of it pleases me very much, but at the same time satisfies me that my first impression was correct—as such first impressions usually are—that I have not the extensive and at the same time minute acquaintance with this part of systematic geology which could alone entitle me to undertake so honourable a duty as reviewing it. I regret it very much, because I am much gratified and flattered by your thinking of me; but I have never practised the ungentle craft of reviewing upon the strength of amateur snatches of knowledge, and I am very sure that others will be found most willing and far better entitled to render you this friendly office. Why should not Sedgewick or Dr. Fitton, who used to write so well in the *Edinburgh?*'

To E. BATTEN, ESQ.

'EDINBURGH, *April 5th*, 1846.

'. . . Your most comfortable letter from the Athenæum aroused all my bachelor sympathies, and abstracting from the existence of our respective wives and three small *et cæteras,* I began to fancy the corner of the library which you had chosen to make your own, for I take it for granted that you choose the library, and the quiet hours from nine till eleven, and I heard in memory the quiet tick of that small chimney clock and saw the dull glare of that most ponderous and ugly chandelier. But you have returned to the realities of life. Let me not arouse in Mrs. Batten the latent jealousy which that "bachelor's pot of mixed tea with dry toast" —the cosy arm-chair which so often woos the literary old gentleman on the north side of the fire-place to sleep—the charms of the pamphlet and magazine table, are apt, too justly, to excite. After all, I will say this, that the still life of a club is more winning than its intellectual life. I am sure I have not made three acquaintances the more by being a member of the Athenæum; and except very intimate friends, such as you and I, who are always glad to meet, a club promotes rather indifference than union. So great is the fear of being bored or the possibility of saddling yourself for ever with a pestilence in your own drawing-room, *i.e.* the club, that men's advances are always very guarded. My associations of pleasure in the Athenæum, I repeat, are much more with things than men—nice books, and cosy chairs, and warm cups of tea, than with sages, wits, and poets. But you, as a London man, may have a different experience, though I do not expect it.'

When the session 1845–46 was over, two months were spent in visiting, with Mrs. Forbes, sundry places on the Borders, and in a tour through the lakes of Cumberland.

July found Forbes once more at Chamounix, and making experiments on the Mer de Glace. All August was spent in those regions, and September brought him back to Mrs. Forbes at Largs, where she had remained during his absence.

The very full correspondence which Forbes has left

will form the best record of his life for most of its future years.

To E. BATTEN, ESQ.

'EDINBURGH, *December 6th*, 1846.

'. . . The old Dean is now a green grass-plot. I looked in the other day—the gateway-bell and all is as it was. How that bell reminded me of you! how often at its sound have I started up to see you walking up the avenue! The avenue and the holly-hedges are there; but, instead of terminating in the tall pile of masonry, it opens on a flat turf soon to be full of graves. Nothing more surely was wanted to point a moral. . . .

'My class, together with the Royal Society, affords me nearly as much work as I can overtake, and sometimes I am very tired; but my class is so attentive and earnest that I cannot but do my best, and I am thankful to feel more vigorous than I have done for some winters. A week ago I had eighty-five papers given in, containing answers to thirteen questions; the week before I had eighty-three: so you see I have some evenings' work with such a class. *Viva voce* examinations on Saturdays make little way. I think of trying six or seven minutes' examination at the beginning of each lecture.'

To the Same.

'EDINBURGH, *Jan. 10th*, 1847.

'. . . You must allow me some voice in deciding whether the glacier question may safely be abandoned to itself. Your repeated assurances that everyone is satisfied are very pleasing, as expressing a sort of public opinion; but I have the very best reasons for knowing that only a small portion of strict men of science, whose opinion must ultimately decide the matter, are convinced, or at least, if convinced, have the candour to allow it. Dr. Whewell has spoken out manfully in the new edition of his "Inductive History," and I hope Lyell will do as much, and then we may expect others to follow. But the result of my reading in the history of science is that in questions of mixed evidence like this a man must work it out to the utmost limit he can, if he means that it shall finally be associated with his name, which is my desire. . . .

'I very much agree in what you say about Popery. It is ludicrous to hear it blackened like Atheism. The truth is they are only tinselled and varnished over, and often distorted, yet very deep and holy truths—you know I have no leaning to the Tractarians, or I would not say so much. But there is sad bigotry and uncharitableness in religious parties. Most heartily also do I agree with you, my dear friend, in what you say about the questionable benefits of intellectual education merely to the masses, which I have hardly seen better stated.'

At the meeting of the British Association at Oxford in June 1847, Forbes was present as usual. Of what there took place, the following letter to Mrs. Forbes preserves some interesting collections :—

'RUGBY, *June 29th*, 1847.

'I really could not manage to scribble a line yesterday, we had such a busy day at Oxford: it is the first time I have missed. I breakfasted as usual at half-past eight, and was busy in the section, having a diagram to draw for Sir D. Brewster, and to listen to an admirable paper of Thomson's, and to Leverrier, who gave us a long and most interesting discourse in French, at the latter part of which Prince Albert "assisted," with the Duke of Saxe Weimar. The moment the Prince was gone, two o'clock, I had to rush home and dress for the luncheon which took place at Exeter College. Our drawing-room was the garden, where a party of the Oxford dons and chief persons of the Association and principal foreigners met Prince Albert, and stood twenty minutes with their hats off; fortunately it was a very mild day, and no hot sun, but quite dry. The Prince was most exceedingly amiable and affable, and perfectly unaffected. Sir R. Inglis did me the honour of introducing me to him particularly, and we had some ordinary talk, whilst the Bishop of

Oxford entertained us with some anecdotes. The luncheon over and the Prince gone, I had an engagement to look over the Natural Philosophy apparatus, and that was hardly seen when dinner hour arrived. I dined at Prof. Powell's, where we had a small but excellent party, including Airy, Adams, and Leverrier. I had resolved not to go to the evening lecture, but to go early home to prepare to be off this morning; but at Powell's we heard that the evening was very fine, and that Neptune might be seen, and accordingly at ten I started for the observatory, in company with Leverrier and Adams. Fancy the interest of seeing the new planet for the first time in company with its two discoverers! Of course there could be no going to bed without seeing it, and accordingly, as it rose late, I did not get back to my college till very nearly two in the morning. I was up at seven punctually, as the coach was to start at eight o'clock, and my packing was not near done. I arrived here at three o'clock, and found all well. After such a day's work as I have described, you will believe that the tranquillity of this place is a most agreeable variety, though I own that I have enjoyed a most delightful treat at Oxford in my intercourse with so eminent men.'

On his return from Oxford, Easter-Tyre in the Strath of Tay became his residence for the summer of 1847. His journal tells of days spent pleasantly reading Agassiz, and Clarke's Travels; studying Airy's Lunar and Planetary theories; writing on the geology of Vesuvius; making excursions in the neighbourhood, and ascending Schihallion.

Autumn brings him visitors from abroad, M. Studer and M. Schintz. With these he undertook an extensive tour through the Highlands, during which they visited Ben Nevis, the parallel roads in Glen Roy, Staffa, Iona, and Arran. These names are suggestive of geology, and accordingly we find him next spring writing geological notes on Ben Nevis, no doubt from observations made during this tour. The great Ben Nevis group of mountains retained for him a peculiar attraction, even after he knew that he would never again be able himself to explore their scarped precipices and snow-filled gorges. He used then to tell what he had once hoped to have done for that brotherhood of Bens, and to speak of the fine field that lay there almost untouched for some future geologist, and the delightful work that might be produced by the combined labours of some skilful man of science and some scholar who knew the Gaelic names which every crag and corrie bears. Science and poetry might combine to read the character which nature and man have together inscribed on the blank folds of those mountains.

He had scarcely returned to Edinburgh to begin the session 1847–48, when he had an attack of severe influenza, from which he hardly recovered all the winter.

The following letters, to the friend who for some years was his chief correspondent, mark the chief family incidents during the College session :—

'*February 9th*, 1848.— . . . We have just returned from consummating the happy event of christening your name-son. . . . It was a very happy moment to me this, consecrating as it were for another generation the memory of the very deep attachment and affection which I have long borne to you.'

'*April 5th*, 1848.—. . . Were I to write for a week, I could not tell you what we have gone through since I last wrote. Our dearest Eliza has been at the very brink of the grave, and has been mercifully restored to us under seemingly hopeless circumstances. All thought of recovery was abandoned, and we waited to see her die; but, rejecting no means of safety, and having our careful and excellent Dr. Smith in the house, we saw her gradually revive, and she is now, I may say, convalescent. You may believe I am thoroughly exhausted with anxiety and watching, and Alicia is not much better; but I have returned to my lectures, which are very heavy work. I have loads of examination papers, which must be my excuse for writing thus briefly.'

When May 1848 set him once more free from College duties, Forbes returned on his track of last autumn, to conclude his survey of Ben Nevis before writing his geological notes on it. The following are some records of his whereabouts :—

'FORT WILLIAM, *Wednesday morning, May 24th*, 1848.

'. . . I am about to start, Swiss fashion, with a hot roast leg of mutton in my knapsack, a guide, and some letters to some charitable folks by the way—for, as to inns, there are none—to circumvent Ben Nevis. There will be two days without a letter; I am very, very sorry.'

'SHOOTING LODGE ON LOCH EILT, *May 24th*, 1848.

'Here I am in as wild and remote a spot as you need wish to see. There is hardly a village nearer than Fort William, which is at least seventeen miles off —which I have just walked over—by the most execrable sharp-stoned footpath you ever trod. A most kind welcome, but nothing else, met me here. There is only a gamekeeper living here, and the stock of provisions not extensive—no milk, no butter, no bread, no eggs, no cheese, no meat, no fish, no fowls, no ham, no anything except oat-cakes, and, strange to say, tea and sugar; so I supped on tea, and dry bread, which I had brought with me, sopped in it, and I am now getting into a bed with no sheets. Good night, my darling! all blessings attend you!'

'SHOOTING LODGE IN GLEN SPEAN, *May 25th.*

'To you, my darling, a shooting-lodge on Loch Eilt and a shooting-lodge in Glen Spean may seem to be very much alike at present; to me they are very unlike. To-night the richest cream, the freshest butter, the best eggs, crimp toast, a block-tin teapot, cocoa-nut candles, and everything corresponding—such is the luck of travelling. You will really require a guide to my wanderings, so I am making you, on the other side, a map of my route. . . .

'Yesterday I went from Fort William to Loch Eilt. B is the place where the two men were found dead last year. This morning I went to the top of a hill marked A by the dotted line, to Loch Treig, and so down here. I am very, very tired, but have been often and often thinking of you and my little ones. I shall probably make to-morrow a short day, and get back to Fort William on Saturday. Does it not give you a good idea of my Ben that I shall have been three days in going round it, and travelled nearly seventy miles? I am greatly surprised how much insight I have got into the geology of this difficult country during the few fine days I have had; but I am quite satisfied that, to do the thing properly, I must sit down before Ben Nevis for two or three months, and use every fine day. Do you think, darling, you would not dislike it very much, coming to this wild place, and under the tempestuous sky, and me rambling about the hills to places where, even with a pony, I would not like my wife to go? I have really been looking what alternatives there would be, and I find none—except living in the inn, which is most comfortable, but in the heart of this poky village; or else to take Glen Nevis House—a shooting quarter just at the foot of the Ben. I was through the house, and it is comfortably furnished; but it is in such a wild, solitary glen, overhung by the mountain, I think you would not like it.'

'FORT WILLIAM, *Sunday evening, May 28th*, 1848.

'I wrote you a long letter only a few hours ago, but I can't bear you to lose a post, and I am going to start early for Ben Nevis, and shall not be back till late. This has been a superb day—Ben Nevis as clear as crystal. I hardly dare hope for so good to-morrow. This is the first day I have had a comfortable dinner, or hot meat, since Saturday week. It is not ten, and I am writing by good daylight. I have been reading all your letters, and am deeply interested in all you tell me. I am so rejoiced with your accounts of the children; they did so warm my heart. Darling things! I often think of them, and I look at all the children on my journey with very different eyes from what I used to do, down even to the infant in arms. . . .

'*Wednesday.*—Monday was a bad day, as I wrote; here is another worse. But yesterday I had such a dose of walking as might count for two or three days. The weather was fine, but mist as usual on the top of the Ben, so I did not attempt, but I circumnavigated him at a very high level, over some of the most frightfully rough ground it is possible to see. My object was to examine, as far as possible, the contact of the granite and black porphyry. There is a very wild and precipitous glen at the back of Ben Nevis, with a sort of pass, at the head of which we can get into Glen Nevis. I shall make you a little drawing; it will please you better than description. . . .

'Our pass was by A, the summit of the Ben being at C. Neither my man nor I had ever been in the said place before, and as I said, mist surrounded the upper part of the Ben, and came down somewhat lower than the ridge. So, as we approached the head of the glen, we got only glimpses of snow-fields and broken rocks above us; and at length we were immersed in the fog, which fortunately was not very deep. We kept on the rock as long as we could, and at length found that there only intervened between us and the ridge a short steep ascent of drifted snow, most truly Alpine. It was too late to think of receding, and it was not far; so, assuming my new mahogany tripod as an Alpine stock, I proceeded foremost to make steps in the most approved Swiss fashion, to the no small edification of my companion, who had never seen such an operation before. The upper few yards were so steep that I actually could not get one foot stuck into the snow before the other, and had to get along sideways.

'We luckily got to the top safe and sound, and found less mist on the opposite side. We came down a most tremendous steep into Glen Nevis, but without any risk, having been eleven hours incessantly on foot; so perhaps I was not very sorry to see rain this morning, to give me a good excuse for staying at home. I have so very many geological notes and specimens to arrange that I have abundant occupation.'

'FORT WILLIAM, *May 29th*, 1848.

'. . . I wrote at once to Mr. Robertson, enclosing Mrs. Batten's letter, telling him that I should leave this for Drumnadrochit on Saturday, that I should spend Sunday in Strathglass, and wished to reach Kintail by Tuesday night to catch the Skye steamer going north on Wednesday morning. He will thus have time to give notice to the people up the Glen, &c.

'I mentioned in my last that your letter to McGregor had been most useful. I completed the circuit of Ben Nevis successfully in three days. There is typhus fever at Kinloch-Leven. Mr. Milner was from home, so that I did not go down there, but contented myself with a bird's-eye view and scramble over the hill to Loch Eilt, where, if I had not all the comforts in the world, I had a truly kind reception from Oliver, the gamekeeper; got a view next day from Scuir Eilt and down the Glen to the head of Loch Treig; slept at Inverlair (Mr. Walker's). A letter from McGregor promised me comforts and luxuries, but I was rather knocked up, and rode and drove into Fort William next day, nor did I feel well enough to deliver your letter to Colonel Ross. I have since been up Ben Nevis, and should have been on the Ben to-day; but though yesterday was magnificent, just because it was Sunday, it came on to rain in the night *à la mode du pays*, and is drizzling away.'

'PORTREE, *June 11th*, 1848.

'. . . I can assure you Mr. Robertson did not belie the character you had given him. He paid all honour to Mrs. Batten's kind introduction. I spent part of Saturday and all Sunday last at his hospitable house, going to church at Erchless Chapel. A very wet Sunday it was, and I had come to a resigned state of mind to go by Glen Affaric unless the weather was really stormy, in which case I should have gone by Dingwall. I had, however, better luck, and Monday was most agreeable, and improved every step of the way, so that when I reached Loch Affaric it was really a lovely afternoon, still quite early, the hill-tops all clear. I spent the evening rambling about, and was as much delighted with the scenery as I had been with the drive from Fasnakyle. I had arranged to go up Mam Suil in the morning, but unfortunately in the early morning the mist fell down on the hills, and I was afterwards glad I did not attempt it, both because the afternoon became extremely wet, and because I was a little unwell. I arrived at Kintail much fatigued, although the walk is not particularly long or difficult, at least if we except the descent from the Balloch. That part of the road and the view from the pass is certainly very striking. I arrived very wet and rather chilled at the manse at an early hour, and was very kindly received by Mr. Morrison, who with true hospitality, finding that I was very anxious to get on by next morning's steamer to Skye, after giving me dinner, kindly sent me in his gig to Balmacarra, where I arrived most wet and weary at 10 P.M., at a small but comfortable inn, and next morning I had just sat down to a capital breakfast when the alarm of the steamer was given and I had to run and gallop on a pony for a mile and a half, and after many adventures and much energy I and my luggage just reached the steamer in time. . . .

'I should have written to you sooner, but I have hardly had a moment, having been almost constantly with my friend Necker, who was very glad indeed to see me, and who talks incessantly to me till past eleven every night. As he never speaks to any person here, you would think he had kept all his conversation of the last three years to devote to me. But it is a pleasure to see him so well and to be permitted to cheer his solitude. I return home, that is, to Bridge of Allan, on Tuesday, and to our real home soon after, where we mean to remain for some time. I forgot to say that I can hardly suppose Mam Suil to be as high as has been reported. Judging by appearances, and the height of my barometer at Affaric — unfortunately broken at the Balloch—I should conclude it to be under 4,000 feet, perhaps considerably. Have you got the height?'

June, July, and August of this summer were spent—a rare thing for him—in Edinburgh.

September found him at Fettercairn, and October at Conway, in North Wales. To these two expeditions the following letters belong:—

'ABERDOUR, FIFESHIRE, *September* 21*st*, 1848.

'. . . We much enjoyed our visit to Fettercairn, and I must say it is very pleasant to see so happy and united a family, so rejoicing in the performance of every mutual and other duty. Their child is indeed a charming girl, and a model of all that I could wish mine to be, especially in gentleness, thoughtfulness, and most child-like simplicity, with charming gay spirits. We spent a day or two at St. Andrew's on our way back, where Alicia had never been, and we enjoyed it very much. The weather has been almost perfect for ten days or a fortnight past. We returned home to-night, and found our little flock all well and happy to see us. On Saturday we are going into Edinburgh to hear Jenny Lind, for the first and probably last time. It is an indulgence which I would not deny Alicia, though I myself am scarcely worthy of it.'

'BIRKENHEAD, LIVERPOOL, *October* 1848.

'. . . I have been on a sudden escapade to North Wales to see the Conway and Britannia bridges. I left home this day week, and proceeded to Manchester, where I visited some engineering establishments. I then went to Conway to see the tube, and by a singular accident I arrived just after the second tube had been floated into its place. But though so far unlucky, I could not complain, as it enabled me to meet Mr. Stephenson, and to discuss the whole matter with him, and receive every explanation from him which I could possibly desire. We next went to the Menai Straits, and saw the stupendous works there. He is indeed a very interesting as well as able man, this Robert Stephenson, and I have formed quite a friendship for him. He appears to be a most liberal man in point of money, no grasping, no pretension, candid, and most affable; often acknowledges mistakes, and is ready to listen to what everyone has to say. I was really delighted with him. We parted on Saturday morning, when I went round Snowdon, passing Sunday at Bethgelert.'

When Forbes met his class at the opening of the session 1848–49, he devoted two lectures to the refutation of a view of knowledge which he regarded as sophistical and mischievous, but which had recently been recommended to an Edinburgh audience by a high authority.

Some time before this Mr., afterwards Lord, Macaulay had come from London, for the special purpose of delivering an oration at the opening of a large public library and lecturing Institution in Edinburgh.

Macaulay took occasion to ventilate one of those startling but not very profound paradoxes which he knew so well how to set forth in pointed and brilliant epigrams. He maintained that 'a little knowledge is not a dangerous thing,' that there is no line by which superficial knowledge is separated from profound. He said that he never could prevail on any one of those persons who talk of the danger of shallow knowledge to let him know what was their standard of profundity. When we talk of deep and shallow, we are comparing human knowledge with the vast mass of truth that is capable of being known, and, tried by this standard, the most learned and profound are almost as shallow as the most ignorant. The sage of three centuries ago is not equal in knowledge to the child of to-day; the intellectual giants of one age become the intellectual pigmies of the next. Homer, and Plato, and Shakespeare, and Bacon must dwindle into insignificance as the world grows older. These views Macaulay sought to establish by pointing to the fact, that any mechanic can now learn in a few days more about the celestial bodies from Herschel's astronomy, than the astronomer of the Middle Ages could learn in a lifetime ; and Faraday taught the little boys and girls in his Christmas-holiday lectures more chemistry than all the alchemists ever knew.

These plausible sophistries were, of course, set off by brilliant rhetoric and by amusing illustrations which all could appreciate. But they were too false in themselves, and too likely to find favour with the ignorant and the self-conceited, for Forbes to allow them to pass unchallenged. He therefore entered into a thorough and elaborate exposure of the falseness of Macaulay's whole position; and, after delivering his views in two successive lectures to his class, he published them in the form of a small book, entitled 'The Danger of Superficial Knowledge.'

Forbes began by showing, that the terms 'great' and 'little' are really unmeaning if used in relation either to the whole amount of truth in itself, or to the whole of what is knowable by man. It is only in reference to the mind that receives knowledge, and to its capacities of receiving it, that knowledge can be called either great or little. It is even repugnant to common sense to speak absolutely of a little knowledge being a dangerous thing. If it be dangerous, it is dangerous to something sentient and capable of inquiry—to the human understanding which receives the knowledge; and thus the effect may be expected to vary, not according to the absolute measure and amount of the knowledge, but according to the capacity, the habits of thought, and the previous attainments of the mind to which it is addressed. And the same relativeness holds not only with regard to individuals, but with regard to the same nation in different ages, and to different nations in the same age.

If, instead of a little knowledge, we speak of knowledge which is superficial in quality and sudden in acquisition, then it may be truly said that such knowledge is dangerous. A legacy of £1,000 bequeathed to a poor labourer whose whole previous income was £30 a year, may very probably prove injurious, or even ruinous, to his habits and character; while to a man who has been accustomed to spend £10,000 a year it becomes merely a small and useful addition. Power, wealth, knowledge, when suddenly possessed by those unprepared to use them, become to them sources of real danger; it is experience, or the wisdom of experience, which renders these gifts safe and fruitful of good.

Such sudden and superficial knowledge is called empirical, and is acquired by memory, or by a very narrow experience, not deduced from extensive study and general principles. There shall be two men who in the amount of facts they know, that is in their useful information, seem to be as nearly as possible on a par, yet the knowledge of the one shall have been gathered from mere memory of things read in books, the knowledge of the other shall have been thought out by himself, the principles on which it rests apprehended and made his own. Between these two kinds of knowledge, though their amount be exactly equal, there is a whole world of difference. It is one of the most cherished fallacies of self-conceit to confuse between the mental power which originates great discoveries and the very commonplace power required to understand and apply these discoveries when made.

That anyone should for a moment confound the state of mind of a person capable of originating the Theory of Gravitation and of composing Newton's Principia, with that of a reader of the same work, even one who completely comprehends every argument it contains, seems hardly conceivable. Yet those who see no distinction of one sort of knowledge from another, except in the facts with which it makes us conversant, evidently fall into this confusion. They forget that facts have nothing great or little in themselves. The greatness is in the ideas they disclose, the power they call forth, the light they spread, the usefulness they generate. It is in the individual mind—in the mental power, which originates and grasps the knowledge—that the greatness truly dwells.

'A man may deduce truth entirely for himself, as Kepler did when he deduced the primary laws of the planetary motions, or he may be led to perceive and apprehend it as an act of his understanding, by receiving a mere indication or hint from another, as Galileo constructed a telescope from a hint of the toy of Jansens. Or he may extend and enlarge the knowledge which another has added to the common stock, as did the great mathematicians of the eighteenth century working upon Newton's theory of gravity. Or he may passively assent as an intelligent reader to the truth of what is set before him, as does every day the competent student of Euclid's Elements. Or finally he may receive on trust the conclusions of another without attempting to analyse the proof or comprehend the reasoning; and this last is the humble acquisition of many of the readers of popular science of our own time. All these are various ways in which the knowledge of the same facts may be attained by the understanding. It were childish to suppose that the effects of this knowledge could be the same, either in kind or degree, in all these cases, or on the character and habits of thought of the person acquiring it.'

After showing how false and arrogant are the pretensions of those who disparage the great thinkers of early time, because they knew not many facts and processes which we now know, and who hold that the intellectual giants of the past become the intellectual pigmies of the present, he points out that this temptation is greatest in sciences especially progressive as are the physical.

Macaulay had said, 'The knowledge of geography which entitled Strabo to be called the Prince of Geographers would now be considered mere shallowness on the part of a girl at a boarding-school.' 'The contrary is the fact,' Forbes replies: 'the knowledge of Strabo was a profound knowledge of geography, and it produced the effects of profound and well-directed knowledge,—it was a knowledge ever increasing, yet ever tempered by the conviction of ignorance; a knowledge which taught his contemporaries to enlarge their acquaintance with the common family of man, to extend commerce, and to preserve human life. Whereas the knowledge of the boarding-school, unless it be tempered with more humility than can be possibly looked for whilst such comparisons are uttered by men of talent on such occasions, will begin in ostentatious displays of memory, and end in pedantry and conceit.'

Again he says, 'The empiric who has acquired a sort of sleight-of-hand acquaintance with a parcel of facts, and uses them ingeniously, may outshine for a time the more profound and practical student, and the latter may even be unable to cope with him in his own narrow field of display. But for the production of fresh knowledge or substantive additions to the capital of the human mind, our superficial aspirant is incapacitated by the very habits which have gained him a popular reputation. We have it on the authority of Newton himself, that if he was in anything superior to other men, it was in the faculty of patient and concentrated thought. "I keep the subject constantly before me," he said, "and wait till the first dawning opens slowly, by little and little, into a clear light;" a sentence which speaks to the experience of all who have accustomed themselves to habits of patient study. But the superficial thinker has no such dawn. Facts are things which admit of no degrees. He knows them or he does not. He is in a blaze of what he calls light, or else he is in total darkness. He evidently never thought in the way that Newton thought; his knowledge, if knowledge it be, comes to him through some different avenue. Facts are not knowledge, any more than books possess understanding.'

By these and such like arguments, less sparkling but more solid than those of the brilliant essayist, Forbes combated Macaulay's paradox. He then concludes by suggesting certain safeguards against the superficial knowledge which puffeth up. Such are, the making every acquisition of knowledge not an isolated effort, but part of a total mental discipline; the not resting in facts, but striving to grasp the laws and generalizations by which they are connected, to appropriate the ideas or theories which bind them together and unite them into a body of science; the habit of combining study with that reflection, divorced from which reading becomes mere cramming—a simple effort of memory or mechanical expertness; the habit of looking before and after—back to the lives and discoveries of the great thinkers of old, forward to the time when all our borrowed learning shall have disappeared, and all that posterity shall call his will be that small residuum which each man by patient thought has made really his own, and discovered for himself and for mankind: lastly the cultivation not of the mere intellect only, but of our whole moral as well as mental nature. 'The discipline and education of the moral sense, the exercises of habitual piety, and the sanctions of revealed religion are truly found, by dearly bought experience, to be necessary for the temperate and useful employment of the reasoning faculty.' The whole discourse breathes the sobriety and truthfulness of a man, who knew what it is to go out in some directions to the confines of the known; who in thought was at home there, and who had striven to carry the limits of knowledge some steps forward into the unknown. And as his experience, so his language about knowledge was altogether unlike that of the rhetorician who, viewing knowledge merely from a distance, has his imagination fired by the achievements of others, but knows nothing of what it is to think out difficult problems for himself, or to wrench from Nature her secrets by patient inquiry.

During the December of the session of 1848-9, Professor and Mrs. Airy visited Forbes, and stayed for some time with him in his house in Park Place. The Astronomer Royal lectured for Forbes in his College class room, and received from the University the degree of LL.D.

The summer home of Professor and Mrs. Forbes and their young family was for the year 1849 Chiefswood, in the Rhymer's Glen, well known to readers of the Life of Scott as the home of the son-in-law and biographer of the poet. Forbes' time there was spent in writing for the *Edinburgh Review* a paper on Greenwich Observatory, in working at a map of the Mer de Glace, reducing his Alpine surveys of 1846, or in a tour among the Border hills. This was varied by a longer excursion in autumn, of which the following letter is a record :—

To M. STUDER, BERNE.

'CHIEFSWOOD, *August* 29*th*, 1849.

'You must have had more "second sight" than myself when, in your kind and most welcome letter of the 14th July, you prophesy a tour in the footsteps of Walter Scott as my summer's occupation. You will not easily believe, perhaps, that, not content with the scenery of Melrose, the glen of Thomas the Rhymer, the Abbey, and the glen of the White Lady of Avenel, I have been on the track of the Pirate, the Fitful Head, the Cathedral of Kirkwall, the stones of Stennis, and the Isle of Hoy. In short, I have returned from a tour of Orkney and Shetland. I was induced thereto by my friend Mr. Airy having projected this excursion. I visited Orkney with him and Mrs. Airy, but I went on to Shetland alone. Caithness, the northernmost part of the mainland of Scotland, and Orkney are chiefly remarkable, geologically, for the great extension of certain members of the old red sandstone formations, which have been ably described by Mr. Hugh Miller, and which contain, you are aware, a host of fishes which characterize that formation, and which are perhaps nowhere found in such perfection, although in Russia the same species have been found. These are the fish named and classified so successfully by Agassiz, such as the Asterolepis, Glyptolepis, &c. I visited some of these fishbeds in Orkney, and I am promised a large series from Caithness; of which I can offer you a part if you choose to have them. Shetland is a long voyage beyond the Orkneys, and as the islands can hardly be explored except by water—the interior of many being covered by almost impenetrable peat bogs—you will understand that one is very much at the mercy of the winds and waves. Being no great sailor myself, you will believe that I scarcely ever expected to visit these remote but interesting islands of Thule; but, on the whole, opportunity favoured me, and I managed to see in a short time, and by seizing every occasion, a good deal both of good geology and of good scenery; and as a Scotchman I feel glad to have visited a country so remarkable in several points, besides being the northern extremity of the kingdom. A steamer conveyed me to Lerwick, the capital of Shetland, but from thence I had to make my way in small boats, and to depend on the hospitality of the native gentry, there being no regular inn from one end of Shetland to the other. Considering my anti-marine propensities aforesaid, I felt rather proud when I stood on the northernmost headland of the Island of Unst, the northernmost of the group, almost in latitude 61°, and looked the North Pole in the face; nor had small pleasure in breaking the chromate of iron and Diallage Rocks and serpentine of Balta Sound, whereof I have specimens for you, if you care to have them. Pure hydrate of magnesia was also found there. This was the station of Biot and Kater in their pendulum researches, and I lived with Mr. Edmonstone their host. The west of Shetland, the least known part, struck me however most. We have there the granitic hill of Boeness, the highest land there; and the phenomena of magnificent granite veins into the adjacent rocks, together with very curious associated claystone porphyries. These various formations produce a grand variety of cliff scenery, which I scarcely remember to have seen equalled except near Staffa. The outmost Island of Foula, far to the west, shows a gigantic precipice of 1,200 feet, rising perpendicularly out of the sea: but the sea and climate are so wild that it is a serious exploit to visit it, and I did not attempt it. I do not think that, even in the Alps, I have seen more terrific single precipices than in Shetland. Altogether I am much pleased with my trip. Do not think that I have forgotten either Ben Nevis or Switzerland. The opportunity of securing so delightful a residence as this is for one summer for my family decided me upon taking it, as I might not have so good an occasion again, but I will not forget the Ben. Again my long-projected visit to London, which I have regarded for years as a kind of duty, I took in place of a foreign excursion; but I very fully intend to make one next year, although I fear that the meeting of the British Association in Edinburgh may occur at an inconvenient season. Shall you not try to come over if it takes place in August?'

The session of 1849–50 found Forbes stronger in health than he had been since his illness in 1843, and with a larger number of students in his class, 203, than in any former year. During this winter he resumed his long-suspended experiments on Radiant Heat, but after spending much time on these, he found his labour fruitless through some defect in the apparatus.

'EDINBURGH, *January 29th*, 1850.

'I have not had such good working health as this winter since before 1843. The test and proof is that I have been able to return to my laborious experiments on Heat without suffering from it, and can at all times apply myself without any significant symptoms which for six winters past have told me of mortality and increasing age. I wish you could see my new little room in College, which is really a charming den, and fitted up with an express view to my Heat experiments; whereas you remember my old arrangement at the case in the great room used to get very comfortless by four o'clock of a winter's gloaming when you used to look in upon me.'

'*March* 17*th*, 1850.

'. . . I am going to read a paper at the Royal Society to-morrow about a wonderful meteor which was seen here on the 19th December last. I will tell you its route as nicely as you know that of the omnibuses in Oxford Street. It moved in a straight line parallel to the horizon of Edinburgh for two or three thousand miles, without any regard to the earth's curvatures. There is an independent gentleman for you! Doesn't even touch his hat to old mother Earth as he draws a tangent to her just to the west of St. Kilda, 88 miles above the Atlantic.'

When May 1850 came, Forbes had looked forward to taking Mrs. Forbes with him to spend a summer among the Alps. This was prevented by the sudden illness of his second daughter. After an anxious interval, his child was so far recovered as to allow him to set forth, and he travelled with his sister as far as Kissingen. Thence he went on alone to Chamounix. Ten days were given to complete his survey of the Mer de Glace, and then he had to return home. This was the last of his Alpine work: for though he passed another summer in Switzerland, he could then only gaze on the Alps from the lake-shores, not assail as of old their heights. It was the meeting of the British Association, held in the August of that year in Edinburgh, which called him reluctant away. Sir D. Brewster was President, Forbes was Vice-President, and presided in Section A, and Mr. and Mrs. Airy were guests at Park Place. Forbes himself was unwell the first two days of the meeting. After it was over there followed tours with Mrs. Forbes, first to the west coast, then to Aberdeenshire. During the months of August and September he lived with her at Phesdo, near Fettercairn, from which place he and Mrs. Forbes made the Aberdeenshire excursion described in the following letter:—

To E. BATTEN, ESQ.

'PHESDO, *September* 17*th*, 1850.

'I am anxious to write to you whilst under the invigorating influence of an eight days' tour into Aberdeenshire, with most beautiful weather—except just at starting, when we had a dense and cold fog. We first lionized Aberdeen very thoroughly, spending nearly two days, one of which we devoted to old Aberdeen, where I wished much for you. I very much recommend you on your next Scotch trip to make up your mind once for all for a sea-voyage—which you have been, I do not know how many years, talking about, but I observe it always evaporates into rail—and go to Aberdeen, which is nearly as quick as to Leith, and as cheap. You will be delighted with the old King's College, unless I am much mistaken, or unless I am even now preparing you a disappointment by saying so much about it. I was most struck by the College chapel, which is the choir of the old one, the remainder being converted with slight change into the library, where the holy water basin at the old entrance is still quite uninjured. The chapel contains the original carved-oak stalls, &c., in most admirable preservation, quite unlike anything else which I recollect in Scotland. It seems that the iconoclastic spirit was never so strong in that country. Crowns, mitres, and bishops' tombs are quite uninjured in the old cathedral, though in other respects it is fearfully transmogrified. The old Gothic bridge of Don is very interesting, and that I recollected well. Leaving Aberdeen—I pretermit the other sights—we went up the Don. I was on the whole much disappointed with the scenery, though it is an eminently industrious, thriving country. The neighbourhood of our former family place, Monymusk, was the chief exception in point of scenery; there it is highly beautiful, with some grand timber. We spent a day at Castle Fraser, an admirable specimen of what is called, I believe, Flemish style, as it prevailed from two to three hundred years ago. We subsequently visited Craigivar, also a fine old turreted castle. We found the Dee at Aboyne, where, and subsequently, we had always difficulty in finding accommodation, every inn being full in that region. We always slept out, not in the open air, but in a supplementary *hospitium*. Thanks to our stars, we escaped "the gathering" at Braemar, which would have made it intolerable. We enjoyed the drive up to Castleton exceedingly. I knew already the neighbourhood of Castleton and Invercauld, but not the stretch of Deeside, which is truly beautiful. The Queen and her mother have made a happy choice. We spent Sunday at Braemar, but did not go to the Queen's church, being nine miles off. Yesterday, Monday, we mounted Lochnagar with the aid of ponies. The morning, misty as usual, cleared up into a splendid day; and, indeed, each day just now is magnificent, and as like the last as can be: the barometer nailed at Set Fair. We scrambled down the hill to Glen Muick, a rough affair, and came out just at the Queen's hut, where, by the merest accident, her Majesty had arrived only an hour before, and was already off to fish in the dark waters of the utterly lonely Loch Muick, and this as evening drew on; we had not time to follow her Majesty or wait her return, but had to make the best of our way to Ballater. We crossed the Cairn over here to-day.'

During the remainder of his time at Phesdo, he was occupied with the theory of probabilities as applied to double stars, which, he says, cost him much thought and labour; with Fourier's theory of heat; and with writing a new course of University lectures. But, before he returned to Edinburgh for the winter, indifferent health forced him to give up severe study.

The journal of the ensuing session notices experiments on the conduction of heat, a visit to Professor (now Sir William) Thomson, who had been appointed to the Natural Philosophy chair in Glasgow University, and last letters which passed between Forbes and his young friend John Mackintosh, who then lay dying at Carnstadt, near Stuttgard.

Mention has already been made of the mutual affection between these two, which, at first arising out of their relation as professor and student, soon passed into warm friendship. It will be remembered how, in a former summer, they had travelled together in the Pyrenees. Ever since they had kept up an affectionate intercourse and correspondence. In the previous summer Forbes had met Mackintosh at Kissingen, and had seen and heard from himself how his health was failing. It is thus Forbes writes of him in February 1851:—

'EDINBURGH, *February 9th*, 1851.

'My dear friend John Mackintosh is dangerously ill. I received a letter about ten days ago, dated from Tübingen, the 21st January, kind and warm-hearted and gentle, as he always was, yet evidently a little alarmed about an increasing affection of the lungs, under which he has suffered ever since his imprudent exposure and fatigue in Italy, in May and June last year. This I learned from him first at Kissingen, in June or July last, but always under the seal of strict confidence not to alarm his family. It was a sad mistake. The disease was commenced and aggravated by imprudence; and with the characteristic unconsciousness of consumptive patients, he had evidently no conception of actual danger even on the 21st January. But on the 28th, his friend Mr. Röstlin writes to a mutual friend, Mr. Nelson of Newport, to break to his family the utterly unsuspected tidings that he is rapidly sinking into the grave. How strange, how appalling to us the ways of God! Is it because he is so pure, so good, so beloved, that he is to be removed from the evil to come? I wrote him a letter to-day, full of unrestrained affection, in sad uncertainty whether his living eye, which so often beamed with regard towards me, will ever glance over it, or even his ear catch the words when read: for I have written to his mother to entreat that it may be read to him, if necessary. I long to have an interest in the thoughts and prayers of that dying saint—for dying I fear he is.'

To JOHN MACKINTOSH, ESQ. (*Tübingen*).

'EDINBURGH, *February* 1851.

'. . . No moderate distance, no dispensible engagement, would have hindered me from going to attend the bedside of one whom I love so well, and who I know would have conferred more benefit on me from that bed of sickness than I could possibly have hoped to communicate to him. Under present circumstances you well know how impossible this is. I must satisfy myself with imagining the pleasure which meeting your mother and sister after so long an absence will give you, and with indulging the hope that your good constitution and placid temper may yet enable you to surmount the attack of disease. Yet neither of us would blind ourselves to the alternative of such a struggle. I have once, as you know, found myself not far removed from those circumstances in which you are now placed: abroad, and attacked by a prolonged illness of doubtful termination, warned by a medical friend at home, to whom Dr. Nasse's opinion had been communicated, to prepare for a longer journey than that to Italy. That was in 1843, and yet I am still here. But the recollections of that solemn period fill my mind with various and opposite emotions when so vividly recalled by the circumstances of your state, dearest Mackintosh.

'In all things, I know, you have greatly the advantage of me. Sincere and absolute will be your resignation. How habitual and deep your repentance for the sins and errors of your past life; how natural and unfeigned your trust in the Saviour whom you have constantly believed in. I know how little, comparatively, I had such grounds for comfort, and how unworthy I now am to offer you spiritual consolation. But as I do feel these things deeply, I see no good reason for suppressing them. And on such subjects your thoughts will most surely be turned with most pleasure when this letter reaches you; and should it be too fatiguing to you to write, do, dearest Mackintosh, send me some friendly message, your love and blessing, by the pen of one of your friends and watchers; and think of me as one who would be present with you in body if he could, but most certainly is so in soul and spirit: and that God Almighty and our Saviour Christ may be with you now and ever, to support and comfort and defend and reward you, is the prayer, dearest Mackintosh,

'Of your affectionate friend,

'JAMES D. FORBES.'

To this letter Mackintosh was able to reply, not however with his own hand, but in words which he dictated to his sister.

'MY VERY DEAR SIR, 'CARNSTADT, *February* 1*st.*

'The kindness of your letter overpowered me; but more, it sustained and comforted me. Since I wrote you I have indeed been brought very low, and think my end must be near. I suffer greatly from pain in the spine, which has been weakened by so much expectoration, so that sometimes I can find rest in no position. Yet the Lord has never yet tried me above what He gave me strength to bear, and His goodness to me is infinite. I think this light affliction over, and then His own presence. My family are a great comfort to me. Your name is written on my heart, and I can never forget all your gentleness and kindness. If it be the Lord's will to take me home, it is my joy to think that we part only for a season. I desire to be most kindly remembered to Mrs. Forbes and the dear children, and ever am,

'Your devotedly attached, and grateful,

'JOHN MACKINTOSH.

'I know I have your daily prayers. The ground of my confidence is the alone merits of my dear Redeemer.'

And so ended the friendship of these two, all of it that was allowed on this side the grave. The record of it is beautiful in itself, and will find a response in many hearts in Scotland which remember one or both of these two friends, so different in character yet united in affection. In a Journal written long after, Forbes made this entry:

'*March* 11*th.*—John Mackintosh died at Carnstadt, near Stuttgard. No event for a long time made such an impression on me: he was a most holy character.'

On the 9th of April the remains of John Mackintosh, having been brought from Germany, were laid by his own last wish in the Grange, as near as might be to the grave of Chalmers. Forbes was one of the friends who attended that funeral.

The letters contained in this chapter reveal a great change that had come over his feelings and character during this last decade of his life. In the autobiographic letter with which the fourth chapter closes there is something too much of self-sufficingness, too great independence of human sympathy, unless it came from the great ones of science. That letter is no doubt a faithful transcript of him as he then was, or at least as he appeared to himself to be. He was then in a great degree self-contained, but perhaps he believed himself to be so more than he was. We may give an exhaustive account of all that we feel in ourselves, and it shall be true as far as it goes. But there is in every man much more, whether of strength or weakness, of good or evil, than the plumb-line of his consciousness can at any one moment get down to: much that lies in him dormant, waiting but the change of circumstances which shall awaken it and bring it to the surface. So it was with Forbes. How strong the family affection was in him we have seen by the way he felt his father's loss. Then came the gradual breaking up of the old home, the scattering of the family, the absorption in scientific discovery, till he, finding himself almost alone, and yet with all his powers in full and delightful exercise, fancied that he could so live always, and that he and science were a match for any two.

But the change of circumstances that was needed to alter all this came in 1843. The peaceful and happy married life which then began, the birth of his children, and the new family ties that grew round him, called out the deep family affection which, during the first active years of manhood, had slept in him unsuspected. Married life, and the home it gave him, unsealed anew those feelings which since his father's death, and the breaking up of his first home, had been without adequate objects: new channels were opened for the long pent-up streams to flow in.

When the summer of 1851 came, Forbes took the house of Craigieburn, a romantic spot on Moffat Water, and there he left Mrs. Forbes and his young children, while he set off on the 21st of June for the Fiords of Norway. As this was the last of his foreign tours, our narrative must turn aside for a while, not only to describe this Norway summer, but to cast a retrospective glance over all the summers he spent abroad from 1835 till 1851.

FORBES' TRAVELS.

CHAPTER VIII.

TRAVELS, 1826 1830.

'If we were to classify travellers in a very general manner, we would consider them as either "travellers" or "tourists;" meaning by the former, persons who visit unknown or distant regions, and by the latter, those who confine their rambles to comparatively civilized countries —as, for instance, to any part of Europe. . . .'

These words are taken from an unpublished article on 'Travel' by James Forbes, written about 1839 or 1840. Its pages, from which space forbids us to quote at greater length, show how well he had studied the philosophy of travel in all its relations; and if, according to his own definitions, the limited geographical range of his journeys places them in the more modest category, it must be allowed that the 'travels' of many men have been less fruitful than the 'tours' of James Forbes.

We cannot do him justice, however, unless we bear in mind that half a century of improvement in locomotion lies between us and the time when he first began to travel. The men of science, the mountaineers and the tourists who now, each returning summer, throng the valleys in which Forbes found himself alone, can hardly realize the difficulties which beset the path of one who pioneered for all three classes, in days when experienced mountain guides did not exist, and railways were unknown.

He laboured, and other men have entered into his labours. The theodolite has been planted on peaks which in his time were deemed inaccessible by travellers, and fatal by guides; for even then, legends of air too rarefied to support life, and of avalanches started by the human voice, still lingered among their crags—the 'trailing skirts' of that departing night of superstition, which had before peopled them with dragons and chimæras. Stories, too, of savage rudeness, if not of crime, hung about many of the less frequented valleys of the Alps. The fact that a pass had never been crossed by man before, was then a reason for avoiding it; it is now an invincible attraction towards it: and Forbes, although by no means the first of English mountaineers, contributed greatly to this change of feeling, by the popularity given by his writings to those Alpine regions which he was among the first to explore.

To attempt a portrait of the traveller himself, is impossible. Many who were the companions of his professorial work, many who knew the tenderness of his friendship, and the deeper tenderness of his domestic life—and all this encased within a shell of cautious and sensitive reserve which left on some the impression of hardness—had seen only half his nature. When on his travels he was another man, and yet the same. To an artist's appreciation of beauty (and he was no contemptible artist), combined with carefully trained powers of observation, there was added the outburst of a sunny and joyous spirit. How he drank in the pure mountain air, how unwearied was his light active step, and how, at the sight of a fresh gleam of sunlight on the landscape, a cloud shadow, a flower even, he would break off his perpetual merry whistle for an exclamation of delight—above all, the chivalrous pleasure in confronting difficulties and dangers, which belonged to his strong and noble character—there are but few now left who can remember. Of the companions and the guides of his journeys some preceded, and some have followed him

'From sunshine to the sunless land,'

leaving the difficult task of attempting to portray him as he was, to those who had seen him only after many long years of shattered health had caused his early vigour and energy to be 'half-forgotten things.' Comparatively few of his letters written from abroad remain, and his well filled note-books cannot supply their place, for they are but dry bones from which to reconstruct the living man. They are records of facts only, for the impressions made on his mind and imagination by the more poetical aspects of nature, he felt, but seldom recorded.[1] And yet, as a dried flower is often more valuable to a botanist than a drawing of the same flower, however artistically coloured, the following sketch of James Forbes' travels, the 'plain unvarnished tale' afforded for the most part by his journals—not written to be read by others[2]—may bring

[1] While urging the moral advantages of mountain travel, 'Their solitude,' he writes, 'is the parent of reflection, and draws forth to daylight the capacities of that dimly seen inward being, which now begins to assert its claim to individuality, but which, amidst the busy turmoil of life, might remain a secret and a puzzle, even to itself. At such times the mind becomes capable of seriously entertaining thoughts which in hours of luxury or business would have been instantly discarded. The young mind in particular seems to discover a link between its powers of conception, and the greatness of the objects to be conceived. The seeds of a poetic temperament usually germinate amidst mountain scenery, and we envy not the man, young or old, to whom the dead silence of sequestered nature does not bring an irresistible sense of awe—an experience which a picturesque writer has thus expressed: "It seems impious to laugh so near Heaven!"' (Unpublished article on 'Travel.')

[2] In 1856 he thus wrote to a friend:—'There is a most serious difficulty in writing travels of many years back from mere notes taken at the time. I have tried it, and found it so. It is twenty-one years since I was in the Pyrenees, nineteen years since I was in the Tyrol and Carinthia, and that minute local fidelity in tracing impressions, which gives value and freshness to a "Tour-book," vanishes, I need not say, in an incomparably shorter period. My mere tourist's notes, in the form of a diary, it would be unbecoming in me to publish, as they are not worth it: while whatever matter was added would run the risk of being either exaggerated or obsolete.'

him more personally before the reader, than if any attempt had been made to present the narrative in a more attractive form.

Forbes' first journey was made in 1826, at the early age of sixteen. He travelled with a large party through France and Germany to Italy, rarely halting for more than a couple of days until they reached Venice, where he was compelled by illness to make a longer stay, after which he continued his journey to Rome. Between this capital and Naples, Forbes spent no less than six months, during which time his remarkable papers on the 'Physical Geography of the Bay of Naples' and other subjects, which were sent anonymously to Sir David Brewster and by him inserted in the *Edinburgh Journal of Science*, had, as the young student found on his return, laid the foundations of his future fame. Shortly after Easter he left Rome, and his journey homewards through Switzerland is rendered memorable by his first visit to Chamounix, a full account of which has been already given in the earlier pages of this book. In 1832 he travelled with his brother Charles to Heidelberg, and afterwards continued his journey alone to Switzerland. He revisited Chamounix, passing on this occasion a week there; and after making the tour of the Italian lakes, was suddenly recalled from Geneva by the news of Sir John Leslie's death. He hastened home to contest successfully the chair of Natural Philosophy at the University of Edinburgh.

It was not until 1835 that Forbes resumed his travels. Three years had now passed since he first occupied a Professor's chair, and the lectures of the still youthful philosopher had proved that he was more than qualified to fill it: but the work had been necessarily severe, and it was full time that he should again have recourse to travel, not as before, for education and amusement only, but also for the relaxation afforded by such a fallow for the mind. 'Those,' he writes, 'who are compelled by circumstances to fill an active public station during three-fourths of the year, can alone appreciate the influence of a temporary seclusion from ordinary habits and impressions. These "hours of idleness," as many would term them, are not lost. As far as physical capacity goes, they serve to store up energy, which, distributed over the other ten months of the year, render every labour not only more tolerable, but also more productive. So, too, of the intellect. During the hours of continuous study many ideas float before the mind of every man, and some are caught as they pass; but he is then too much occupied in the nerve-labour of research, to have much of the superfluous energy required for invention. It is during these hours of relaxation that thoughts of diviner mould arise unbidden, and that the mind, like a magic mirror which converts the distorted elements of a landscape into a congruous whole, reflects from its untarnished surface the scattered impressions derived from experience and meditation, into one bright focus.'[1] During the previous session he had been delivering a course of lectures on heat; and as the hot springs of the Pyrenean valleys promised new and interesting observations on this his favourite subject, he thither betook himself early in the May of 1835.

After some days spent in Paris, he was, as he expresses it, 'safely delivered by the diligence in its portmanteau fashion at Orleans,' whence, by way of Tours and Bordeaux, he reached Pau the beginning of July.

Journal, July 7.

'I left Bordeaux at 7 A.M. by diligence for Pau, and was agreeably surprised by the beauty of the country, for as far as Langon we followed the course of the Garonne through a perfect garden of vines and fruit-trees. . . . The approach to Pau is delightful, through a richly-wooded country, with very fine chestnut trees and oaks, while the box-tree begins to appear in the hedges. To-day the mountains were shrouded in mist, all but their bases, and yet I thought the view from the upper

[1] Unpublished article on 'Travel.'

part of the town, even under these conditions, one of the finest I had seen, especially that from the promenade which belongs to the chateau, but is open to the public.'

Journal, July 9.

'A splendid morning. . . . The whole range of the Hautes Pyrenées was now uncovered, and presented as noble and rugged an outline as I ever saw. I hastened to the promenade to enjoy this superb view, which renders it, I should think, the finest of its kind in Europe. After walking an hour on the promenade, and learning the names of the principal mountains, by the aid of a map and of Brugière's Outlines of the Pyrenees, I sought a new point of view by crossing the Gave de Pau and mounting the exquisitely rich côte on the other side—going out by the road to Lourdes, and returning by that to the Eaux-Bonnes. As I expected, I found in the midst of rich scenery some magnificent points of view, especially that of the Pic de Midi de Pau, looking up the Eaux-Bonnes road, which is bordered by poplars of great height. . . . I left at 4 P.M. for Eaux-Bonnes. . . . . The springs of Eaux-Bonnes enjoy a high reputation, especially when taken inwardly. . . . It is very singular that of two of the springs, which are about 200 yards from each other, one should be a hot spring, and the other cold, or nearly so: and that with nearly the same constitution, as Dr. Darrall assures me; the cold one containing, however, more sulphur. The costumes are very picturesque: the women wear showy red hoods, flat topped and peaked; and the men, scarlet sashes of silk or worsted, which contrast admirably with the dark brown garments of undyed wool with which they are clothed from head to foot.'

Eaux-Bonnes was succeeded by Eaux-Chaudes, with its splendid gorge, to which Forbes had seen nothing superior, except the Via Mala. 'If ever there was a valley of disruption,' he writes, 'it is this one, though I do not pretend always to decide. This confirms Dr. Daubeny's theory of hot springs, especially as these waters issue just at the junction of the granite and limestone. The limestone rises always to the granite, more as it approaches it, and at last is elevated in horizontal strata on the top of it—at least so far as I could judge from a very imperfect examination. What confirms the view of the granite being the upheaving agent is that the valley of disruption is perpendicular in direction to the strata (Hopkins' theory). It is remarkable that there are marks of water wearing (obviously not weathering) on the rocks at a great height above the torrent.'

The scenery appeared to him more and more beautiful as he passed on by Argeles to Luz, a little village lying on the brink of a brawling stream, embosomed in rich meadows, and crowned by a quaint old castle, the chapel of which is a remarkable specimen of Romanesque architecture. The hotels, even there, were delightful compared with those of many more pretentious localities, and for four weeks Forbes made this charming spot his head-quarters.

Journal, July 14.

'Following the custom of the place, I hired a pony and rode to Barège . . . . As unfortunately the supply of the Barège water is quite inadequate to the demand, it is constantly husbanded in cisterns, where it must necessarily undergo a change of temperature, depending more or less on the season. For this reason I took the temperature of only three of the principal springs as they flowed from their cisterns. In these reservoirs the *barégine* or fatty matter is copiously deposited: I procured some with difficulty, the bath-keeper, whose perquisite it is, trying as usual to mystify me. The country people have great faith in its application as a poultice, but I believe this is not sanctioned by the faculty, who, however, seem to be in almost profound ignorance of the mode in which these waters operate, for it is sometimes as unaccountably energetic as it is at others the reverse. Certain it is that the greatest care is necessary in the use of them, and in some cases they have been confessedly fatal. Barège is a perfect hospital, and one is stared at for not having crutches. In the evening I rode up the magnificent pass of Les Echelles, towards Gavarnie.'

To Miss E. Forbes.

'Luz, *July* 22.

'. . . . I am still enjoying myself amazingly, and, although I have been here ten days, I have seldom spent ten days more agreeably. I have been singularly fortunate in my quarters, a fact which makes me the more unwilling to leave them and try my fortune in some other place. I have a delightful room, and the people are attentive in the extreme. They at once got into my ways—and my ways are pretty irregular except that I always dine at four, which gives me a long evening to walk or ride, and in the morning I am up at five or half-past. . . . I have been twice at the top of a mountain called the Pic de Bergons—which is 4,501 feet above my bedroom—and it gave me one of the finest views in the Pyrenees. From it only (as yet) have I seen the famous "cirque" of Gavarnie, with its cascade 1,400 feet high, although it is distant no more than a three hours' ride. This may surprise you, but I do not go post haste to the sights. . . . The Pic de Bergons is charming, and although about 7,000 feet above the sea, it is enamelled to the top with the most beautiful flowers, while on one side it is clothed with watered and mown grass slopes, almost to the summit . . .

'I went to spend last Sunday in a very wild valley, the Vallée d'Héas, in the upper part of which there is a chapel, resorted to by "all the world" on the Eve of the Assumption. It was no fête-day when I was there, so I spent the hours among grand and savage scenery, with only the apparatus of "a prayer-book and a conscience" (as Dr. C—— would say), and I spent them much to my mind. Nor did I think the occasion desecrated because my geological hammer accompanied me. . . .

'I like the people hereabouts better than any I have seen; they appear to me to have really a "bon naturel," and not to be spoiled by strangers, of whom there are very few, except invalids. As I was gargling my throat this morning (having got it relaxed), the good, worthy *femme de chambre* bolted into my room, thinking that I was in the "dead thraws!"'

Journal, July 18.

'I ascended the Pic de Bergons in magnificent weather. I left Luz at about six, and though I ascended very gently, I reached the summit, not by the easiest path, in exactly three hours. This surprised me, as its height, according to Reboul and Vidal, is about 730 toises, or nearly 4,700 feet above Luz. Fifty minutes from the top I found a copious spring with a temperature of 40°·5, and as I walked pretty uniformly, its height above Luz may be 3,400 feet nearly; but I had not my barometer. At the top, a little to the south, and in the shade, I swung my needles very carefully, but the observation was almost entirely vitiated by the discovery of my geological compass in my pocket, after all was over, and after I had previously searched for it in vain when I did want it.

'The view was very magnificent: it was my first sight of Gavarnie. . . . I descended a very steep and fatiguing path towards St. Laurent (which is not visible from the summit), and was well rewarded by a most exquisite view of the last-named village and the valley of Luz, which, from the extreme steepness of the mountain, seems to lie under one's very feet. This, with the stream of the Gave, mapped out as it were upon it, in all its windings, formed the most exquisite picture of the kind I have seen. I must not forget to commemorate a visit from two immense eagles, as I sat alone on the top. One of these, which I had for some time watched making his circuit, seemed suddenly attracted by my appearance, and with the utmost coolness made towards me. He came so near that I absolutely thought he was going to pounce upon my straw hat. But after coming quite close, and making two circles of reconnaissance around my head, (*qu.* a good omen?) he solemnly flew off with his mate towards the Vielles Niéges. I estimate the span of his wings at three and a half or four feet at least. The geological structure of the Pic de Bergons is exactly what I anticipated from the examination of its base as seen from the road to Gavarnie : alternating traps, slates, and limestone, the latter extremely slaty at the summit.'

To Miss Forbes.

'Argeles, Hautes Pyrenées, *August* 5.

'. . . I write to you from a place which, from its situation, may be considered a paradise. It is just where the mountains are lowering themselves into the plains, and where a southern climate is combined with the freshness and variety of mountain scenery. All the little hills are covered with chestnut trees, and trellised with vines, while the cottage gardens are filled with beautiful peach and fig trees, the latter covered with fruit. . . . This morning I have come on fast from Cauterets, one of the most beautiful and (unfortunately) fashionable of the Pyrenean watering-places, the environs being most romantic, and more in the style of the Alps than any part of the Pyrenees with which I am acquainted. The pine grows in characteristic abundance, and in this I confess I have a great delight, for it is a far finer mountain tree than the oak, or beech, or even chestnut, which abound elsewhere. There is some most savage scenery near Cauterets, and splendid waterfalls—but all this profits nothing so long as "Mordecai the Jew sits at the gate," and one is subjected to all the misery of fashionable infestation, much discomfort, and exorbitant charges. . . . From what I have seen, which is yet but a small part, I am disposed to say (although with Dr. Chalmers I will not give a "final deliverance") that anyone who has seen the Alps well, need not die of despair if he does not see the Pyrenees (which otherwise of course he must do). But observe, I include the Italian Alps, which I think among the noblest, and without which he could not have a conception of the mixture of Alpine and Italian scenery which surrounds me here. For example, it seems to me that the Pyrenees have even more peculiarities than the Italian Alps, and some advantages over them, especially in the character of the people. Here, almost everything is peculiar, and the flowers are almost all different from ours.'

Journal, July 29.

'I left Luz on foot at half-past five, for Gavarnie. . . . Beyond Gèdre, where we breakfasted, I found, in accordance with the accurate account of Charpentier, magnificent crystallized limestone in the granite: the granite here, like that of Héas, is altogether a strange rock, and often passes into a slaty structure, becoming gneiss, or mica slate, besides enclosing masses of these rocks, and forming an absolute breccia. But where crystalline, it seems to me never stratified: the fissures and the flat surfaces they leave can never be traced to any distance, and when they disappear, are as often at right angles to the former as not. The appearance of vertical stratification is often caused by the rain-courses, as in the valley of the Reuss at St. Gothard, to some parts of which this valley has a great resemblance. . . .

'*August* 11.—I left Gavarnie at five, on horseback, and in an hour arrived at the Cirque . . . attaining the Brèche de Roland a little before ten. Notwithstanding that my guide had pronounced the glaciers "assez méchants," I found, as I expected, that it only required a little patience to make good each step with an axe, and that real danger there was none. . . . Until the new snow has melted, it is all straight walking. . . .

'I was so little fatigued, that I walked to Gavarnie by preference, riding to Gèdre in the afternoon.

'I never enjoyed an excursion more. The Brèche itself quite exceeded my expectations, and well repaid my second ascent. I found that the first time I had not been more than thirty minutes' easy walking from the top. The view of the Cirque from Saradetz is magnificent, as from this point it loses the confined look it has from below.'

Forbes' next resting-place was Bagnières de Luchon, and while there he made an excursion into Spain, which, although rapid, was not devoid of small adventures.

Journal, August 16.

'I started for an excursion into Spain, that is, for the Vallée d'Aran, the upper part of the valley of the Garonne, which has been very awkwardly separated from France. I left Luchon at about seven, with Bertrand Lafond, the old guide of my friend M. de Raffetot, who, in recommending him to me, remarked, with great simplicity, "Si le bon Dieu le trouve aussi brave que moi, il sera bien heureux!" and I am disposed to say as much of him myself. The walk to Boussos, or Bosost, is charming. A gently rising, richly-wooded valley leads to the Port de Portillon, from whence we descended to Bosost, where the fête of the Assumption is chiefly celebrated.'

To MISS FORBES.

'BAGNIÈRES DE LUCHON, *Aug.* 17.

'I have just returned from an excursion of two days into Spain, where I slept last night. Yesterday I started with a fine morning from hence, and in three hours was in Bosost, where I found men gaudily dressed, with showy embroidered silk handkerchiefs, worn as ladies do, dancing the Bolero; but while strolling about I fell in with a far more interesting sight, a chapel seated amongst rocks on the banks of the Garonne, where mass was being said. All the neighbouring rocks were covered with men, women, and children in the attitude of prayer, all in the most gaudy dresses—quite Spanish—while several pilgrims with staff and escalop completed the scene, *à la* Wilkie. Having stayed some hours, I might have returned here with the other less enterprising curious who went to Bosost, but I had arranged to sleep at La Vielle, the capital of the Vallée d'Aran, and to return a different way by the mountains. . . .'

Journal (*continued*).

'I left Bosost for Viella at half-past three, and in an hour and a half reached Las Bordas, where I was rudely stopped by some ragamuffins of the Spanish National Guards (Urbanos). My permission from the Maire of Luchon to go to Viella and return by the gorge of Artigues de Lin and the Port de Venasque was examined, and then my baggage (which had been already examined) and myself were forwarded, under the charge of an Urbano, to the Governor of Viella, to whom I carried a letter from M. Barron, of Luchon. Of course I thought my troubles were ended, but the Governor proved a genuine Spaniard, hasty and caustic, and ordered my immediate return, under guard, to Las Bordas, treating me with scant courtesy. I at once resolved to descend at night to Bosost, where I could lodge better than at the *posada* at Las Bordas, and this I effected at the risk of travelling in Spain at ten o'clock at night, having performed the journey from Viella with all convenient speed. At last I arrived at the excellent establishment—not quite *posada*, and almost private house—which I had left in the morning, and where I had been lodged in the most cleanly and comfortable manner, with the delicacy and attention of a private family. The dancing was still going on, so I remained until half-past eleven, witnessing the Bolero and Fandango, which were kept up with great spirit chiefly by the men. I was escorted to bed by the two masters and the mistress of the house, and slept so admirably that I could not hesitate to pay a charge of 8 fr. next morning, four of which, at least, might well be ascribed to the quantities of bows and civilities I had received. . . . . Having slept a night in Spain, I left it Spanish fashion, escorted by a troop of dancers, who danced me out of the village. I forgot to mention that a priest, a brother of mine host, joined in the dances with all the alacrity of the rest. It was quite a domestic performance, and there were no strangers but myself. . . .

'*August* 18.—Started from Luchon about six for the Port de Venasque. . . . . Its opening on Spain was magnificent, but the Maledetta was much clouded: fortunately, however, the mists cleared away so far as to let me have a complete view of this giant mountain, with its prodigious glaciers, which seem to me to rival those of Mont Blanc, and to vie with the ice-fields of Grindelwald. The valley of Venasque is a fitting foreground for such a scene: bare rocks with only scattered pines, savage in the extreme. Altogether this is far the finest thing I have seen in the Pyrenees.

'From the Port de Venasque I descended into Spain for an hour and a half to see the Trou du Taureau, a most singular spot, which I wonder is not more known. A river as large as the water of Leith, but incomparably more rapid and deep, is lost at once in a limestone gulf with precipitous banks. The river partly flows under the solid rock, and is partly carried in vortices through openings in the sand which has accumulated at the bottom of the crater-shaped gulf. The river reappears in the valley of Artigues de Lin, communicating with the Vallée d'Aran, and I should have seen its new source but for the incivility of the Governor. There can be no doubt of the identity of the streams. After much rain (as my guide informed me, stating that he had seen it himself) the water rises to a great height in the crater from want of sufficient outlet; and when this is the case, the stream reappears in the valley of Venasque at a spot which I saw, not very far from the Trou du Taureau, being forced out of its subterranean bed, no doubt, by the hydraulic pressure in the crater.'

Quitting Luchon with regret, he entered the department of the Arriège at St. Beat, and travelled by St. Girons to Ax, from which town he dates the following letter to his sister:—

To Miss E. Forbes.

'Ax, Département de l'Arriège, *August* 26.

'. . . Here I am, in quite a new country, perhaps less frequented by strangers than almost any part of Europe furnished with good roads. . . . But notwithstanding all this, and although the mineral waters are stated to be more abundant here than anywhere else, and equal in quality to those of Barèges, nobody comes here unless from Toulouse and the neighbourhood. Altogether, I am very glad that I came here, not only because I have seen a great deal of good scenery on my way from Luchon, but also because there is perhaps no other part of France where you find the French quite pure and unmixed with strangers; and moreover, being a place of considerable local resort, there is a good hotel. The abundance of hot water here is something extraordinary. There are sixty or seventy springs in the town, and one of them, in the market-place, of the great temperature of 168° (scalding hot), constantly sends forth a stream as thick as your arm, to the great convenience of the inhabitants, who use it for washing and every sort of purpose. This is by far the hottest spring I have seen, except the baths of Nero, and perhaps the most abundant. It is marvellous to think where all this comes from, but the quantity of the Pyrenean waters is indeed surprising. . . .

'I was dreaming, the night before last, of my arrival at home. I am not tired of my journey; on the contrary, I have enjoyed it increasingly. I am more and more convinced of the value of such a fallow for the mind, connected with the most favourable reaction upon it of a vigorous bodily state. Yet it is curious how in travelling my thoughts almost always recur to old times, and to the sacred associations of childhood, which, about my time of life, begin to be forcibly felt. Circumstances the most minute present themselves in fresh and glowing colours both in my daily walks and nightly dreams; and many and many a time, when wandering through the beautiful forests of box-trees, have I recollected the days of our feasts, for which it used to be "my favourite green"—as poor Mr. Shand says—and recalled the pattern with which I used to decorate the table, my favourite employment. Yes, the *coup d'œil* of a vast series of those sumptuous entertainments in all the various parts of poor Collinton are presented to my imagination. Pleasing reveries, which I feel it a privilege to be able to recall, and which people the solitary hills and valleys of a distant land (to the eye of one who would be still solitary though all the villages below were to turn out their throngs) with the images of country, home, and kindred—and of some faces which could never be seen again but for the aid of that same happy spirit of association; for—

> "She can give us back the dead,
> E'en in the loveliest looks they wore."

* * * * * *

I did not think to have been led from the hot waters of Ax to writing sentiment, but writing as I do to you, my dear E.—to endeavour to impart a feeling, not to affect an unfelt emotion—I would wish to convey some notion of the trains of thought with which travelling after my fashion (which the Frenchmen whom I meet in parties of from six to twenty or thirty pronounce "bien triste!") inspires me. But this I think one of its greatest advantages.

'The inns in these regions are not in general very bad, but the attendance is execrable, and affords a lesson of patience which one might go through the worst parts of Britain without learning. To get hot water to make tea has cost me more running up and down stairs, more strength of lungs (not a bell in the Pyrenees), and more twirling of thumbs, than all the conduct of an English journey. I have acquired an intimate acquaintance with the French cuisine in the same way, and marvellous it is to see the dinners they turn out with one or two saucepans, a few bits of live charcoal, a gridiron, and an infinite number of earthen pots, which in this country are perfect salamanders. Another of the French arts of which I have always intended to speak is that of the *coiffure* of the women in the S.W. of France, between Bordeaux and Barèges. It consists simply of a handkerchief,—very pretty, certainly, but adjusted with extraordinary skill. . . . The art appeared to me so well worth learning, that I served an apprenticeship to it, and actually "coiffed" a demoiselle in a shop at Barèges, with the greatest success ; so that I hope to be able to impart it, and, like Brummel and his neck-cloth, to leave an imperishable legacy to my country. . . .'

After another descent into Spain for the purpose of examining the springs of Las Escaldas and its neighbourhood, Forbes passed on by Veronet and Arles to Perpignan, where he finally took leave of the Pyrenees, and, travelling by diligence to Toulouse, entered the volcanic districts of Auvergne, and established his head-quarters at Clermont.

To Miss Forbes.

'Clermont, *September* 22.

'. . . . Clermont seems to me the most pleasantly situated of all the large towns of France which I have seen. It is perched upon an eminence rising from an extensive and fertile plain, which forms a sort of bay amongst the hills which surround it for two-thirds of the horizon. These hills are, for the most part, connected with a plateau or table-land of granite, from which the volcanoes rise; and of these, the Puy de Dôme, the highest and most noted, is conspicuous from Clermont, being about the distance of the nearest Pentlands from Edinburgh, and greatly resembling some of them in shape.

'You are not to suppose that the word volcano, as applied to most of these hills, has anything problematical in it, for they are as obviously volcanic as Vesuvius, and their vast torrents of uncultivated lava occupy the valleys with their cindery masses like those of Vesuvius or Ischia. The craters whence the lava has proceeded are equally visible, and often so red as to look as if they were still glowing hot. Besides these unequivocal volcanoes, there are others of diverse character which serve to connect the former with the origin of mountains, like Arthur's Seat and the Pentlands—whence the importance of Auvergne for geology. . . . The Puy de Dôme was the first mountain up which a barometer was carried, at the suggestion of the famous Pascal, and I ascended it fully as much in reverence for his memory as on any other account. He was a most remarkable man; and as he was a native of this place, I hoped to have obtained some new particulars about him, but in this I have failed.'

Journal, Sept. 16.

'I left Clermont on horseback for Mont Dor. The morning was fine, but the Puy de Dôme was clouded over, and we ascended the granite plateau from which the chain of Puys arise, until the Puy de la Vache and its neighbouring volcanic orifices appeared, when we crossed the lava streams to which they give rise, and which are quite similar to those of Vesuvius, with only a stunted vegetation of birches and willows.

'The craters of La Vache and Lassolas are well preserved, and completely thrown open on the S.W. side, to which they have directed their lava streams. . . . There is quite a circle of craters here, among which Mont Jughat and Mont Chat are conspicuous. But the rain soon began to fall and rendered more uninteresting an uninteresting country, until I reached the Mont Dor, when it poured torrents, so that I was soon chilled and wet. During the descent the weather moderated, and I was able to admire the very pretty scenery of the valley. . . . I was, however, truly glad to arrive at my destination.'

Journal, Sept. 19.

'A cloudy morning, as usual at Mont Dor, but it cleared up, and I started about half-past eight to return to Clermont by a circuitous route. I found my way with some difficulty to the rock of La Thuillière, which with the Roche Sanadoire forms one of Elie de Beaumont's centres of *soulèvement*. The road, although marked as such in Desmarest's admirable maps, does not exist in some places; but notwithstanding this, when I asked a peasant on entering it whether I could lose my way, I got the usual answer: "Oh, mon Dieu! non." In a fog it would be impracticable without a guide, but the absolute stupidity of the peasantry in giving local indications is quite characteristic of them. . . . Stopping at Malvialle, I climbed to the top of La Thuillière and enjoyed a superb view. La Roche Sanadoire is a noble object, and the hills round this supposed centre of elevation have remarkably the form of an amphitheatre. Everyone calls these rocks phonolite—I should rather have named them compact felspar, as I did not observe the sonorous character, nor the vitreous lustre. But it little matters.

'The Sanadoire is beautifully columnar, and the Thuillière, although also columnar, splits readily into tabular heaps. A little to the north of these rocks, on the other side of the valley, I found granite not indicated by Beaumont or Desmarest, and the situation of which is remarkable. . . . I passed close to the Puy de Dôme and descended to Clermont by La Barraque, reaching it about five.

'*September* 20.—. . . Ascended to La Barraque, and collected the fine basalt with olivine which occasionally occurs in a very singular manner: having all the characters of imbedded masses of granite, when examined it turns out to be only olivine, with some reddish cretaceous and clayey matter. May not olivine be formed from roasted granite? . . . I continued to ascend to the Fontaine du Berger on the great road to Limoges, then struck off to the Puy de Pariou (whose lava current I had for some time been following), and reached its base in two and a half hours' slow walking from Clermont.

'I ascended to its summit, but was sorry to find its lovely bowl-shaped crater covered with grass and heath, instead of being in its native beauty. Thence I went slowly on to the Puy de Dôme, which I ascended from the Petit Puy. . . .

'*Mem.*—Decomposed granite is sometimes hardly distinguishable from tuff, and domite has nearly the same ingredients: I have no doubt that it is an altered granite. . . . The view from the top, which is easy of ascent, and not so high above the plateau as it looks, is very noble, and I admire more and more the situation of Clermont. The form of the mountain is that of a parabola squared down at the vertex, and its top has somewhat of a crater aspect, but it has obviously been artificially dealt with. A chapel built of quarried basalt and scoriæ formerly existed.

'I descended by the valley of Royat, which I was rather too fatigued to enjoy, as I had not recovered from my cold and the fatigue of Mont Dor. I reached Clermont at five.'

Journal, Sept. 22.

'I left Pontgibaud under the guidance of my innkeeper, to see the Fontaine d'Oule, a hollow in the great lava stream which contains ice in summer, and none in winter. A guide was most necessary, not only to find the place amidst that scene of desolation, but when found to find the ice, the season being so far advanced that scarcely any remained; but a piece "large enough to swear by" was at length discovered. My theory of this is simple. There are notably many parts of the *cheire*, or lava stream, where snow lies in the hollows during most of the year. The *cheire* being composed here almost entirely of detached blocks, I presume that the melting of the said masses of snow produces currents of water at 32°, which, dropping through this grotto at a spot entirely sheltered from the rays of the sun, may readily by its own evaporation be sufficiently cool to render it solid. It is obvious that this process can only go on while the ice is melting, *i.e.* in spring and summer; in winter, as the source is frozen, there is no further accumulation of ice.

'From thence I took the road to Volvic, passing to the west of the Puy de Louchadière (*i.e.* La Chadière, or, in *patois*, the arm-chair), whence comes the lava which is crystallized near Pechadoire. After a pleasant ride of about two hours, I reached the lava quarries above Volvic, situated in the *coulée* of the Puy de la Nugère. The lava was remarkably fissile in the direction of its length, most in a horizontal, then in a vertical plane; and the cavities, which are numerous, are in the same direction. Near the surface, and also, as the workmen told me, near the bottom of the *coulée*, this structure is lost. It becomes compact like basalt, and is no longer able to be readily split,—so far is it from being true that lavas are always most porous near the surface.

'From Volvic to Clermont I followed an agreeable road near the base of the granite table-land, and passed several *coulées* of recent lava. I reached Clermont in about two hours from Volvic.'

Forbes made no stay at Clermont, but started at once for Paris, journeying without intermission through the rich country watered by the Allier and Loire. At Paris he conversed with MM. Arago, Libri, and Fresnel, and obtained from the former the temperature of the hot springs observed by him, making at the same time a careful comparison of the thermometers actually used, with his own. By M. Melloni, Forbes was introduced to Baron Humboldt, 'whom I found,' he says, 'very much what I expected, but I had little time to form a judgment of him. He is very conversational, especially on subjects connected with his own labours, and speaks English fluently. His face is not very striking, and does not at first sight greatly convey the idea of intellectual power.' Forbes only spent two days in Paris, and then travelled back straight to Edinburgh.

The following sketch of Forbes's tour in 1837 is contributed by Mr. E. C. Batten, who was his travelling companion on that occasion. The latter had been for some time his pupil, and between them a deep and affectionate friendship had grown up. Shortly before, Forbes, in his official capacity as Dean of the Faculty of Arts, had presented him, together with the other successful candidates, for admission to their academical degrees, and at the close of the session proposed to him that they should travel to Germany together.

Mr. Batten's *Narrative.*

'Early in the May of 1837, Professor Forbes and I started from London for Antwerp, the first stage of the longest tour ever made by him in one year.

'His first object in visiting Germany on this occasion was to acquire thoroughly the German language, and for this purpose he proposed passing six weeks at Bonn; we accordingly travelled thither by Antwerp and Brussels (where we received a warm welcome from Professor Quetelet), Aix-la-Chapelle, and Cologne.

'Very happy were the six weeks we passed at Bonn. The unusual lateness of the spring caused us to be in time for the first glorious burst of green leaves in the woods which surround Godesberg, and clothe the sides of many of the Siebengebirge. Our mornings were taken up by German lessons, and the lectures of the Professors; and after a mid-day dinner at a *table d'hôte*, our afternoons were devoted to exploring in every direction the lovely scenery which surrounded us. . . .

'The simple society of the Professors of the University was always open to us. . . . A pleasant country walk, preceded by a tea, or followed by a supper, formed the staple of their modest, yet kindly hospitality; but this enabled us to enjoy still more the intellectual and well-sustained conversations arising out of such social intercourse.

'Among the names of those whose acquaintanceship we enjoyed, and whose lectures we attended during our sojourn at Bonn, I may mention Dr. Klausen, who lectured on mythology; Professor Moggerath, Dr. Sach, and Professor Walter, whose subjects were moral philosophy and law; Professor Goldfuss, Professor Mendelssohn (grandson of the composer), and Schlegel: above all, Baron von Buch, the geologist, with whose vigorous mind and enormous attainments Forbes seems to have been particularly struck.'

The following account of their meeting is taken from Forbes's own journals:—

Journal, June 15.

'Finding Von Buch here, I left my letter from Professor Jameson for him, and in the course of an hour he called. I was instantly charmed with the kindness and simplicity of his manner; and as I could see at once that his character for singularity was not one which I had any reason to fear, we were immediately excellent friends.

'*June 16th.*—Von Buch having given me last evening his paper on craters of elevation, I left for him one of mine on the same subject, as well as another on the Pyrenean springs, and this morning I had a visit from him, and a long conversation on the latter subject. He expressed himself as being much struck with my paper, and thought my remarks on the influence of granite, at its junction with other rocks, on springs, original. . . . Altogether he was much pleased with it. As to the history of the elevation theory, he states that he first pointed out four elevations belonging to different epochs —all of which are well seen in Grimm's map of Germany —chiefly by tracing the river-courses. The mountains about the Elbe form one system, the Táunus mountains another, the Schwartz-Wald a third, and the Alps a fourth; . . . pointing out, as evidence of the comparatively recent elevation of the latter, the inclination of the tertiary strata. He gives Elie de Beaumont the credit of finding exact criteria for determining epochs of elevation, but attached little probability to the theory of parellelism on a large scale, considering the systems of chains rather overdone. . . . He considers that carbonated springs are, as a rule, higher in temperature than others, and spoke of the large number of them which are found in Nassau and in other countries, the deeper valleys of which contain hot springs. . . . He also spoke of a theory of variations of atmospheric pressure in summer and winter, depending on the fact that the superior current will sooner find its way to the surface of the earth in winter than in summer. . . .

'In the afternoon I observed the dip of the magnetic needle at Poppelsdorf, with very concordant results, and afterwards called on Professor Gustav Bischoff, who showed me the sheets of his work on the temperature of the earth. He had made experiments on springs, with a view to getting a law of decrease of temperature in ascending; but he was "bothered" by the influence of the proper temperature of the different depths from which they take their origin. He does not consider that carbonated springs are, generally speaking, hotter than others. . . . He has also made experiments on the cooling of spheres of melted basalt, having a diameter of twenty-seven inches, and others of nine inches. He finds for thermometers at different depths in arithmetical progression, the temperature increase geometrically, and the velocity of cooling is as the diameter simply, and refers to Newton's experiment, which it confirms, and which seems curiously mangled in quotation by Dulong and Petit. The specific gravity of glassy slag is less than that of slowly cooled basalt. . . . Von Buch came in as we were examining these.

'Bischoff also took a piece of basalt, split in two, and tied together with twine (the use of the two pieces was to test the completeness of the fusion), fixed it in the axis of a cylindrical mould, and poured molten iron round it. This latter in cooling (according to him) expanded and squeezed the basalt. Certainly, many dykes of basalt in iron appeared, very prettily, but I doubt the squeezing, especially from the large cavities left in the iron, a fact which was also noticed by Von Buch: whether there would be squeezing or not, would be somewhat difficult to predict. . . .

'*June* 17*th*.—I went to the Drachenfels by steamer, in order to make my preparations for a course of hourly meteorological observations at which I purposed assisting, when who should I find at the steamer but Von Buch, on his way to Paris, to which he intends going on foot from Mayence, although he had not said a syllable about it last night. He received me with his usual cordiality, and we had so pleasant a chat that I regretted our arrival at Königswinter. Speaking, amongst other things, of mathematics, he remarked that their invention requires to be stimulated by an appeal to nature; original mathematicians being more or less physicists. "En effet," he continued, "la nature a plus d'esprit que nous. And therefore," he added, "it is a false principle to build up complicated explanations to account for one or two observed facts. Let facts multiply themselves until they explain one another. If this does not happen, I would renounce a theory which should require me to argue, instead of to interpret nature."'

MR. BATTEN'S *Narrative resumed.*

'We made two or three excursions from Bonn, each having for its object something of scientific interest; and after a short absence in the country of the Upper Rhine, during which Forbes had remained at Bonn, we met at Königswinter for the purpose of carrying out a plan, into which he warmly entered, of continuing on the heights of the Drachenfels a series of meteorological observations, which had been taken by a society of the students of the Natural Philosophy class in Edinburgh for some time past. He had never attended our meetings, but had encouraged the formation and progress of the society, and now joined me in taking the observations hourly for two days and a night. It was this hearty zest with which he encouraged and entered into the efforts of his students, whose love of science had been first roused by himself, that forged so firm and lasting a link between him and them; . . . and now, after five lustra spent in law chambers and law courts, I can look back with undiminished interest to that watch on the "castled crag of Drachenfels," beside one of the purest votaries of Science, to whom she was unfolding every day more and more her wonders and her blessings.

'On the 26th of June we had finished our studies at Bonn, and proceeded by Göttingen and the Hartz mountains to Berlin. At Göttingen our stay was short, but we visited the Professors, especially Blumenbach, and Wöhler, the Professor of Chemistry; while Forbes interested himself greatly in the system of management and mode of education followed at the University. We then walked through the forests of the Hartz, examining the geology of that district, sleeping on the Brocken and testing our thermometers in its snowdrifts, and after an interesting visit to the hæmatite iron mines of Backberg, we reached Berlin early in July.

'At Berlin we remained for a fortnight together, preserving pretty much the same course of study as we had done at Bonn. Our German lessons were renewed, and we attended lectures on various branches of science, while many of the Professors, among whom I may name Gustav Rosé, Weis, Ehrenberg, and Encke, were most kind and attentive in making us acquainted with all the objects of scientific interest which Berlin could afford. . . . With Professor Encke Forbes was principally charmed. His astronomical and his magnetic observatory —the latter a pretty cottage in a garden, charmingly situated at some little distance from the centre of the town—supplied my fellow-traveller with two inexhaustible sources of pleasure: a splendid telescope by Fraunhofer, and a spot for swinging his magnetic needles, and continuing the series of delicate observations on the direction and intensity of magnetic force, which he had commenced at Bonn.

'Always more thoughtful for others than for himself, my dear friend chafed under the notion of keeping me within the walls of a city during the dog-days, and accordingly persuaded me to go before him into Saxon Switzerland. He soon after joined me at Dresden, but his extra sojourn in the Prussian capital was not without its effect on his health, and he did not recover his usual vigour until we reached Carlsbad.

'At Dresden I had the advantage of his long experience and consummate taste in Italian pictures as a guide through its galleries, after which we gladly exchanged the heat and dust of cities for further geological explorations, in the course of which we visited the pitchstone rocks of the Tribisch Thal, and the remarkable contact of granite with chalk in the neighbourhood of Meissen. After examining the springs of Töplitz, where Forbes was still feeble and unwell, we reached Carlsbad, and lingered there until his health was completely restored. He was equally attracted by the scenery and the society of Carlsbad, and greatly impressed by its remarkable geological position. "The great tortoise-like shell," he writes, "of recent formation, incrusted over the source of both its groups of springs, now bearing much of the weight of the town, and having within an enormous pressure of gas, is one of the most singular of natural artifices conceivable."

'On the 10th of August we started in a private diligence for Prague, and thence posted across to Budweis, taking the tramway to Linz. . . . Travelling southward, we were now daily getting nearer to the mountains, which I then saw for the first time, though both the Alps and the Pyrenees were already familiar to him; and I well recollect his quiet consciousness of the mighty secrets which lay before us, when he introduced to me a bluish, indistinct line rising above the horizon, as "The Alps!"

'The few days we spent in the Saltzkammergut, passing by Ischl and Hallstadt to the magnificent scenery of the upper Gosau lakes, and thence by steamer to Saltzburg, were days of exquisite enjoyment. I never again trod the mountain side in company with my friend until after his trying illness in 1843. Now, on those towering heights, commanding the most enchanting Alpine scenery, he was full of vigour, his foot planted firmly, his eye ever keen, and his whole tone and air that of a thorough mountaineer.

'Soon, too soon, we reached Saltzburg; and as I was shortly due in England, he persuaded me to leave him to his more detailed examination of the Berchtesgaden, urging me, however, in words which he repeated in a letter written from Windisch-Matrei, to return home as soon as possible, "because then I shall expect you the sooner in our old-fashioned capital, and welcome my dear friend all the sooner to my old-fashioned house." And thus we parted, to meet again in Scotland.'

After the departure of his friend, Forbes, having comforted himself by a series of tough magnetic observations, proceeded to visit the ancient springs of Wildbad Gastein; and having completed the circuit of the Gross Glockner by crossing the beautiful pass of Malnitz, the Matrayer Jock, and the Velber-Tauern Col, he arrived at Innspruck, and soon passed on to Trafoi and the Stelvio.

Journal, August 30th.

'I meant to have crossed from the Saltzburg country by the Velber-Tauern; but, the morning being very bad, I remained patiently at Windisch-Matrei all day, in the afternoon walking up the great lateral valley of the Isel-Thal, a well-cultivated and well-peopled, but most remote spot. A stranger was a real lion; and even at Windisch-Matrei, the *chef-lieu* of a district, my visit was a wonder. The apparition of an Englishman (Lord Auckland) in 1819, and of another in 1832, were still fresh in the recollection of mine host. . . . Next day I started for my Col, on a promising morning, with a guide who also promised fair, and had been strongly recommended; but his request for a morning draught of brandy did not augur well. Very well he went, however, as far as the Tauernhaus, or inn at the foot of the Col, which required fully three hours to cross, and afforded some fine scenery. But as we ascended his pace slackened, and he soon became quite as incapable as my drunken guide of last

Tuesday. On this state of things a thick mist descended, which completely enveloped us; whereupon he insisted on sitting down every five minutes, and drinking water when he could not get brandy, for he had no private store, and my flask was soon exhausted. At length he evinced symptoms of perplexity about the way, which made me suspect he was feigning in order to gain time, and was in the pleasant predicament of doubting between my guide's honesty and his fallibility, when a happy opening in the mist showed the path far below us, just when we were about to be cut off from it by precipices. As it was, we had a difficult scramble over a snow-bed of extreme steepness.

'The Vellar-Thal, into which we now descended, was very wild and precipitous, forming lower down a tremendous horseshoe of precipices—like the Oules of the Pyrenees—over which innumerable cascades poured their waters, and it was late before we reached Mittersill, where I slept. . . .

'Landeck is beautifully situated. The limestone crags which surround it are very steep and bare; but in the midst of them, and on the wrong side of the Alps too, fruit-trees and Indian corn abound, and even the vine grows hardily. When noon sounded to-day from the church at Obsteig as I passed, I saw the field labourers take off their hats—one knelt—and remain in an attitude of prayer until the bell ceased. At the same place I was greeted by an old woman with the extraordinary salutation, "Hübsches Gesichts," which, I presume, may be translated "good dinners!" But the most common form of greeting is "Gelobt sei Jesus Christus," to which the answer is, "In Ewigkeit!" . . .

'I left Mals early, to cross to Bormio by the pass of Santa Maria, through the Grisons, purposing to return by the Stelvio—a charming expedition. The road to Santa Maria lies through a pleasant valley, where the Romance language still lingers, passing Taufers and Münster, the latter in all the brilliancy of a fête day; and then a steep but good horse track carries it on, always through the same ravine, so that it is impossible to miss the path, which eventually joins the great Stelvio road at Santa Maria Sopra. After dining and undergoing passport formalities, I descended to the baths of San Martino, near Bormio, in about two hours. . . . I was exceedingly struck by the appearance presented by one of the snowy peaks in the neighbourhood — perhaps Monte Cristallo, not, I think, the Orteler Spitze—as seen from the elevation immediately above the Cantoniera di Santa Maria. I saw it under singularly favourable circumstances as regards light, and it had the appearance of being formed of solid ice. It seemed absolutely translucent. . . . Next day I left the baths of Bormio, on a lovely morning, to retrace my steps to Santa Maria. I must confess that the engineers have done their work well, for the ascent on the Italian side to so great a height is certainly wonderfully easy. The view towards the south is very wild and grand—wide, undulating snow-fields of great beauty; and immediately on leaving the refuge at the top, the prospect of the Tyrol opens and enlarges for some hundred yards. The day was unusually favourable, and it was perhaps the grandest scene I ever saw. The remainder of the road seems as though it were planned on purpose to show off to the best effect the Orteler Spitze, with the superb glaciers which descend from it, and the magnificent brotherhood of peaks of which it is one. Their forms are beautifully varied, and the stupendous accumulation of snow on their summits is such as materially to affect the characteristic form of their outlines. In fact, under certain lights, as I observed yesterday, some of their summits have the appearance of pure icebergs. . . . I think the advantages of mounting from the Italian side very great; for, although the cream of the view comes upon one at first, it is all the more striking, while, at the same time, its details are successively developed in such a manner as to keep the interest on a stretch during the whole of the descent.'

Forbes now began a chase after his knapsack, which had been mis-sent; and as he had gone to the Stelvio in the lightest of marching order, with full pockets only, he was glad to capture the fugitive some days after, at Botzen. Having picked up a welcome budget of letters at Trent, he entered the 'Dolomite country,' then so little known, now the paradise of artistic tourists, and proceeded in the direction of one of its finest peaks, the Marmolata.

Journal, Sept. 14th.

'This morning, whilst warming myself dreamily during the rain at the kitchen fire at Cavalese, I got an excellent geological lesson from the boiling of certain thin porridge made of Indian corn (polenta) and milk. The air-bubbles disengaged during the process formed the most beautiful elevation craters, often with little interior ones, formed by the immediate sequence of another bubble at the same point. . . . I walked from Cavalese to Vigo, in order to understand fully Von Buch's section of the adjacent beds of rock, and in this I succeeded beyond my expectations. But the strata of limestone and sandstone do not all dip towards the axis, as he has represented, but some from it, and with evident disturbance. I went in search of minerals to a dealer's, but only found some very indifferent specimens.

'The scenery from Vigo to Campedello, and from that to Canazea, is in the highest degree striking—in fact, I know nothing of its kind to compare with it: one is entirely surrounded by jagged peaks of dolomite. The amazing crags of the Lang Kofel come into view near Campedello, and afterwards the magnificent outworks of the Marmolata and Sasso di Val Fredda are seen, in all their majestic beauty. As I wished to examine the Lang Kofel more minutely, as well as Von Buch's sections between Gröden and Colfusco in the Abteier-Thal, I availed myself of a tolerable horse track which wound through the forest, and, after passing close under the precipices of the Lang Kofel, entered the Grödener-Thal. . . . I then proceeded to scale the range dividing Fassa from Gröden, but before we had ascended for an hour and a half, my guide exhibited the usual symptoms of knocking up; in fact, although he had stood yesterday's hard walking very tolerably, it was now apparent that a substitute must be found, and I was lucky enough to pick up a day-labourer at Plan who accompanied me to Colfusco. . . .

'The ascent of this Col was extremely fine; at each side magnificent dolomite cliffs, and behind, the towering pinnacles of the Lang Kofel. . . .

'Quitting the neighbourhood of the great dolomite peaks at Niedersdorf, I started one charming morning, with all my goods on my back—so sick was I of guides! . . .

'I had yesterday passed the source of the Drave, and to-day I descended the valley in which its course lies. The scenery is very pleasing, at least so it seemed to me, for I have seldom enjoyed a walk so much; and as it becomes constantly deeper and more striking, it is on that account finer to descend than the reverse. . . . Its forests also give to its scenery the peculiar character of the Carinthian valleys, as they consist of a variety of spruce fir, exceedingly taper and cypress-like; and although the individual trees are seldom or never fine ones, the change from the ordinary Alpine pine-forest is agreeable. The whole way down, on the right, the valley was overhung with savage dolomite peaks and crags, until I drew near to Lienz, where the valley of the Drave suddenly opened with great beauty, showing a wide extent of cultivation, dotted with neat villages and churches. . . . I took an "Einspänniger" from Lienz to Greifenburg, where I passed the night, as I had resolved next day to leave the post road, and cross to Villach by the Weissen See.'

While Forbes now zigzagged in an easterly direction, southwards to Laibach, and again northwards to Gratz and Vienna on his homeward journey, he does not appear to have found human nature improve as he went. 'Manners,' he writes, 'seem to degenerate as I go eastward. In the Tyrol they are far from polite, but here they are even less so: and the intolerable and perpetual "Was schaffen sie?" (what do you want?) and "Wohin?" (where are you going?) reach even to the acme of Transatlantic curiosity. . . . To-day I was much struck with the disagreeable appearance and air of the country people. The Krainisch, especially, are a piggish looking race, and they seem quite in their element when herding that animal—their frequent occupation. The form of their heads is as peculiar as their features, and their costume is characterized by peaked hats, long leather boots, and general dirt.' Nor yet did the guides of Southern Austria find favour in his eyes, to judge from this leaf of his notebook:—

'Synopsis of guides engaged by me for the longest and shortest distances in Saltzburg and the Tyrol:—Out of seventeen, five were good. Three also good, but arrived exhausted. Four stopped, or did not start. One left behind. One good, but cheated. This explains why I was generally my own carrier!'

Next year Forbes again visited Germany, accompanied by his sisters; but their tour did not extend beyond the baths of Kissingen. In 1839 he resumed his travels, and started for Paris early in May, on his way to the volcanic districts of Auvergne.

To Miss Forbes.

'Lyons, *May* 20.

'. . . Of all my Paris reminiscences, by far the most interesting is having seen Daguerre's pictures, a matter which is now of some difficulty; but M. Arago kindly arranged it for a large party of us, the day before I left Paris. I went certainly prepared for a disappointment, yet I have no hesitation in saying that I was pleased beyond my most sanguine expectations . . . in short, it baffles belief. All sorts of objects were represented: furniture, plaster casts, curtains, &c., all fixed by the camera obscura; open-air views at different hours of the day fixed in three or four minutes, including the most exquisitely delicate objects, and giving the shadows proper to the time of day in a way no artist can do. . . . . Arago repeated several times Herschel's remark on seeing them: "Nous ne sommes que des barbouilleurs —on n'aurait pas pû s'en former une idée. C'est admirable! c'est une miracle!"

'Afterwards I spent some hours with M. Babinet and his friend M. Moigno, and was agreeably surprised to find him well acquainted with my experiments on heat, as well as to find that he had formed a just estimation of Melloni's claims to his discoveries in polarization. I showed him my method of making polarizing plates of mica, a process which I afterwards explained to Lerebours the optician, leaving with him some specimens, and then visited M. Cauchy, who received me with great politeness and explained very clearly his views on the laws of reflection of light on metals. . . . . I gave him a reference to MacCullogh's memorandum, which he did not know, and he expressed great interest in the subject of heat, seeming much pleased by the account I gave him of my own experiments.'

Journal, May 17.

'M. Elie de Beaumont called in the morning, when I gave him my results of two years' computations of the conductivity of soils. . . . I afterwards called on M. Isidore Niepce, who had the week before showed me the first specimens of the Daguerre process, as discovered by his father M. Niepce, and whom I found a most gentlemanly person. He explained the process circumstantially, and I have no doubt correctly.

'I examined the pictures very minutely. The substance is acknowledged to be silver, plated on copper, and Arago suspected that the metals produced some galvanic effect, but it is evident that the surface has most to do with it. . . . . The pictures of both Niepce and Daguerre have this very singular peculiarity, that the parts which are dark, and therefore unaltered, have an aspect simply metallic; and in the case of those on glass, the glass appears wholly uncoated. The sensitive surface which is first applied must therefore be either wholly transparent like varnish, and whiten slightly on exposure to light, like chilled varnish, or, what is more probable, the unchanged portions of the composition are subsequently removed by some chemical action. The consequence of this is, that if you look along any of these plates, so as to get a specular reflection, the picture is reversed, the lights becoming dark, as in Talbot's process.

'I am also inclined to think that this composition is easily soluble, at least in Niepce's process; for he showed me on one of his pictures a blot made by a touch of the finger, where a bunch of flowers should have been. . . . In our presence M. Arago promised Daguerre to advise the Ministry to give him a pension, and, if they declined, he proposed to bring the matter himself before the Chamber. The same evening I left Paris by diligence for Dijon, *en route* for Chalons and Lyons.'

To Miss Forbes.

'Lyons, *May* 20.

'. . . . I arrived here a few days ago, and am astonished by the magnificence of the panoramic views which Lyons commands; I do not think I have ever seen anything to equal them. . . . I am on my way to the South of France—probably first to the Velay, right bank of the Rhone, then to Avignon, where please write, if you have occasion, as I am quite at sea as to future arrangements. I came down the Saône to-day in a steamer from Chalons, a blissful change from a diligence, and yesterday travelled from Dijon through a country filled with imperishable names. The man whose patriotism would not warm on the plain of Marathon would be less inhuman than he whose blood does not circulate more freely in the environs of Pomard, Volnay, and Chambertin, and whose ideas are not elevated by the proximity of the "clos de Vougeot"—although it would puzzle Johnson himself to make that sonorous.'

R 2

To the Same.

'MONTPEZAT, DÉPARTEMENT DE L'ARDÊCHE, *June* 2.

'. . . . I arrived from Le Puy on Friday, and am astonished at the beauty and interest of this country, which rivals even the Italian Alps and the Pyrenees. In a geological sense it is interesting in the highest degree, and I have four or five craters, just as well-defined and as recent-looking as Vesuvius, within a day's walk. But what chiefly delights me is the exquisite beauty of the scenery. There are deep granitic valleys divided by serrated mountains, through which here and there a cindery volcano thrusts his roasted head, while the valleys are clothed with chestnut and mulberry trees in the most exquisite manner. Then the climate, instead of the cold of the Haute Loire, has an Italian feel, tempered however by Alpine freshness. I came here intending to spend half a day, and I shall end by spending four or five, for happily, as I wrote to Jane, things do not press. This is an admirable country for sketching: Jane would never be idle here; you are as sure to find lava in a valley as water, and Giant's Causeways are innumerable.

'I was very much interested to discover accidentally yesterday that a very few years ago a meteoric stone fell in open day in the immediate neighbourhood. The noise was heard by hundreds of people, many of whom have described it to me, and I have to-day been fortunate enough to procure a piece of it. I will try to get more.'

Journal, July 5.

'I went in search of the spot where the meteoric stone was said to have fallen, at Libounez, between Juvinas and St. Pierre Colombier, where I found that everybody in the neighbourhood spoke of it as a thing of yesterday, and never to be forgotten. The field was immediately shown to me, a small enclosure just below the village; and I inquired for the actual spectators of the fall. Dolmaas, who had been mentioned to me, was dead, but with some difficulty I found two brothers named Serre, who were working with some others in a potato field when the stone fell among them. One of these men gave me in his *patois* a most animated account of the scene and of their extreme terror, and with the aid of an interpreter I extracted the following particulars, in which those present whom I questioned agreed.

'The stone fell on the 15th June, 1821, at half-past four in the afternoon, whilst the sky was clear, and the wind blowing from the north. A long rolling noise was first heard; then an explosion like the discharge of a cannon, which occurred five minutes before the stone fell, touching the ground within a few feet of the terrified bystanders and penetrating it to a depth of seven palms (about 5½ feet) in a vertical direction. No lightning accompanied the noise or the fall, but the stone burnt the ground, reducing it to a cindery condition.

'The men were frightened, but not stunned, though the noise was heard at a great distance—one man present declaring that he had heard it at Argentière, distant five French leagues in a straight line. The hole was the same size as the stone. There was no scattering of earth, and it had so wedged itself between two others that it could not be removed without breaking it. The people of Libounez thought it was the devil that had fallen! and did not venture to dig up the stone for seven days, when it was sprinkled with holy water by the priest! It weighed 220 pounds, as I was told by the man who weighed it, and it was sold for six francs; but the fragments have been so dispersed that I with difficulty obtained one or two morsels, although I inquired for them in all the surrounding valleys.

'Such was the circumstantial and apparently authentic narrative which I gathered from the spectators of this most curious occurrence. They are corroborated by a manuscript account, or *procès-verbal*, drawn up by the Maire of the Commune of Juvinas, and forwarded by him to the Prefecture of Privas, where I subsequently discovered and copied it. . . . The meteor of Juvinas is one of the very few which have fallen so near to intelligent spectators as actually to endanger their lives.

'In these regions the appearance of the people and villages is quite Italian, only that the people are more honest, and excessively simple. The accommodation is homely enough, but I have found an inn which contains within a very untempting exterior a clean bed and most civil people: and this, I own, has had some influence in postponing my departure. . . .

'All this is within one day's journey of the Rhone, where "Milords Anglais" are rattling along at this moment, ignorant that on their right hand lies one of the finest countries in Europe! Never mind, let them remain so. A traveller is quite a curiosity, and there is no tradition of an Englishman having ever sojourned here before.

'I arrived at Thuez, quite astonished by the exquisite beauty of the valley, which is far finer than that of Montpezat, and has a more southern aspect. I found excellent quarters at a small inn, but unpleasant, in so far that as the profession of innkeeper was there of secondary importance, there was a disposition to consider the traveller as an equal, and the obliged person. After dinner I went to examine the lava stream, which, proceeding from the crater of Mouleyres, has filled up the whole valley to a depth of 200 feet, and, abutting against the granite mountains opposite, once formed a lake by damming up the torrent of the Ardêche. I descended by the Goule d'Enfer, a ravine formed at the junction of lava and granite, and found myself in a wide gorge, the scenery of which was perfectly magnificent. A lofty wall of basaltic pillars stretches away for more than a mile, in some places 180 feet in height; and while its columnar structure is in the lower part vertical and regular, the upper portion is composed of irregularly placed nascent columns, but the whole is evidently the result of one eruption. . . .

'I am more and more charmed with the scenery of Thuez. . . . Above the lava stream the valley expands and the river runs in a wider bed. At this point, above the southern bank of the river, rise the picturesque ruins of a castle, and here the scenery reminded me of some of the best-wooded of our Highland glens, a resemblance which was increased by hearing a peasant chanting with much taste a wild, plaintive, irregular air, exactly resembling some of our Gaelic dirges. . . . Wild fig-trees grew plentifully around, and were covered with fruit; but towards Mayras the valley became wilder, and the vegetation more scanty. . . .

'Next day, after finishing my examination of the lava cliffs of the Ardêche, I started on foot for Jaujac, determined to ascend the volcano of Neyrac on my way . . . and soon came in sight of the extensive colonnade of basalt which occupies the southern bank of the Alignon for more than four miles. It is more beautifully columnar but far less lofty and grand than that of Thuez; and after scrambling along the north bank with some difficulty for two miles to Jaujac, I was somewhat disappointed with the latter. The Coupe de Jaujac is a low, strong (*sic*), elliptical crater, which has burst at the end of its longer axis, and from the firmness and dimensions of its lava walls I presume it was once lofty—as indeed it must have been to have contained any part of the prodigious flow of lava which proceeds from it, and which was evidently the result of one eruption.

'I waited an hour and a half for the courier for Aubenas, and almost laughed when he approached in a heavy, six-seated, antediluvian vehicle, drawn at a foot's pace along an excellent road, by the most miserable of broken-down ponies. When we got within some three miles of our destination, the courier, in order to make more speed, shouldered his bags, and walked on! a proceeding which I imitated. . . . While settling with the postmistress, I had a new proof of what has struck me very much during this journey—the universality of the fame of Scott. When asked my country, I generally said *Ecossais,—Anglais* being too generally associated abroad with pride, wealth, and extravagance —and I scarcely ever named myself such, in any place, however remote, without the name of Scott being immediately quoted as that of my great "compatriote," who had interested everyone in my country. Even at Jaujac, where I had, as usual, given myself out for a Scotchman, a pettifogging *notaire*, whom I met at dinner, immediately quoted "Le Tay" and "La Clyde" as classic streams. This is a sort of fame to which I scarcely think Scott attains in our country, and surely it is the highest of any. Even among the wild lava streams of the Vivarais he has made his country known and its natives honoured.'

A few days more, and Forbes had quitted, with regret, the beautiful valleys of the Ardêche, and, after a short stay at Marseilles, began to make his way northwards towards the Monte Viso, by one of whose passes it was his intention to descend into the valleys of Piedmont.

Journal, June 29.

'I provided myself with a guide from Meyronnes, and started to cross the Col de Maurin to St. Veran—a most picturesque and striking route, especially at one point above the village of St. Paul, where the river rushes through a narrow gateway in the limestone rock. . . . Higher up the valley is choked by a mass of *dêbris*, a fallen mountain, which descended, it is said, in the fifteenth century, and has evidently caused the formation of the Lac de Paroi. There, the cultivation of corn ends; then comes a narrow pass, which is succeeded by a more open pastoral valley, extending several miles to the Col de Louget, leading by an extremely gentle ascent into Piedmont. In this pastoral valley I saw several masses of serpentine interposed in limestone, and about to be worked for ornamental purposes.

'Having reached the head of the valley, instead of crossing the Col de Louget, I turned to the left, and proceeded to ascend abruptly to the Col de la Cula, leading to St. Veran. The ascent was not very agreeable, for after very steep grass slopes abounding in marmots' holes, came patches of soft snow, into which we sunk knee-deep. and then beds of soft mud, from which the snow had lately melted ; so that a course of loose sliding stones which succeeded was a positive relief. From a point a little to the right of the Col, we had a most magnificent view of Monte Viso and its surrounding chains, together with the Alps of Dauphiné. The descent to St. Veran is steep but not difficult, and on the way I observed the height of the highest level at which I found corn grown in that valley. The village of St. Veran lies at a height of 6,591 feet above the sea. Barley and rye are here cultivated up to 7,000 feet. The appearance of the village of St. Veran—the highest in Europe, except that of Soglio in the Tyrol—is most singular: it is a mass of wooden spars, more like ships' masts and rigging than dwelling-houses, the latter being built of logs, with extensive verandahs on one or more sides, filled with spars for the purpose of exposing to the sun their ill-ripened harvest, which is sown in July, and reaped, ripe or not, in August of the ensuing year, when it is hung up to dry or ripen as it can. . . .

'The only bed at the inn had so disgusted me that, at the risk of hurting the feelings of the inmates, I went off to the Curé and begged a lodging—a request which he courteously and promptly granted. He was somewhat of a man of the world, and expressed his anxiety to be useful to mankind in general, and to me in particular—a feeling for which his order does not often get credit, and for the absence of which they are sometimes unjustly abused. He had a good house, the only stone-built structure in the village except the church, and he treated me very hospitably. . . .

'Next day I had a charming walk to Abries, where I presented my circular letter from the Directeur des Douanes at Paris, to the captain in command here, who immediately sent me one of his men as a guide for the Col de Viso ; and having made all necessary preparations, we started about five for the chalets of La Tronchet, where I had resolved to pass the night. The walk was most beautiful, and on arriving at our destination we proceeded to cook our tea in a saucepan, and then retired to rest on some very clean straw spread on the clay floor, with my knapsack for a pillow, placed against the solid rock which formed the fourth wall of the house. There was, of course, no glass in the windows, but the shepherd had considerately stuffed them up, so that by keeping up a good fire the night passed very tolerably.

'Next morning my *chasseur-douanier*, by name Rey, called me at three, and at half-past we were already on our way to the Col de la Traversette. The ascent was over steep grass slopes as far as the chalets of Monte Viso—miserable places, in every way worse than those of La Tronchet—and from the summit the view was superb, stretching away to the hills above Genoa, and, of course, only to be seen in the early morning. Unfortunately the Alps in the direction of Savoie cannot be seen from the Col, and I neglected to ascend a point on the left, which commands them. . . . The precipitous way in which the Col appears to overhang the plain is extraordinary, so much so that, feeling no desire for a closer acquaintance with Grissola and Pæsana, considering also the value of so good a guide, I proposed to him on the spot to attempt the circuit of Monte Viso, returning to sleep at the chalets. To this he assented, and having observed the barometer and dismissed my porter, whom I had brought from Meyronnes, to await us at the chalet, we at once started. . . . We sought the entrance of the ancient gallery[1] on the Piedmontese side, but were unable to discover it, as it was still covered with snow. There is no question, however, that it is still accessible from that side, and even nearly complete. My guide had entered and examined it, and had observed an iron hook for suspending a lamp; the Curé of St. Veran also

[1] This remarkable gallery, which pierces the Col de la Traversette at about 300 feet below the crest of the ridge, and at a height of at least 9,500 feet above the sea, was cut through the mountain by Ludovico II., Marquis of Saluzzo, in order to facilitate the intercourse between his territory and the adjoining valleys of Dauphiné. Many years after Forbes's visit it was cleared out, and again used; but in some cold seasons it remains choked with snow, until the summer is far advanced.

had been there two years ago; so there appears to be no doubt that the gallery is nearly complete, although the entrance is small and obscure, and the cold and wet make it very disagreeable to traverse.

'Our walk soon became ticklish enough. Large fields of snow lying at steep angles, and affording very bad footing, had to be crossed, some precipices climbed, and immense tracts of loose stones scrambled over: but at length we reached the Lago de Vallante, which is held to be the proper source of the Po. The lake had a circumference of about two miles, and was still frozen over.

'Hitherto our course had been clear; but now, how to get round the back of the mountain, which Rey had never visited, was more doubtful, for a rocky spur descended to the plain, presenting an edge very like a cock's comb. . . . Luckily we found a gap which let us pass, and we thought we were about to complete the circuit by finding a Col to conduct us back to France, when suddenly we fell upon a valley, perhaps 2,500 feet deep; it was the valley which has Ponte at its head, and into it we had no alternative but to descend. Here we had the greatest difficulty in obtaining any information from the inhabitants of the chalets. . . . At last, however, we met with an old man, who resolutely refused to believe that we had crossed the mountain from Abries, but directed us correctly to the descent of the lofty Col de Vallante; so, after dining on cold chamois, we proceeded with as much heart as we could raise to regain the high level from which we had descended. The Col de Vallante joins the Monte Viso on one side, as La Traversette does on the other, and both on the top and during the descent on the French side we found enormous snow-beds, down which we scrambled and slid as best we might, sometimes sinking to our thighs, always hearing the water flowing under us. I made the guide go first, but often made a plunge where he escaped. This side of the Monte Viso is awfully precipitous. Rey, though he had been on one part of it, had evidently no idea of the extent and savage wildness.

. . . During our descent into the valley of Ponte I had wrenched off the heel of one of my shoes, a small matter elsewhere, but serious here ; and although greatly fatigued and annoyed by this, I judged it better not to stop at the chalets of La Tronchet, but to go on to La Monta, where there was, so to speak, an inn. We had been out altogether sixteen hours, with short halts, so in spite of the accommodation I managed to sleep well.'

Next day, notwithstanding his injured shoe, he crossed into Piedmont, by the Col de la Croix, to Bobbio and La Tour, where he made the acquaintance of M. Revel, the Principal of the Protestant College, who subsequently accompanied him to the defile of Pra del Tor, famous in Vaudois history.

After an interval of bad weather, which exercises, as a rule, a disturbing influence on the temper, and may possibly be accountable for the fact that Forbes alludes to the inhabitants of Pignerol as 'odious,' he quitted this flourishing town 'de bon cœur,' and proceeded in the direction of the Pelvoux group, by way of La Perouse and Cesanne. At Monestier, however, 'the most primitively nasty place' he ever saw, he met with unexpected difficulties.

'After visiting the springs of Monestier,' he writes, 'and getting tea with great difficulty amidst an uproarious group of Sunday revellers, a gendarme, who had asked for my passport, very coolly and distinctly informed me, with evident marks of satisfaction, though civilly enough, that I could proceed no further, but must remain at Monestier for at least several days. On my remonstrating, he pointed out that the *signalement* was wholly wanting in my passport. It was one of Lord Granville's, and I am afraid rather a slovenly document. However, after a very short discussion as to the non-importance of this part of the solemn farce, I walked upstairs, and produced a letter from the Directeur des Douanes at Paris, requiring all his agents on the frontier to assist and aid me to the utmost of their power. Then the little gendarme—who was a sharp, Napoleon-like fellow, and, I suspect, had some little grudge against the English—saw that his game was up. He fought, however, as long as he could, verified my signature, and started difficulties about the fact of my letter bearing a date subsequent to the day of my departure from Paris; but all this I answered by producing the letters of M. Griterin and M. Arago, with their envelopes addressed to me at Avignon, which I had fortunately preserved. The evidence, in fact, was so complete, that "Napoleon" could only say that he was satisfied, and that I might proceed. I conducted myself all through with great *sang-froid*, but I confess that I did not sleep any the sounder for the interruption.'

Having overcome this little difficulty, he started for Bourg d'Oysans, which he reached after a long day's walk over the Col de Lauteret, and then advanced into the heart of the Pelvoux group, to St. Cristophe and La Bérarde. 'The scenery,' he writes, 'is stupendous, unparalleled, perhaps, except in the neighbourhood of Mont Blanc or the Orteler Spitz. . . . It is far finer than the Val Romanche, and Vénos is quite exquisite.'

Forbes then went to Grenoble, and by Chambery to Geneva, where he found his brother John established with his family at the Ecu. In company with them he paid a visit to Chamounix, where they made several of the smaller excursions, amongst others that to the Jardin; 'the scenery of which,' he writes, 'was not less fine to me than I recollect at the commencement of my Alpine journeys, and at this I greatly rejoice.'

Towards the end of July Forbes went round by the Col du Bonhomme to Aosta, where he sought about for a man to guide him towards the then unknown valleys, above which rise the Grivola and Grand Paradis. He succeeded with difficulty in finding a guide—'a short, thick-set man, afflicted with goitre, but a good-natured, honest fellow enough, and who walked wonderfully well. He was married to his *third* wife!'—and after examining the iron mines in the neighbourhood of Cogne, he proceeded to cross the chain of the Grand Paradis to Cuorgne, *en route* for Turin.

Journal, August 1.

'. . . I left Cogne with a guide recommended to me by Dr. Grappin, for Ponte. There were four different paths to choose from, and that recommended by Brockedon is the only mule path, I believe, the others passing over glaciers. The one I selected, which is called the Col de la Nuova, is considered the shortest, at least in point of time. I found the glacier part of it by no means difficult, and on the whole the pass resembled the Col de Traversette on Monte Viso : but on the Piedmontese side the descent is most precipitous. The chain of Mont Blanc is finely seen from the summit—indeed from several points in the valley of Cogne ; but in order to see Monte Rosa, I was obliged to climb a rock which rises on the right of the pass, and from this I obtained a fine view : but the northern side of the valley of Cogne conceals too much, and on the whole, judging from Brockedon's description, the view from the Col de Reale must be preferable. . . . After descending to the village of Campiglie, the scenery becomes charming, and so it continues without interruption to Ponte, which is beautifully situated, but the landlady of the inn was drunk, and the inhabitants were disgusting—so I went on to Cuorgne, where I found better quarters. My guide, who yesterday had been very compassionate, inviting me to sit down, and advising me not to look over precipices, began to find that I was made of tougher metal than he thought. I don't think I sat down quite enough for him, for he was completely knocked up before we got to Cuorgne.

'*August 2nd.*—I left Cuorgne by diligence for Turin. . . . As the weather was very hot, I rested after my journey, but in the afternoon I saw Sig. Plana, and found him, as usual in such cases, overwhelmed with official engagements—what lost time compared to the petty consequence or the emoluments of an official position! He spoke of Libri's book on the History of Science as wasted time, which would better be employed by enlarging the scope of science; and he spoke in the same strain of the popular writings of Herschel and others . . . but does he do better himself? . . .

'*August 3rd.*—I called on M. Sismonda, Professor of Geology, and saw his carefully arranged and catalogued collection, as well as his geological colouring of Maggi's map of Piedmont: I also visited the Egyptian Museum. Sig. Plana kindly gave me a copy of Taylor's very rare "Methodus Incrementorum."'

After leaving Turin, Forbes proceeded on foot to Lanslebourg, and next day crossed the Col d'Iseran, which he describes as 'savage and wild in aspect, and planted with numerous crosses, commemorating not only accidents, but murders.' This brought him to the 'abominable' chalets of Lignes, and eventually to Bourg St. Maurice, Chambery, and Lyons. He thence returned home by Paris, and early in August was again in Scotland.

CHAPTER IX.

ALPINE TRAVELS.

FORBES spent the summer of 1840 at home, but in the autumn of that year an event occurred which sent him back to the Alps the following summer with renewed interest. At the meeting of the British Association held at Glasgow, he fell in with M. Agassiz, who, already well known as a naturalist, had turned his attention to glacial phenomena, and had recently published his 'Etudes sur les Glaciers des Alpes.' Forbes was already familiar with the general aspect of glaciers, and had been much struck by what he saw of them. M. Agassiz spoke of a series of observations which he proposed to make on the Lauter-Aar Gletscher, of such a nature as to entail a residence of some days in a hut on the glacier itself, and invited Forbes to share his bivouac. The invitation was gladly accepted, and Forbes accordingly started for the Alps early in July 1841, accompanied by his friend and pupil John Mackintosh.

After a short stay in Paris, they descended the Rhone to Valence, and thence drove to Thuyetz, where Forbes found himself once more among the ancient volcanoes of the Vivarais. To these he devoted himself until his rendezvous with Agassiz at the Grimsel, which was fixed for the 8th of August.

The six weeks which intervened were well spent. 'During my previous visit in 1839,' he writes, 'I had found in the Vivarais a united attraction of scenery and geology, together with that isolation and remoteness which lends a peculiar, though doubtless a selfish, charm to a prize which we imagine that others have in some degree overlooked. This caused me to fix my quarters in the very first village I reached, and again two years later (1841) to revisit every point of geological interest, to extend my notes, and to prepare a map and drawings of the volcanic phenomena.' The unwonted presence of travellers appears to have excited considerable curiosity among the somewhat morose inhabitants of this remote district. At Thuyets the Professor was taken for a tax-gatherer, and roundly abused by an old woman on that supposition. Again, at La Bastide, beside which rise the ruins of a castle once belonging to the Count d'Autraigues, a victim to popular fury during the French Revolution, the consciences of the men whose fathers had sacked his chateau and divided his lands smote them, and they said, 'The grandsons of the old Count are come!'

After lingering for a few days at Thuyets, 'sketching, geologizing, and bathing in the limpid Ardêche,' they turned their steps towards the district of Le Puy, and, although foiled by bad weather in many intended excursions, they explored much interesting country. But time pressed, and they soon turned eastward towards Agassiz' trysting-place, which it was Forbes's intention to reach by the Little St. Bernard, the Col Ferret, and the valley of the Rhone, having first, however, revisited the Alps of Dauphiné, and crossed, if possible, some of the glacier passes of the Pelvoux group. In company with Mr. Heath of Trinity College, Cambridge, who had joined him at Grenoble, he succeeded in making two interesting passes: the first of which, the Col de Sais, had probably never been previously made by a native, the second, the Col du Célar, certainly never by a traveller.

To MISS FORBES.

'LA GRAVE, VAL ROMANCHE, DAUPHINÉ,
A wet Sunday, August 1*st*, 1841.

'. . . In my letter to John I described our passage of a Col which no strangers, so far as we could hear, ever crossed before, and how we descended into a valley blocked up by glaciers at its head, which testified its astonishment by the mouth of the Mayor of the village, who asked for our passports, and told us that *bâtons ferrés* were prohibited in France. The gendarmes were secretly sent for, from the nearest town, in order to inspect the strange strangers. After a day of repose, we jumped out of the valley, as we had jumped into it, across a glacier, which proved more difficult than the last; and the precipices we had to descend altered, I confess considerably, the sense which I had always attached to the word inaccessible. The inns are so horrible, that we lodged with the *curés*, and met with an amicable reception and clean beds, an immense luxury. Our time is limited by our rendezvous with Agassiz at the Grimsel on the 8th, otherwise we should have spent more time in these valleys; but we have seen enough to give us a fair notion of their general as well as of their wildest features. Never in Switzerland elsewhere have I gone through such places. We went to the valley of Arnieux, Felix Neff's parish, and spent an afternoon very pleasantly with the Protestant pastor, a rough zealous man of the Covenanter character, and well fitted, I daresay, for his arduous and ill-paid situation. . . .'

They reached the hospice of the Grimsel on the evening of the 8th, and found that M. Agassiz had arrived a few days before, the *avant-courrier* of the numerous party who were to share his hut on the Unter-Aar Gletscher. Dr. Voght the naturalist was also there, and M. Zippach the keeper of the hospice, while two days after their party was increased by General Pfuyl, with MM. Studer, Escher Von der Linth, and Desor. On the 9th M. Agassiz, with Forbes and Mr. Heath, made trial of the hut as a sleeping-place, only to be driven back next day by heavy snow. But although thus foiled in their first attack, they did not return empty-handed; for then for the first time what is called the veined or ribbon structure of glacier ice—which, although previously remarked, had been passed over as superficial and unimportant—was observed by Forbes, and at once pointed out to the others as a phenomenon of the utmost significance and importance. 'It was fully three hours' good walking,' he writes to Professor Jameson, 'on the ice or moraine, from the lower extremity of the glacier to the huge block of stone under whose friendly shelter we were to encamp; and in the course of this walk I noticed in some parts of the ice an appearance which I cannot more accurately describe than by calling it a "ribboned structure," formed by thin and delicate blue and bluish-white bands or strata, which appeared to traverse the ice in a vertical direction, or rather which, by their apposition, formed the entire mass of the ice. The direction of these bands was parallel to the length of the glacier, and, of course, being vertical, they cropped out at the surface; and wherever that surface was intersected and smoothed by superficial watercourses, their structure appeared with the beauty and sharpness of a delicately-veined chalcedony. I was surprised, on remarking it to M. Agassiz as a thing which must be familiar to him, to find that he had not distinctly noticed it before; at least, if he had, that he had considered it as a superficial phenomenon, wholly unconnected with the general structure of the ice. But we had not completed our walk before my suspicion that it was a permanent and deeply-seated structure was fully confirmed. . . . We did not sleep that night until we had traced it in all directions, even far above the site of our cabin, and quite from side to side across the spacious glacier of the Finster-Aar.'

Next day they returned to the hospice, but on the 11th the whole party of eight went to the hut, where they slept—'a close fit!'

The scene of their encampment lay among the grandest and wildest scenery of the Bernese Oberland. The two great glacier streams which have their origin in the neighbourhood of the Finster Aarhorn and Schreckhorn unite in their downward progress at a rocky promontory, which bears the name of the Abswung, and their combined streams form the Unter-Aar Gletscher, which travels for five miles from the Abswung before it gives birth at its lower extremity to the waters of the Aar. On the medial moraine, which, taking its rise from the point of junction of the two branches, stretches down the centre of the main glacier, may still be seen the remains of the huge rock which was now to be their only shelter for some days, although the ice in its patient march has since then borne them many hundred yards in the direction of the Grimsel.[1]

The following letter to Miss Forbes gives an animated picture of their bivouac life :—

'*From* M. AGASSIZ' CABIN ON THE AAR GLACIER,
'8,100 *feet above the Sea. August* 10*th*, 1841.

'You never, I am sure, got a letter written under circumstances like the present. Here are M. Agassiz, Heath, and I stretched out on beds of hay under a huge detached rock, lying on the ice of one of the greatest glaciers of Switzerland, in the very midst of which we have taken up our abode for some days. The nearest house is the hospice of Grimsel, five hours' walk nearly. But add to all this, that we have passed the night in a storm of thunder, lightning, and snow—the most tremendous thunder I ever heard, mixed every now and then with the fall of an avalanche or the rolling down of stones from our own bed—the ice. If you will turn to Plate 14 of Agassiz' Atlas, . . . you will find a most exact picture of our residence. It is on what is called the moraine of the glacier of the Aar, the immense heap of stones which goes along the centre of the glacier, and which is not all stones to the bottom as you might suppose. It is really ice protected from melting by the covering of stones, so that one side of our parlour and bedroom is a wall of ice, which supports the huge block which makes the roof of our house ; which wall is an object of study by the crevasses which form themselves in it, increased

[1] When Forbes revisited the spot in 1846, the block was found broken in two.

by the heat of our bodies. The kitchen is without, under the same rock, where strong decoctions of tea, coffee, and soup from time to time are made. It is astonishing how we accustom ourselves to anything. We have our wardrobe nearly all on our backs, and of course we did not undress very much; but this morning, by Agassiz' advice, I sponged myself over from a pail of iced water in the open air with great refreshment. To complete the novelty, I have just smoked a cigar for the first time in my life, with no disagreeable consequences, but the reverse. If this weather lasts, we go down the Grimsel to await better. Agassiz has just asked the guides without how the weather is? And the answer, "It is raining white," is not consolatory: the snow is eight or ten inches deep. We have brought plenty of clothes, books, and instruments, not to mention victuals and wine, which keep mind and body in a most healthy state. We are fortunate in being alone with him, the other companions of his journey being not yet arrived. Agassiz has brought the *Athenæum* and *Literary Gazette* down to the 24th July; so we are quite in the world, and yet out of it. . . .'

Journal, August 13th.

'Occupied much time in actinometrical observations. In the course of the day M. Chambrier of Neuchâtel joined us with three friends, and left us before evening. We went down to the east side of the Lauter-Aar Gletscher, to examine a contact of polished rock and ice, of which M. Agassiz took a section. Some wine we had poured into a hole in the ice (with a temperature of 28°) had wholly disappeared this evening. . . . There slept this night in the *cabane* MM. Agassiz, Voght, Girard, Robertson of Elgin, Burckhardt the artist, Heath, and myself.'

Two young English noblemen, Lord C—— and Lord N——, had arrived during the day, seeking quarters for chamois shooting: but our philosophers were there for business, not for amusement, and they seem to have found the party of chasseurs—as the 'Red Prince' found 'our own correspondents'—somewhat in the way. The hut was full, and Agassiz was inflexible, so they were sent to sleep with the guides.

'*August* 14.—To-day we had a visit from Lord Enniskillen and the Hon. John Cole; the former remained. There arrived, also, M. Martins (of the French Spitzbergen expedition) and M. Canson. We walked up to the very foot of the Finster Aarhorn. The glacier becomes very flat, and has a most imposing appearance of breadth, opposite the entrance of the valley leading to the Strahleck. . . . The view from the flat part of the glacier is most imposing; the precipices of the Finster and Ober-Aarhorns rising sheer from the flat surface.

'We ascended part of the *Névé* or *Firn* leading to the Strahleck, where the texture of the snow is soft and entirely granulated. The blocks appear to rise through it by their action in melting the snow, and forming a pillar of ice upon which the block is gradually raised. This is M. Agassiz' idea. In returning we observed that the vertical structure of the ice must be of a permanent character, for we traced it across old crevasses for considerable distances, well marked by the vertical stratification, and showing a shift, from the lateral moraine, at each crevasse in descending order. We slept this evening, Agassiz, Lord Enniskillen, M. Martins, Canson, Heath, and myself. . . .

'*August* 15.—We had intended to-day to cross the pass of the Gauli Gletscher, but Jacob did not consider the morning sufficiently favourable to call us: and as the sky soon became cloudy and threatening, we resolved to descend to the Grimsel, completing, on our way, our examination of the structure of the glacier, by a careful scrutiny of its lower part. We traced, without difficulty, the vertical stratification on both sides of the medial moraine, but for a considerable space found it not truly vertical, as it leans towards the lower part of the glacier surface, at an angle of about 70°. Towards the lower extremity the ice becomes distinctly granulated and composed of fragments, re-cemented as Agassiz describes, and this structure may be traced at the very edges, but not generally throughout the glacier. The stratification becomes less apparent below, but is still clearly traceable. It rises towards the wall of the valley, whilst the whole middle, from leaning or falling forward, appears to give a central horizontal stratification; but in reality these strata, as I believe, dip inwards at a considerable angle.

'In the upper part of the glacier yesterday we traced the vertical structure quite into the valley of the Strahleck, near the lateral moraines, *i.e.* wherever the ice is well formed, so that I ascribe its disappearance or very imperfect formation in the centre, not to the want of structure in the half-formed glacier or *névé*, but to its incapacity of demonstrating it, there being no formed strata of blue ice. We also traced something of it in the glacier descending from the Finster Aarhorn.'

Monday was a *dies non*, for the rain fell in torrents, but Agassiz and Forbes faced the storm and walked down the valley to Rosenlaui, devoting the next day to a careful examination of the grooved and striated rocks in the neighbourhood of its beautiful glacier. M. Agassiz had effaced the polished surface with a hammer within a triangular space close to the ice, in order to see whether the advancing glacier would renew its presumed work; but they found that the ice had retreated at this point—so the experiment, with which Forbes appears to have been much struck, was as yet a failure. They then walked back to Imboden, where they fell in with Professor Guyot, of Neuchâtel, and, as they were somewhat fatigued, passed the night at Guttannen. M. Guyot was an intelligent companion, and a great tea-drinker, while M. Agassiz gave them a most interesting history of his studies, and the development of his intellectual powers. So the evening passed pleasantly away.

Journal, August 18th.

'A charming walk from Guttannen to the Grimsel. We carefully examined the polished rocks as we passed, rising as they do 1,500 or 2,000 feet above the valley. They are not all polished, and even for large spaces there is not a trace, but I observed two facts which I think nearly general. (1) The polish is decidedly greatest on the side of every projecting rock looking up the valley, against which, therefore, the ice would abut. (2) The inclination of the striæ does not depend, as Agassiz supposes, on the combination of the vertical motion of the block in the ice with the longitudinal motion of the ice (which would be an insensible quantity), but the striæ rise before a contraction of the valley, and fall after such a contraction: the ice rises to surmount an obstacle; having surmounted it, it falls. . . .

'I cannot for one moment doubt that the configuration of the rocks in the valley of Hasli is entirely superficial, and independent of crystalline or other structure, and that it is not performed by water. The continuity of the strokes, the depth of the furrows relatively to the hardness of the rocks, the immense height above the valley to which this polishing action extends, seem conclusively to negative the idea of water action. . . .'

In the afternoon Agassiz went back to the Aar Gletscher, while Forbes and Mr. Heath ascended the Sidelhorn, finding, on their return, that Mr. Faraday had arrived at the Grimsel. The next day again saw them inmates of the hut.

'In the evening we had a splendid display of the phenomenon of the shadow of mountains projected on clouds, which I once saw by moonlight at the Mer de Glace at Chamouni. It was about 6 o'clock, when the sun was near setting. The peaks of the Lauter-Aar and Hugi-hörner were grandly depicted on a cloud a short way above them. This evening there slept at the *cabane* MM. Agassiz, Nicholson, Burckhardt, Heath, and myself.'

This was their last night at the *cabane*. During an ascent of the Schneebighorn and an exploration of the Gauli glacier, next day, Forbes was partially engulfed in a crevasse, and, falling forward, severely strained both back and legs: the accident did not prevent him from pursuing his observations during the remainder of the day, but he did not again accompany M. Agassiz to the Aar Gletscher, remaining for the whole of the next week at the Grimsel hospice. Part of this detention was caused by a burst of bad weather which drove Agassiz from his glacier—'our glacier' as they always called it—and the belated party, 'MM. Agassiz, Voght, Desor, Burckhardt, Heath, Mr. and Mrs. Trevelyan, and myself,' must often have remembered those pleasant days, when the logs crackled on the great open hearthstone, while the storm raged without; and when tired of social chat M. Agassiz lectured on glaciers, or Forbes on the chemistry of heat, until they came to the conclusion embodied in an old German proverb, 'Besser schlechtes Wetter als gar keins!'

A week afterwards, while suffering from snow blindness brought on by his ascent of the Jungfrau, Forbes dictated to a friend the following sketch of the party:—

'*August 26th, Thursday.*—The weather being a little better, M. Agassiz went up to the glacier with M. du Chatelier and M. de Pury of Neuchâtel, who arrived yesterday in the midst of a torrent of rain, which Dr. Voght predicted would prevent the arrival of anyone but a hot-headed Englishman. The morning was tolerably fine, but the snow had fallen quite down to the level of the hospice, and this was all against the projected expedition across the Ober-Aar Joch, for which we had now waited several days of dreary bad weather.

'We scarcely regret, however, on recollecting the sociable hours which we spent in this place, to which so many travellers come with no other wish than to leave it as soon as possible. During the worst weather our little society was almost confined to the following members:—M. Agassiz, cheerful, kind, and frank, not much disposed to active exertion within doors, but always ready to contribute to the cheerful companionship of a party, the chief of which he justly considers himself. Dr. Voght, a laughter-loving, and withal shrewd young man of twenty-three, a true German in complexion, phlegm, and habits. During the excursions of his more adventurous companions his chief home was near the snuggest corner of the smoke-inviting wood-fire which blazed at the end of the long, low hall of the hospice; here he would remain almost as motionless as the red molecules of doubtful origin which it was his chief delight to watch through a powerful microscope in the coloured snow, the gathering of which was the object of an almost daily walk to the Sidelhorn. M. Desor, a Frenchman by birth, a friend of M. Agassiz, and the journalist of the expedition: like all journalists, placed in rather a difficult and dubious position, from which even a certain share of natural quickness and French vivacity did not serve wholly to extricate him. M. Burckhardt, the artist of our Agassian Club, a shrewd sensible man, with a sly smile and some dry humour, often successfully used in rebutting the sallies of M. Desor. Mr. and Mrs. Trevelyan formed latterly part of our daily circle: the one quiet, gentlemanlike, and well-informed; the other quick, rather erudite, and fond of art. They usually spent their mornings with us, occupying themselves by drying flowers or finishing sketches. Zippach, our landlord, was an excellent manager—all went on smoothly, cheerfully, and without annoyance or a single unkind or hasty word. Everyone was well served, and that although the situation rendered it impossible to avoid sometimes a tumultuous overflow of guests. By 8 or 9 A.M. all passing travellers were gone, and a still repose reigned until three or four (we usually dined at one or two), when travellers began to drop in so rapidly that at seven supper was laid for thirty or forty—perhaps even sixty persons—who had all to be lodged in clean dormitories of the establishment. At this hour we generally took tea, in which the whole Grimsel party above named often joined us. Many and pleasant were the evenings we thus spent!'

On the 27th the weather cleared, and they accordingly started for the Ober-Aar Joch, a pass which at that time had been only traversed twice before, with the intention of afterwards attempting the ascent of the Jungfrau. A full account of these expeditions forms an appendix to Forbes's work on 'Norway and its Glaciers,' published in 1853.

The following letter conveyed to his sister a sketch of this, his first considerable ascent in the Alps :—

To MISS FORBES.

'BRIEG, *29th August*, 1841.

'You will probably be surprised when I tell you that I have just returned from the summit of the Jungfrau, where M. Agassiz and I and two Frenchmen, with four guides, arrived yesterday afternoon at 4 o'clock. You thought me wiser than to undertake any such expedition, and I own that but for Agassiz, who had an intense anxiety to ascend the Virgin Alp, so often assailed in vain, the idea would never have occurred to me. Since it succeeded, I am, you may believe, very glad to have made the attempt; not only for the thing itself, but because it gives me a perfect idea of what I have long wished to know, the real amount and kind of difficulty of the most formidable Alpine ascents, for I suppose that there is no doubt that the ascent of Mont Blanc, though longer, is by far less formidable. We commenced on the 27th August by a very interesting walk from the Grimsel, by the vast mass of glaciers which fill up the Oberland, or rather the space between it and the Vallais (as you will see by the map). We ascended the upper glacier of the Aar, and, passing a Col nearly 11,000 feet high, descended with much labour, but without much danger, the Glacier de Viesch for about five hours, until we reached some chalets on the neighbouring glacier of Aletsch, where we slept after a very fatiguing walk of twelve hours over glaciers in every form. Yesterday morning we started at six, being prevented accidentally from going sooner, for the Jungfrau, preceded by the rather ambiguous emblems of axe, rope, and ladder. We had four and a half hours' rapid and not difficult walking over the glacier of Aletsch before we began to mount considerably, and then we had much snow to cross and concealed crevasses to avoid, going cautiously over them tied together by the rope. When quite at the foot of the steep part, we had on one hand nothing but precipices of snow and ice before us, and on the other of rock. The former alone appeared assailable. We crossed a great crevasse, opening nearly vertically, by means of the ladder, and then had to make our way up a wall of snow which overhung it in such a way that if our footing failed we must have gone right into it. The risk was, however, rather apparent than real; for the snow being moderately soft and yet consistent, we could dig in our feet so as to keep a good hold, and thus we gained first a more moderate slope, and finally a Col which separates the glacier of Aletsch (see Keller's map) from the valley of Lauterbrunnen. But the worst was to come, for we had yet 800 or 900 feet to ascend of the final peak, composed of a surface highly inclined and terminating below in precipices some thousand feet high, on the one side in the Roth-Thal (near Lauterbrunnen), on the other side in the Aletsch valley. Had this surface been of snow, as before, it would not have troubled us much, but it was almost all solid ice. Every step was made with a hatchet, and every foot secured before the other could be moved, so that we were two hours in ascending. When we reached the top, we found it a point on which one man could not stand before the snow was beaten down! Of six travellers and seven guides who started four of each reached the top. We were obliged to stand upon it in succession, and planted a flag. The view was clear in some directions and very magnificent. On one side we had only clouds, but such clouds! I noticed one which seemed to extend from the very valley of Grindelwald to 2,000 or 3,000 feet above us. A single cloud of 12,000 feet high was a thing I never saw before. The thermometer was 25°, and we soon descended. We were obliged to come all the way down the icy slope backwards, taking every step (about 700) as we had come up. By the admirable provisions of our guides (of whom the chief, Leuthold, is by far the best in the Alps) this was much less nervous work. . . .'

'VAL TRAVERS, NEAR NEUCHÂTEL, 21*st Sept.* 1841.

' . . . . I wrote to John a few supplementary notes on the Jungfrau business, with one or two scraps in explanation. I had a sharp inflammation of the eyes after I wrote, which you can but too well sympathise with; and I am sure you would think the ascent of the Jungfrau dearly purchased by any permanent injury. Fortunately my cure was rapid and complete, and I took the advice of an English Indian doctor, who was very obliging, and who advised me to avoid glaciers for a time. It is very curious that the glaciers properly so called never affected me in the least degree: it was the intense whiteness of the snow and the force of the sun above 10,000 feet. I then spent nearly a week at the north foot of Monte Rosa:

> "At the foot of Parnassus contented to stop,
> For fear of a kick from the nag on the top!"'

At Zermatt Forbes lost the companion of his previous journeys, parting from Mr. Heath on the summit of the St. Theodule pass, which the latter soon after crossed on his way to Courmayeur and Paris. Forbes lingered a few days longer at Zermatt, visited the Riffelberg, and explored the glaciers of Gorner and Zmutt, after which he retraced his steps down the Rhone valley, devoting a day to the examination of the *soi-disant* polished rocks and moraines in the neighbourhood of Vernayaz. The evidence he there found, seems to have removed from his mind all doubt as to the glacial origin both of *roches moutonnées* and striated surfaces.

'I took a return car to Vernayaz, and ascended by some slate quarries to the village of Salvan, on the south side of the valley of the Pissevache. This is on the junction of the slaty limestone (here quarried as roofing slate), with the equivocal gneiss and granite rocks of this district. I climbed a rock which immediately commands the valley of the Pissevache, and also that of the Rhone, which is itself extensively striated. From this I had a glorious view. The ravine of the Pissevache itself is magnificent, while the Rhone valley stretches away 2,000 feet below. But I was most struck by the stupendous wall of rock which forms the north side of the Pissevache ravine, striated and polished nearly from top to bottom, often in the most distinct and beautiful manner. The striæ are generally horizontal, but some rise and fall a little, depending on the direction in which the rock has offered least resistance. To talk of such striæ having been produced by water is ridiculous. The whole valley of the Trient which is commanded from this point, is composed of *roches moutonnées;* and from the direction of the striæ near the Pissevache, they seem to be due to glaciers descending from the Tête Noire or Dent du Midi, which swept round the Pissevache and descended to St. Maurice; not to a great glacier descending the valley of the Rhone.

'On the shelves about Salvan are many blocks, a considerable portion of which appear to have travelled. Some of them, I think, are the Valorsine conglomerates. These belong to the "moraines" of Agassiz and Charpentier, but this part of their theory is difficult to follow; we may admit it, as harmonizing with other evidence, but scarcely as being a proof itself—at least as far as I have seen. . . .'

Forbes had looked forward to meeting M. Charpentier at Bex, but the latter had gone to Italy, so he went on to Geneva.

Journal, September 11th.

'M. de la Rive called on me and took me to the observatory. . . . I also made the acquaintance of M. Plantamour, the observer, and of M. Emile Gauthier, nephew of the Professor. I saw Professor Gauthier soon after, and he received me with an affectionate warmth which touched me. He is all heart. . . . Every evening since I came, Mont Blanc has been beautifully clear, and the sunset fine; but the recoloration has never been well marked. This shows that it depends on peculiar atmospheric conditions, and favours De la Rive's hypothesis. In *Galignani* of the 11th is a most laughable account of our ascent of the Jungfrau. We are made to have been conducted by a shepherd 80 years old, who ascended for the third time! and to have engraved our names and the date on the flag we planted!'

Much impressed by his observations at Vernayaz, he hastened to Neuchâtel to discuss them with Agassiz.

'My first walk with M. Agassiz was to the polished rocks near the lake, a mile from Neuchâtel, in the direction of Bienne. Quite satisfactory. Glaciers only could have done this; but whence came they?

'Took tea with Agassiz, Mr. and Mrs. Trevelyan, M. Desor, and M. Burckhardt—quite a Grimsel party. Disputed about glaciers the entire evening. . . . Next day I walked with Agassiz and Desor to the "Pierre à Bot," which exceeded my expectations. It is a boulder 50 feet long, the largest in the Jura, composed of granite which Von Buch recognizes as that of the Val de Bagnes. At this level, about 850 French feet, is a sort of ledge, on which the boulders are abundantly heaped, those of one kind usually together. That these blocks were transported by torrents seems incredible. They must have been broken into a thousand pieces!' . . . 'How came they thus to alight on the steep, and there remain? What force transported them, and, when transported, lodged them high and dry 500 feet above the plain?

We reply, a glacier *might* do this. What other inanimate agent could do it we know not.'

It was now the end of September, and Forbes turned his steps homewards, pausing at Berne to discuss with M. Studer the origin of *blocs perchés* and striæ; but although the latter had seen striæ at Bagnes produced by water, which caused him to doubt, Forbes's convictions as to their glacial origin remained unshaken. He soon returned to England by the Rhine and Belgium, crossing for the first time from Ostend to Dover, a route with which he was greatly pleased. 'There were about ten passengers,' he writes, 'and everything so smoothly managed at the Custom-house by a commissionnaire, who will tell as many lies as you like for two shillings!'

Forbes had now fairly served his apprenticeship in glacier observation. During his residence with M. Agassiz, on the Lauter-Aar Gletscher, he had acquired an intimate acquaintance with the varying features of the Alpine ice-world. 'M. Agassiz,' he writes, 'had lately published his interesting work on Glaciers, in which he had embodied the bold reasonings of Venetz and De Charpentier with the results of his own observations; and, guided by this, as well as the ready illustrations by means of example on the spot, which M. Agassiz was as willing to afford as I was desirous to learn from, I soon found that a multitude of interesting facts had hitherto been overlooked by me, although I was already tolerably familiar with Alpine scenes, and with glaciers in particular. Animated and always friendly discussions were the result, and, without admitting in every case the deductions of my zealous and energetic instructor, I readily allowed their ingenuity.'[1]

That glaciers moved, was unquestionable. That in doing so they grooved and polished the rocks which formed their beds, as they passed on, bearing the enormous transported blocks which had for so long puzzled geologists, Forbes was now convinced. But the rate of

[1] 'Norway and its Glaciers,' p. 297.

this motion and its character, whether uniform or intermittent, and whether differential or otherwise—above all, the origin of this motion—were still moot points; and these points, he urged, could only be settled by treating the mechanism of glaciers as a question of pure physics, and obtaining precise and quantitative measures as the only basis of accurate investigation. He resolved, therefore, that he would open a campaign by himself the following summer, and, taking with him sufficient instruments for a complete trigonometrical survey of some convenient glacier, grapple with the problem in a way which had not before been attempted. His previous visits had convinced him that the Mer de Glace of Chamounix would be found in every way suitable for his researches. Although less extensive than the Lauter-Aar Gletscher, this very fact placed its phenomena more within the grasp of a single observer; while the three glacier basins, the snows of which combine at the same point to form its ice-stream, presented every variety of horizontal angle and lateral pressure to which a glacier could be subjected. Next year, accordingly, he hastened thither, and in a few weeks succeeded in placing the facts of glacier motion for the first time on a firm basis.

Journal, Paris, June 7th.

'I left London with Mr. C. Bentinck on the 27th of May, arriving here by Dover and Boulogne on the 29th. On the 30th I was elected a corresponding member of the Institute of France—quite unexpectedly indeed, for I went to the meeting on that day; and had I not been recognized and told to withdraw, I might have been placed in an awkward position. . . .

'The most interesting thing which I have seen in Paris, I saw to-day at M. Libri's. He has acquired from an ecclesiastic a box of Napoleon's juvenile papers, which he had obtained from Cardinal Fesch, during whose lifetime they had remained sealed up. Among them is his commission in the army, signed by Louis XVI., and dated August 30th, 1792; essays historical and political; romances, one on an English subject, the Earl of Essex; letters on Corsica; a private diary, containing a curious hint of an intention to commit suicide—at the age of 17! last, not least however, there was a geographical note-book, the last words of which, from their extraordinary coincidence with his future fate, it was actually startling to read in Napoleon's boyish hand. It concludes in the middle of a page with this fragment of an unfinished sentence: "St. Hélène. Petite Isle."—Surely this beats romance!'

Travelling by Besançon and Pontarlier, he found time to visit M. Studer, at Berne, and M. Charpentier, at Bex, where he was both surprised and delighted by 'The blocks of Monthey,' a series of remarkable boulders transported thither by glacial action. From Geneva he travelled by diligence to St. Gervais, and next day crossed the Col de Forclaz on foot, carefully noting by the way the distribution of the numerous blocks severally deposited during their ancient extension, by each of the glaciers which descend into the valley of Chamounix.

Journal, Friday, June 24th.

'Arrived at Chamounix at a quarter to twelve. Recommended myself to M. Lanvers, the curé, in the name of M. Charpentier, and met with a most kind and friendly reception. He immediately hastened to procure for me a guide who could remain with me at the Montanvert, and, having found one, he obtained permission for him to accompany me out of his turn. His name was Auguste Balmat, and the same night we slept at the Montanvert.'

To MISS FORBES.

'MONTANVERT, *Sunday, June 26th.*

'. . . I have been fortunate enough to make the acquaintance of the Curé of Chamounix, and he has recommended a guide to me, who really promises to turn out

THE MER DE GLACE DE CHAMOUNIX.

a second James Lindsay[1]—in short an admirable assistant. The weather, too, became delightful just as I arrived here, and it is a real pleasure to work amidst such scenes. Indeed, I have much to be grateful for; and many doubts which I had about bringing my plans for the summer to bear, are in a great measure solved. . . .

'I spent this Sunday morning quietly at home until noon, just strolling down to the edge of the glacier, where I read the morning service basking in the sun. What a cathedral! with all the aiguilles for pinnacles, the glacier for a pavement. "He giveth forth His ice like morsels: who is able to abide His frost?" . . .'

The following details of the commencement of his work are taken from his 'Travels through the Alps of Savoy,' p. 129:—

'I resolved to commence my experiments with the very simple and obvious one of selecting some point on the surface of the ice, and *determining its position with respect to three fixed co-ordinates*, having reference to the fixed objects around. . . One day, the 25th, was devoted to a general reconnaissance of the Mer de Glace throughout a good part of its length, with a view to fixing permanent stations; and the next I proceeded to that part of the glacier which lies opposite to the rocky promontory on its western side, called l'Angle. The

[1] James Lindsay, here alluded to, fulfilled the duties of mechanical assistant to the Professor of Natural Philosophy at the Edinburgh University for more than half a century, and has recently (1872) retired from his post. He was then in his 73rd year, still hale and vigorous, and made a most admirable reply, on the presentation to him of a valuable testimonial, by the students of the class. He was in the service of Sir John Leslie from 1814, until Sir John was transferred from the chair of Mathematics to that of Natural Philosophy, in 1819. Since that time he was mechanical assistant under Leslie for fourteen years, Forbes for twenty-seven years, and Tait for twelve years. He had acquired wonderful dexterity in many difficult experimental processes; and was, amongst other things, a most skilful glass-blower. He constructed with his own hands almost all of Sir John Leslie's original thermometrical and photometrical apparatus.

latter presented a solid wall of rock in contact with the ice, and upon this, as upon a fixed wall or dial, might be marked the progress of the glacier as it slipped by.

'The instrument destined for all these observations was a small astronomical circle, or 4½-inch theodolite, supported on a portable tripod. A point of the ice whose motion was to be observed, was fixed by a hole pierced by means of a common blasting iron or *jumper* to the depth of about two feet. An accurate vertical hole being made, the theodolite was nicely centered on it by means of a plumbline, and levelled. A level run directly to the vertical face of the rock, gave at once the co-ordinate for the *vertical* direction, or height of the surface of the glacier. The next element was the position or co-ordinate parallel to the length, or direction of motion, of the glacier. This was obtained by directing the telescope on a distant object, nearly in the direction of the declivity of the glacier, which object was the S.E. angle of the house at the Montanvert, distant 5,000 feet. The telescope was then moved in azimuth exactly 100° to the left, and thus pointed against the rocky wall of the glacier, which was here very smooth and nearly perpendicular. My assistant, Balmat, was stationed there with a piece of white paper, held with its edge vertical, which I directed him by signs to move along the surface of the rock, until it coincided with the vertical wire of the telescope. Its position was then marked on the stone with a chisel, and the mark painted red with oil paint, and the date affixed. These marks, it is believed, will remain for years. . . .

'It was with no small curiosity that I returned to the station of the "Angle" on the 27th, the day following the first observation. The instrument being pointed and adjusted as described, and stationed above the hole pierced in the ice the day before, when the telescope was turned upon the rock, the red mark was left far above; the new position of the glacier being 16·5 inches lower—that is, more in advance—than it had been twenty-six hours previously! Though the result could not be called unexpected, it filled me with the most lively pleasure. The diurnal motion of a glacier was determined, as I believe, for the first time from observation, and the method employed left no doubt of its being accurately determined. . . .'

It is not necessary to describe at length the observations by means of which during the next few days he arrived at the following conclusions, which are found in his notes of June 30th :—

1. That glacier motion is approximately regular.
2. That it is nearly as great during the night as during the day.
3. That an increase of motion observed on the 28th, 29th, and 30th, was due to the heat of the weather.

And to these, next day, July 1st, having examined his stations arranged for the purpose, he adds—

4. That the centre of the glacier moves quicker than the sides. These conclusions he communicated to Professor Jameson in his 'First letter on glaciers,' dated July 4th.

To E. C. BATTEN, ESQ.

'MONTANVERT, CHAMOUNIX, *June 30th.*

'When I tell you that I have been out thirteen hours on the glacier to-day, and that I have done not much less for several days, you will understand that, what with active exercise and observations and taking notes of what concerns my present inquiry, it is with the utmost difficulty I can do anything else except eat and sleep! I rejoice to tell you that everything relative to my glacier observation goes on in the most delightful and favourable manner. I may literally say,

"How happily the days of Thalaba went by!"

Hitherto, the movements of glaciers have been reckoned by years. Some thought they started on, moving at some seasons or hours, and stopping at others; but these six days of hard work have enabled me to establish, beyond a doubt, the regular and constant flow of the glacier, not merely from day to day, but absolutely from hour to hour. You may believe how much this result has delighted me. . . . I am sorry to interrupt my successful operations, but the eclipse of the sun calls me to Turin, and I take the occasion of the trip to visit the glaciers of the Allée Blanche, and spend some days at Courmayeur, in order to make some meteorological experiments. But I hope to return to Chamounix before the 20th of July, and spend three weeks more in this chalet. . . .'

On the 1st of July he started for Turin, in order to observe the total eclipse of the sun which was to take place on the 8th, and was accompanied across the Col du Bonhomme to Courmayeur by his guide, Auguste Balmat, between whom and himself a strong feeling of esteem, and even friendship, was already growing up. At Courmayeur he fell in with Mr. and Mrs. Airy, who were bound on the same errand as himself, and they had a pleasant journey to Turin together.

The daybreak of the 8th found them all at their posts, Mr. Airy on the Superga, and Forbes at the Observatory, where he had arranged to remain with Sig. Plana. On the previous day the atmospheric disturbance had been considerable, and their hopes of favourable weather for the morrow were mingled with fears: the Abbé Baruffi, however, who was one of their party, refused to admit a doubt on the subject; he declared it to be '*impossible* that the eclipse should not be seen. The clouds *must* clear away—it would be "troppo sceleragine" if they did not!' Alas! the good Abbé's ideas of possibility were destined to be enlarged.

Journal, Friday, July 8th.

'Slept very little last night. Often at the window. The dim misty veil grew thicker by degrees, and about 1 A.M. the sky fairly clouded over, leaving scarcely a hope of seeing the eclipse, which was to begin at 5.14. Sig.

Plana, with Sig. Giulio, called for me at four, and we went very hopelessly to the Observatory, arriving about half-past four. The sun showed himself a little, with very rough edges—indicating, as Sig. Plana remarked, "fonctions diaboliques" in the theory of refraction. He then waded through clouds, which cleared for a while, but, as the eclipse proceeded and the sun lost power, became thick and motionless. I caught the commencement between the clouds, but we then saw no more of it until the moon was more than half off—and not even the end. The period of total darkness was, however, sufficiently striking. It came on literally with frightful rapidity, so as to give one the idea of a great movement taking place in the universe, such as might well be supposed to intimate the Day of Judgment. The shadow plainly came from the west and *rushed* to the east, which gave no doubt this feeling of motion—of something passing with portentous velocity, and scattering darkness as it flew. From our station I could distinctly see it sweep over the great plain of Lombardy, with an actual velocity, according to Carlini's memoir, of 1½ Italian miles per second. The effect was appalling. The suddenness of the transition surpassed all expectation, and I think the darkness also. . . . It went off with equal rapidity, and the rush of dawn moving over the plain from the west with a velocity of 6,000 miles an hour was grand!

'Mr. and Mrs. Airy were more fortunate at the Superga. They saw the sun during the whole of the totality, the mass of clouds which hid it from us being above their heads. They saw a ring round the moon of pearly lustre, ⅛th of her diameter in breadth. After the eclipse I went to bed in sadness; but no one's disappointment was comparable to that of poor kind Sig. Plana, who had taken infinite trouble beforehand, and was greatly interested in receiving and entertaining us.

'I spent part of the evening in conversation with him. . . . Speaking amongst other things of Fresnel's wave-surfaces, he observed how limited our powers of observation are! There is a surface of only the fourth degree, of which we can form no general geometrical idea—we describe it by points; and thus conical refraction was discovered, many years after the equation was known. "Quand je veux imaginer Dieu," he went on, "j'imagine un Être qui voit des surfaces de toute ordre." . . . Yes, everyone has his own εἴδωλον of God!—formed according to the extension which they find it easiest to give to their ideas of power and omniscience. The Brahmin gives his Deity a hundred hands—and the mathematician can only arrive at a generalization: which, however feeble, is at least, I think, more happy.'

Returning immediately to Courmayeur, Forbes resumed his glacier studies, and passed a fortnight in the investigation of the Allée Blanche.

A long and interesting day was spent on the summit of the Cramont with Chanoine Carrel of Aosta, whose acquaintance Forbes had made on his way to Turin, and who was as great an enthusiast on the subject of mountains and glaciers as himself; and they afterwards succeeded in tracing the curious geological section of the chain of Mont Blanc, in which the limestone formations appear to dip under the granite, or protogine, which composes the higher peaks, on Mont Saxe, and through the Val Ferrex. Forbes also succeeded in obtaining some authentic statistics as to the former advance and retreat of the Glacier de Brenva, in the shape of the builder's estimate for the repairs of the little chapel of Notre Dame de la Guérison, which nestles among the cliffs overhanging the lower extremity of the glacier, as well as a certificate from the Syndic of Courmayeur containing a statement taken from the archives of the commune, how that in 1818 the Madonna was removed from her chapel, which was then 'écroulée par l'accroissement du dit glacier, qui était monté au niveau de la dite chapelle,' and carried processionally to Courmayeur, where she remained until the glacier retreated, and her chapel was repaired in 1822.

To E. C. Batten, Esq.

'Courmayeur, *July* 21*st*.

'. . . . I have been hanging off and on for some days, waiting for fine weather, with the intention of crossing the Col du Géant to Chamounix, and I meant to have written from the other side. Of my movements, I have only to say that, with the exception of some patience required in the matter of weather, my summer life goes on admirably. I soon expect a visit from M. Studer, at the Montanvert, where I propose to remain until about the 12th of August, and then to take a journey to Monte Rosa and its neighbouring glaciers, returning once more to Chamounix in September. As I wrote to you before, the precision of my observations on the motion of glaciers, gives me great hopes of constructing a sound mechanical theory, the bases of which must be accurate numerical measurements, and not *à priori* speculations about this or that structure of the ice producing such and such consequences. Far from such line and plummet investigations diminishing my picturesque admiration of these mountain masses, I find that each day brings with it a fresh sense of wonder, not unmixed with awe, at the unknown fundamental laws of the mighty movements of their glaciers, from a real knowledge of which, I seem as far removed as ever. . . .'

On the 23rd of July, Forbes returned to Chamounix by the Col du Géant, a route which possessed, at that time, an interest—an interest above all to him personally, which made this one of his golden days. It is now, literally as regards the south side, a beaten path, while the descent of the Glacier du Géant on the northern side has been for some five-and-twenty years 'slowly broadening down from precedent to precedent,' until among the series of alternative routes which can be chosen, according as the weather or the condition of the glacier renders advisable, there is now little hesitation, and no delay. In 1842 it was otherwise. Although at that time more than half a century had elapsed since De

Saussure had traced 'la route nouvellement découverte,' from Chamounix to Courmayeur, and passed sixteen days on the summit of the pass at a height of 11,000 feet, travellers had been few, and the last but one who had crossed this pass was M. Elie de Beaumont, seven years before. Forbes's admiration for De Saussure was great; and in the account of his own passage of the Col du Géant,[1] he devotes no less than five pages to a summary of the work done by the Humboldt of the Alps, on this memorable spot. 'No system of connected physical observations at a great height in the atmosphere,' he writes, 'has ever been undertaken, which can compare with that of De Saussure. At any time such self-denial and perseverance would be admirable; but if we look to the small acquaintance which philosophers of sixty years ago had with the dangers of the higher Alps, and the consequently exaggerated colouring which was given to them, it must be pronounced heroic.'

To E. C. BATTEN, ESQ.

'CHAMOUNIX, *July* 31*st*.

'I write to you, firstly because the bad weather tempts one to the fireside, and therefore to the discharge of one's debts of correspondence; and secondly, because although I know not whether you are certainly aware of the nature of the Col du Géant, yet I may, without undue self-love, suppose that you would not be sorry to learn that I was extricated in safety from its rocks and glaciers. Be it known to you, then, that the very night succeeding my last despatch I set forth at 1.30 A.M., in a fine clear, cold moonlight, accompanied by two guides, to mount the Col du Géant, a pass which has not been crossed for several years, and which is 11,300 feet high, beset with precipices on the south side and with glaciers on the north. A very steep but not at all dangerous climb brought us in six hours of good walking to the top, from whence, in the calm cold light of a splendid and perfectly clear morning, I enjoyed one of

[1] 'Travels through the Alps of Savoy,' chap. xii.

the most magnificent and certainly the clearest panorama I have ever witnessed, extending from Monte Rosa to the Alps, above Grenoble. Mont Blanc lies very near, and apparently not at all at a great height above, while towards the north stretch away the vast ice-fields which it took us the greater part of the day to descend. We reached a part of the glacier so excessively crevassed from side to side, and bounded by precipitous rocks, that it seemed impossible to pass. We were actually foiled in several attempts, and had to retrace our steps, which we took care always to make in such a way as to allow an escape backwards. At last by the perseverance and address of Joseph Couttet, my Chamounix guide (for I had one from each side of the pass), we gradually extricated ourselves, and arrived in the well-known latitudes of the Mer de Glace, which I had often traversed. We reached the Montanvert at 4 P.M., where I once more took possession of my little chamber, and all my properties were at once established in their old places.

'Next day was fine, but I was obliged to confine myself on account of my eyes, which threatened inflammation, for we had passed all the hottest hours exposed to the insupportable glare of very fresh and highly-crystallized snow. I have changed the skin of my face completely since, and the guides suffered still more, but the whole journey was so interesting and beautiful, and we were so favoured by weather, that I never felt so little fatigued by so long a walk, and next morning was as fresh as a lark, excepting only my eyes. I attribute this in a great measure to my having drunk nothing but cold tea by the way, whilst my guides, who drank brandy, were tormented by a raging thirst; since then I have not been able to work much, owing to very changeable weather, but still I have made some progress with my investigations. To-day, having come down to Chamounix on account of the weather, I have the pleasure of a visit from M. Studer, of Berne, who has come here to see me, and with whom I may perhaps travel later in the season. I have been fortunate in my occasional companionships, which, without incommoding me, have completely removed the feeling of solitude which for a continuance of three months might have been oppressive: and even when weather-bound at the Montanvert, I see, as you may well believe, plenty of "monde" of one kind or another, and scarcely a day passes there, without my meeting some one who knows about me or I about them. If the weather permits, I am out from morning till night, and see no one but a stray traveller from the Jardin, whom I meet like a chance chamois on the glacier: by the way I had two within shot—chamois, not travellers—the day before yesterday. . . .'

Notwithstanding the inflammation of his eyes produced by the snow plains of the Col du Géant, Forbes at once resumed his observations on the Mer de Glace. 'On the evening of the 24th of July,' he writes, 'the day following my descent from the Col du Géant, I walked up the hill of Charmoz to a height of 600 or 700 feet above the Montanvert, or about 1,000 feet above the glacier. The tints of sunset were cast in a glorious manner over the distant mountains, while the glacier was thrown into comparative shadow, a condition of half illumination which is far more proper for distinguishing feeble shades of colour on a white surface like that of a glacier, than the broad day. Accordingly, whilst revolving in my mind during this evening's stroll the singular problems of the ice-world, my eye was caught by a very peculiar appearance of the surface of the ice, which I was certain that I then saw for the first time. It consisted of a series of nearly hyperbolic brownish bands on the glacier, the curves pointing downwards, and the two branches mingling indiscriminately with the moraines.' 'They were marked by merely a faint change in the colour of the ice, and were almost imperceptible when examined closely . . . but I clearly saw that they were not properly diversions of the true moraines, as I at first supposed,

THE "DIRT BANDS" OF THE MER DE GLACE.

but merely bands of a peculiar icy structure, which retained and exhibited the dirt. This led me to a very careful examination of the veined structure of the ice, which I found to follow exactly these hyperbolic lines, and to turn round their planes of cleavage very sharply at the bottom of the bend, while a longitudinal section of the glacier would show a dip *inwards* of 45°. This is very important as furnishing the link between the *vertical, parallel,* and *clamshell* stratification.'[1]

A short interval of bad weather obliged him to descend to Chamounix. But soon after M. Studer and sunshine arrived together, and it is easy to imagine with what pleasure Forbes conducted his friend to the scene of his labours, and explained to him their results. 'Having conversed much,' he writes, 'with M. Studer on the history of the glacier question, I made the following memorandum of what I claim as original in my present investigations, and submitted it to him. He considers it historically accurate. The points I claim are these:

'1. The treatment of glacier motion as a problem of mechanical forces, and its examination as such, from exact data observed.

'2. Two experiments for distinguishing between the theories of De Saussure and De Charpentier—*i.e.* those of gravitation and dilatation. One being the exact measurement of a space along the ice, to be measured again after a certain time in order to ascertain whether any expansion has occurred; the other, the determination of the lineal velocity of a number of points along the glacier, in order to ascertain whether their motion is

[1] These remarkable bands are here represented as they occur on the lower portion of the Mer de Glace. The photograph was taken expressly for this work from the summit of the Trélaporte, a buttress of the Aiguille de Charmoz, which Forbes used as one of his surveying stations (G*) in 1844. It rises to a height of nearly 2,000 feet above the glacier, and he thus speaks of it as a point of view: 'The ascent was somewhat difficult, but the view fully repaid us. It commands the whole sweep of the glacier from the chapeau to the Col du Géant, and to the foot of the Grandes Jorasses, as does no other point I have visited. From it the dirt bands on the glacier are beautifully seen.' —*Journal*, August 20, 1844.

uniform, or bearing a ratio to their several distances from the source of the glacier. The former experiment, I have since learned, had been suggested by M. Studer to M. Escher last year, who attempted to put it in practice—unsuccessfully, however—on the Lauter-Aar Gletscher.

'3. The determination of the diurnal motion of a glacier by reference to three co-ordinates.'

After having arranged to meet Forbes at the convent of the Great St. Bernard on the 12th of August, M. Studer proceeded to make an excursion into the Tarentaise, while Forbes vigorously prosecuted his survey, often sleeping *al fresco* among the rocks of the Tacul, but occasionally driven back to the friendly shelter of the Montanvert by the weather, which was stormy and broken. On the 11th he crossed the Tête Noire to Martigny, and next day the two friends exchanged hearty greetings before the hospitable fire which, even in August, is the chiefest luxury in the domicile of the worthy fathers of the Great St. Bernard.

To E. C. BATTEN, ESQ.

'ORSIÈRES, NEAR GREAT ST. BERNARD, *Aug. 13th.*

'I am about to plunge into wild regions where

"Posts never come—which come to all beside—"

'I talk of the summer as if it was to come; and here it is slipping through our fingers! I have been already two months in Switzerland, and now scarcely a month of true summer remains.

'I have left the Montanvert for the present, hoping soon to return for the purpose of making a tour with Professor Studer, of Berne, through the wild valleys between the great St. Bernard and Monte Rosa: we purpose making the tour of Monte Rosa together, and I then return straight to Chamounix.

'My experiments have been laborious, but successful. I have such a quantity of unreduced observations, that I am obliged to crave a general amnesty from my correspondents for poor shabby letters; and as I have a prospect of sleeping in chalets for some time, pen and ink will enjoy a sinecure. It has been a project of a year's standing, this tour with Studer, and I have no doubt that it will prove most interesting, as he is a most desirable companion from his local knowledge, his scientific attainments, and his pleasant, simple manners. He has, very generously, taken my part firmly and strongly in the recent disputes, notwithstanding his older friendship with Agassiz. . . . We spent last night at the Great St. Bernard, which was fixed on as our rendezvous, and whither we punctually arrived from opposite sides of the Alps. We are to ascend the Val de Bagnes, cross the Col de Fenêtre into Piedmont, returning back to the Swiss valley of Evolena, by the Col d'Aosta. I shall then, perhaps, attempt a glacier pass of vaguely mysterious difficulty from Evolena to Zermatt, then cross the St. Theodule, and work round the southern slopes of the chain of Monte Rosa. . . .'

To the Same.

'ZERMATT, NEAR FOOT OF MONTE ROSA, *Aug. 25th.*

'Here I am locked in by a sore foot, which has detained me already three or four days. There seems to be some fatality against my passing the St. Theodule, for last year I was in like manner at its foot, but my sore eyes kept me there. Now, you must take a very good map—or rather the least bad there is—and trace our progress from Orsières, where I wrote you a hurried line ten days ago.

'We ascended the valley of Bagnes, celebrated for the terrible *débacle* of 1818, and saw the mischief-making glacier, after which we passed our first night in a chalet. Next day we crossed the Col de Fenêtre, and descended to the entrance of the valley of Valpellina, which we followed up its whole length, and having again slept in a hayloft, crossed a pass which has never yet got into the way of tourists, and has a fabulous sort of existence in some books: it is called the Col de Collon, and brought us to the Val d'Erin, the capital of which is Evolena.

'It is a long and wild glacier pass, and while descending its northern snow slopes, we were shocked to find the body of a man, in his clothes, which were in good preservation, but the body was terribly disfigured. He had lain there since last autumn! His purse and clothes untouched. We took his purse, which contained small coins amounting to about three francs, and left it with the Curé of Evolena to be given to his friends. . . . Lower down, we found the remains of another man, of much older date; then the bones of two chamois, and finally, the bones of a third man at some distance. Such a thing never occurred to me before, and you may be sure that we resumed our journey silent and awestruck.

'The valley of Evolena is a populous one—only a few hours from Sion, the capital of the Vallais—but a more wretched place for an unlucky traveller to fall into, you never met with; no sort of comfort, cleanliness, or good fare—even the curé was reduced to, if not below, the level of his parishioners. Having some notion of the sort of places we were going through, we had fortunately taken a provision of meat, rice, tea, and coffee, as if we were travelling in Siberia!'

Journal, August 17th.

'Our reception at Evolena was none of the pleasantest. The curé, a timid worldly man, gave us to understand that, owing to his mother's illness, we could not lodge in his house; and as we had no possible alternative but to eat there, we were obliged to submit to the rudeness of his sister, who appeared to be constantly drunk, and was evidently the terror of this ill-regulated household. At length it was announced that one bed was to be had in the village. We inspected it; room for two was out of the question. We drew lots for it. I won!—hard, and knotty, and short it was, but tolerably clean. M. Studer, who slept I know not where, came to me in the morning with a rueful countenance, declaring that he would not stay another hour in Evolena, and he accordingly went off to the Val d'Anniviers.'

Continuation of previous letter.

'M. Studer, whose philosophy was not proof against the discomforts of Evolena, went round by the Val d'Anniviers and the valley of St. Nicholas in order to join me here, while I, accompanied by Victor Tairraz my Chamounix guide, and two men named Biona and Pralong, natives of these valleys, crossed the mountains from Evolena to Zermatt by a pass which has a great character in this country for its height and length, in both of which it exceeds the celebrated Col du Géant—by no means falling short of it in difficulty. There were only three men, I believe, in the valley of Erin who knew of the existence of the pass, and but one who had crossed it; one of the former—Pralong—I had been fortunate enough to procure.

'We crossed most successfully, and I was rewarded by perhaps the grandest view which I have ever met with in the Alps. . . . So you will see from the map that we have pretty well zig-zagged through this—perhaps the least known part of the Alpine chain, but undoubtedly one of the grandest. During all this time I have been walking with a sore foot, which was hurt at Chamounix, and which has now, during the time I spent here awaiting M. Studer's arrival by the valley, begun to take its revenge. There is nothing for it but to lay up; but after superb weather the rain has at last come to assuage my affliction. . . . How the summer flies! It certainly will not have been for me either unprofitable or uninteresting. As to my experiments and conclusions, I am tolerably well satisfied with them, so far as they go; as to scenery and adventure, I have been more than satisfied.'

On the 26th, Forbes at last succeeded in passing the St. Theodule, and crossed the southern spurs of Monte Rosa from valley to valley, exploring their glaciers as he went. The weather was broken; and when they had got as far as the valley of Gressoney, it proved too much for the patience of M. Studer, who fled away by the Col de Valdobbia. Forbes, however, lingered a few days longer, and was rewarded by better weather, so that he was able to make an excursion on the Lys Gletscher in company with M. Zumstein—whose name is borne by one of the peaks of Monte Rosa, which he was the first to ascend—and whose intimate knowledge of these valleys made his companionship most valuable. He then crossed the Col Turlo to Macugnaga.

To E. C. BATTEN, ESQ.

'CHAMOUNIX, *September* 14*th*.

'. . . After my last letter the weather soon mended, and became very fine indeed, so I had a delightful day on the glacier of Macugnaga, and a splendid passage of the Monte Moro. I then descended quietly to Martigny, and from that spent two days on the way here, visiting the glaciers of Trient and Argentière. . . . I have been detained here by a sore hand and arm, but my imprisonment has been rendered agreeable by the society of three Cambridge men, one of whom, Lord F——, a most engaging young man, I knew slightly before. We managed to pass two rainy days drawing and talking very agreeably. It was two months since I had had anything like a chat with an Englishman. . . .

'I am still shut up by weather and my sore hand; my Cambridge friends have left me, so I have time enough for musing; but I expect to-morrow to shoulder my pole, barometer, telescope, and sketch-book, and to be myself again!'

It was indeed time that he resumed his work, for the winter had already set in, and so much remained still to be done, that not an hour of fine weather could be spared. Having measured a base line for his survey of nearly 1,000 yards in length, he proceeded to examine the effects of the first cold of winter on the motion and condition of the glacier, which he had found on his return enormously collapsed and wasted by the autumn sun. Its motion, he found—contrary to the usual belief—sensibly diminished during the cold, while on the return of milder weather, which cleared the glacier of snow, and left it soakingly wet, it resumed its rapid motion. But the first snow-storms of winter came on with more and more violence, and rendered further work impossible for that year. The following account of one of his last days on the glacier, will give an idea of the difficulties he had then to contend with.

Journal, September 25th.

'I had resolved, if to-day were fine, by no means to miss the opportunity of going to the head of the glacier and examining the motion of its tributaries from stations [E] and [K], which I had fixed on the glaciers of Léchaud and of the Géant respectively; but the quantity of snow, independently of the chances of weather, rendered this a serious undertaking. Under David Couttet's advice, however, whom I took with me as well as Balmat, we started in lowering weather at 6.30 A.M. In less than an hour it began to snow, and so continued the whole day; at times very heavily, with drifting wind, but not intense cold. The snow was frequently knee-deep, and it required all Balmat's intimate knowledge of the glacier to keep us clear of bad crevasses by skirting the moraines as much as possible. While passing the icefall of the Glacier de Taléfre, the weather became still worse; and when we were fairly on the upper Léchaud, the wind blew so strongly from the Grandes Jorasses, and the snow had so completely obscured our landmarks for exact observations, that we were fain to crouch behind a stone on the lateral moraine, and make an uncomfortable breakfast at about ten o'clock. At length the snow abated a little, and disclosed the Pierre à Beranger, which was one of my check-marks for the movement of the ice. Hope revived, and a little wine and brandy restored some of Balmat's confidence: we therefore advanced up the glacier, but the storm again increased in violence; and as we got to the foot of the rock on which my station [E] stood, Couttet, who had been hitherto the chief encourager of the expedition, said very quietly, "Nous allons faire une bêtise!" . . .

'The wind was now driving the snow in our faces, but I begged that we might at least stop and take shelter as before, in hopes of a better moment which indeed was not long coming. We then ascended and planted the instrument, but the Pierre à Beranger, although quite visible to the eye, was lost in the telescope owing to the intervening snow-storm. When I was beginning to despair, however, it cleared, and became nearly fair—at least so far as to allow a sight of the important mark, whereupon Balmat descended to the glacier, while Couttet held up an umbrella to defend the theodolite. I thus determined the motion of station [E 1] nearest the eastern edge of the Glacier de Léchaud :—

	Feet.	Inches.
'29th July—8th August . . .	10	11
8th August—25th September .	45	1
Total (48 days) . . .	56	0

	In.
Daily motion	11·3

'This observation was carefully repeated, and notwithstanding the unfavourable weather, is as trustworthy as any I have made. One thing struck me as being exceedingly remarkable. The staff [E 1], immersed to a small depth in the ice, with twelve or fifteen inches of snow above, having been exposed to a great cold for some days—the thermometer at the Montanvert having stood at 20°—and the glacier having been covered with snow for probably a fortnight, was yet standing in water, and quite loose in its hole.

'The surface of the glacier generally was dry—not a rill of water in the *Moulins* or elsewhere, yet this congelation appeared to be superficial, and to have penetrated no great way into the glacier.

'Both Couttet and Balmat were afraid all the way of breaking through the snow into a water-hole, a thing I thought very unlikely, but which, nevertheless, happened to both of them during the day : hence it seems probable that, as Couttet expressed it, "The snow keeps the glacier warm, as it does the ground," and that even in the depth of winter, cold penetrates very slowly. It becomes most important for Charpentier's theory to ascertain the effect on the motion of the glacier of the complete freezing of its surface-water—which ought, according to him, to be attended with a great dilatation and movement of the glacier. . . . In endeavouring to sound the great *Moulin* opposite the Glacier de Taléfre on the Glacier de Léchaud, the line broke, and deprived me of my geological hammer, which I had used as a plummet. I describe it, in case of the improbable circumstance of its ever appearing again lower down. It is of this form, with a globular and a cutting end, weighing about $2\frac{1}{2}$ pounds; the handle about sixteen inches long. The *moulin* was stopped with snow.

'[*June 22nd*, 1858.—By a singular accident Mr. Alfred Wills, accompanied by Auguste Balmat, found this hammer "not far below the Tacul" in 1857, when it was at once recognized by Balmat. See Mr. Wills' letter of Dec. 3rd, 1857.]

'In returning down the glacier, I was struck by the exquisite blue colour of the snow, by transmitted light, when pierced by a stick to the depth of six inches or more. It was, I think, quite as intense as I have ever seen in ice of the same thickness. The snow was quite fresh, having fallen within the last few days, and was neither very dry, nor yet very moist. I think it is not the same in lower regions. Near this point also—the Tacul—I observed this blue colour in pools of water of trifling depth on the glacier. . . . We reached the Montanvert at half-past three, for going up the Glacier du Géant was out of the question.'

A couple of days more spent on the glacier, a few final observations to render the triangulation on which his map was based more complete, and Forbes's summer work was over. On the 28th he descended from the Montanvert for the last time. 'It was a beautiful evening,' he writes, 'the Dru and the Verte were magnificently coloured by the setting sun, and I rejoiced at leaving Chamounix with a favourable impression, after the long bad weather we have had. Next day I quitted with great regret that charming spot where I have spent so many weeks, and which I have now seen in every phase of summer and winter. Undoubtedly it deserves its reputation, and I have every reason to congratulate myself on having made it my head-quarters.'

After spending a week at Geneva, he returned to England by the same route he had followed the previous year, and while visiting M. Studer at Berne, *en route*, he addressed the following letter to his sister:—

To MISS FORBES.

'BERNE, *October* 9*th*, 1842.

'. . . I spent the month of September, which was variable, chiefly at Chamounix, having a determinate task to perform, which I had very considerable anxiety about accomplishing within the time to which I was limited; and, in doing so, had to submit to considerable inconveniences. The snow lay to some depth below the Montanvert, the thermometer fell to 20°—in my bedroom to 39°—and I made an excursion over the Mer de Glace in the midst of a snow-storm, which gave me a good idea of an Alpine winter; but these difficulties added something to the value of the results, and, on the whole, I have reason to be satisfied with what I have done.

'I have determined the velocity of twelve points on the glacier, during an interval of more than three months—in fact almost all the accessible part of the season. These include more than sixty measurements, and besides all other things, I have a complete survey of the glacier and its tributaries, for constructing a map. I am not sorry that the world should have an opportunity of comparing the results of Agassiz' mode of working with those of mine. His force consisted of a paid surveyor, a paid draughtsman, a chemist, a geologist, a trumpeter—I know not whom besides—in all nine masters, and nine guides who all lived on the glacier. I avoided society, and the disturbances attendant on notoriety; and with one assistant only, except on rare occasions, have performed all my summer's work—my ordinary expenses being eleven francs a day! . . .'

CHAPTER X.

ALPINE TRAVELS, 1843—1851.

DURING the winter months of 1842, Forbes was busily employed in the reduction of his observations, and in putting his notes into form for publication. The result was his 'Travels through the Alps of Savoy,' one of the most charming books ever written on Alpine travel, and not unworthy of its confessed prototype, the travels of De Saussure. It was published in the summer of 1843; and almost at the same time another important event in his life, his marriage, took place. Early in July he started with his young bride to spend his honeymoon among the Alps.

It had been Forbes's intention to revisit his many friends at Bonn, and then to travel by the Rhine to Switzerland. But this was not to be. The hard work of the previous session told upon him now, and for the first time in his life he broke down. One Monday afternoon they travelled from Antwerp to Liège in weather so sultry and hot, that Forbes could not bear the close railway carriage, and at the first stoppage they took their places in an open one. It was then near sunset, and the sudden and dangerous change from afternoon heat to evening chill soon came. Next day, when they arrived at Bonn, he was seriously ill—at intervals scarcely conscious; but he would not allow that he was so, and when several of his old friends visited him the same evening, he exerted himself so far, that they did not observe anything amiss: when they had left him, however, he became much worse; and when his anxious wife succeeded in getting a doctor sent for, his malady was pronounced to be gastric fever, complicated by severe pulmonary inflammation, and slight hopes were held out of his recovery. She never left him night or day, but for ten days he did not recognize her. It was a long and anxious time—how long, those of our readers only who have passed through such a trial, can say. At last the crisis came, and the doctor was of opinion that he would not live through the night. She spent that painful night—like all the rest—alone with him, and in the morning could thank God that he still lived, but it was many days before she could realize the fact that he was recovering. The fever lingered about him still, and at last, about the middle of August, a yearning for the cool breath of the glaciers came upon him, and he told the doctor that he felt he could never get well unless he left the Rhineland, and travelled at all hazards to the Alps.

The doctor stood aghast at the idea, and his prophesies of evil seemed likely to be fulfilled, for during the slow journey from the Rhine to the valley of the Rhone, Forbes's health did not improve. But once at Bex, with glaciers within sight, and his old friends De Charpentier and Von Fellenberg to talk to about them, his life and spirits came to him again; and he was, for the time, almost his former self—sketching the blocks of Monthey, 'and no less charmed with them,' he writes, 'than last year'—planning how much work he was still justified in attempting under the change of plan necessitated by his illness, and arguing with Charpentier as energetically as ever.

On the 4th of September they crossed the Col de Balme, and some time before reaching Chamounix they were met by a crowd of guides and friendly people of the place, who were anxious to welcome 'Monsieur For-bés,' and show him that his kindly, genial manners and undoubted intrepidity had gained their esteem—an esteem with which his name is still mentioned there.

The first week removed all doubt that the remedy which Forbes had prescribed for himself, was the right one. 'I have visited the Flégère, the Chapeau, and the Col de Voza,' he writes, 'and my cough has gradually disappeared. The climate agrees with me far better than the soft air of Bex, and the hot sunshine is alone unfavourable.'

On the 11th he once more established himself at the Montanvert, where he hourly gained strength and health, and resumed his observations, accompanied by his faithful guide, Auguste Balmat, 'who,' he writes, 'to his honour be it said, refused all remuneration for his trouble last winter in watching the glacier, as well as for the days spent with me last summer.'

Forbes's chief purpose there was to ascertain the progress which the glaciers had made during the winter, by measuring the spaces passed over by certain blocks which he had marked, and which Balmat had from time to time observed, during the spring and summer. In the case of one of these, near the lower extremity of the Mer de Glace, its movements had been too irregular to give satisfactory results, owing to the broken character of the ice on which it rested, and the fact of its having been gradually shunted on to the edge of the moraine, where it now lay stranded. Enough, however, could be gathered to prove that, contrary to received opinions, the glacier moved with considerable velocity even in winter.

Journal, September 12th.

'I walked up the glacier with Balmat and my small theodolite. . . . The "Pierre Platte" was in every way an unexceptionable landmark, and I resolved to cross the Mer de Glace for the purpose of observing it—an exertion which I should hardly have ventured for a less interesting result. It was easy to find the motion of the "Pierre Platte" [c] since the visual line was fixed. From

June 27th, 1842, to September 12th, 1843 = 442 days, it had moved 320 feet, or 8·7 inches per day.

'I then ascended the *couvercle* to the edge of the Glacier de Taléfre opposite the Jardin, and corrected some details of my map. Being anxious to obtain a good bird's-eye view of the Mer de Glace, and more particularly its moraines, I mounted a rock at the south-west foot of the Aiguille du Moine, which commands the whole glacier magnificently as far as the Montanvert, and was fortunate enough to make from there a very interesting observation.

'The dirt-bands were beautifully traced before me, as I saw them last year: but above the Trélaporte (beyond which on that occasion I was prevented by a fall of snow from tracing them), I counted six new ones on the lower Glacier du Géant, after which succeeded a space of the glacier which might contain three more, but these latter were not distinguishable. Above this, again, came a striking and beautiful appearance which was quite new to me. The immense fall of snow of the previous winter had not been entirely melted during the whole summer, and still lay in all the hollows where it could accumulate, and accordingly the higher part of the Glacier du Géant was marked by broad snow bands, evidently filling hollows in the surface of the glacier; and these bands corresponded precisely in form and intervals to the dirt-bands. These wrinkles on the ice resembled those on a cow's horn, and I am not sure whether they have not one point in common—the indication of annual growth. . . . Counting from the nine dirt-band spaces previously observed above the Trélaporte, I counted ten of these snowy wrinkles, which I considered to form a continuation of the former—altogether nineteen dirt-bands, terminating opposite the Aig. de Blaitière, and, as the distance is about 9,000 feet, averaging 474 feet apart. . . . Auguste and I managed to scramble down a steep *couloir* from my station on the rocks of the Aig. du Moine: the forms of the neighbouring ravines were savage in the extreme, the rock of this district being peculiarly hard and rough, but we regained the glacier from Les Egralets, and returned to the Montanvert. . . . In passing the great Moulin—this year particularly fine—I found a piece of wood, which on examination appeared to have once formed part of a ladder, a very stout one, as it had holes for the insertion of rungs. This may probably have been a portion of De Saussure's ladder, much chafed by the glacier. It was almost at the same spot that I found similar fragments in 1832.

'*September* 13*th.*—I found myself rather breathless from yesterday's fatigues, but not otherwise unwell. . . .' After having fixed the position of numerous blocks on the glacier, with a view to the future observation of their winter progress, and determined the position of true north and the variation of the compass on the map of the Mer de Glace, 'I descended to the source of the Arveiron. While returning from thence to Chamounix, below the chalet of Planaz, I was struck by a singular thought: What is the origin of the mass of blocks which we ride over every day on the ascent to Planaz, while on our way to the Montanvert? I was struck with them from having to consider the course of the torrents of Grepon and Fouilli, or Blaitière, which descend into the valley of Chamounix, and which both struggle with difficulty and by a crooked course behind the wall of *débris* which lies along the foot of the mountain, until they join the Arve in the plain below. What are these *débris?* are they not the lateral moraine, corresponding to that of Lavanchi? I sent Auguste early next morning to break a number of the blocks at random, and he brought me twelve or fifteen specimens, all undistinguishable from one another, and from the granitic blocks of Lavanchi.'

Forbes's time at Chamounix was cruelly short, three days only on the glacier: and then, with health at least improved, he recrossed the Tête Noire to Martigny, where he was joined by Mr. E. C. Batten, the companion of his sojourn at Bonn in 1837, who was travelling with his bride and her sister. They all travelled together to Leukerbad and the Gemmi, on the top of which they parted, and Forbes was able to revisit the glaciers of

Grindelwald before the winter cold set regularly in. 'I there found,' he writes to Lord Cathcart,[1] 'an exact confirmation of the views which I have published respecting the origin and structure of glaciers. I found the forms, simple and compound, of the ribbon structure to be such as I have described, including the gradual rise of what I have called the "frontal dip," from a very low angle near the lower end of the glacier, up to 75° towards its origin, especially on the glacier of the Strahleck, above the Zäsemberg, where it forms the "Mer de Glace" of Grindelwald. The other principal affluent of that icy sea, viz. the glacier descending from the Viescher Hörner, exhibits the same wrinkles exactly as those which I have just described upon the Glacier du Géant—perhaps even better marked. As the lower glacier of Grindelwald furnished an excellent example of all the modifications which I have elsewhere shown to belong to the canal-shaped glacier, with branches, so the upper glacier is an exact representative, in its lower part, of the oval glacier, of which I have taken the glaciers of the Rhone as a type; whilst many of the tributary glaciers of Grindelwald and the Jungfrau bear ample testimony to the general fact that the structure of glaciers is developed during their progression, and after their primitive stratification has been annihilated by their being projected in avalanches over appalling precipices. . . .'

On September 26th the winter snow began, and after being delayed a day by it at Rosenlaui, 'we passed,' he writes, 'some starving days at Thun.' They then travelled through Berne and Zürich, greeted by old friends at both places, and at length arrived at Bonn with the Battens, whom they had met at Mayence.

The pure mountain air had served Forbes as a walking-stick, but it could not serve him as a crutch: and as the winter still came on, and he still was moving northwards, it became a serious question whether he could continue his journey to Edinburgh, and attempt to undertake his

[1] Fifth Letter on Glaciers.

winter's work there. Dr. Nassé, who had attended him in the former attack at Bonn, and whom he again consulted on this subject, gave a decided answer in the negative; and Forbes, with a heavy heart, addressed a letter to the Lord Provost of Edinburgh, requesting leave of absence for the winter.

His request was immediately granted. 'I have received,' he writes to Mr. Batten, 'the most gratifying letters from Scotland. Kelland and Henderson have done everything with the best grace, and in the kindest manner. The Principal writes to me that both he and the senators have cordially endorsed my conduct; the Lord Provost and Council have been conciliatory and accommodating; and last, not least, my own family decidedly approve of my remaining abroad, without, however, any unreasonable apprehensions. All this comforts me in the highest degree. . . . I doubt not that all will go on well under Kelland's care, and that James Lindsay will be very useful and in high favour. Even the grim *Scotsman* has relaxed into a very kindly paragraph in my favour. . . .'

They turned their steps once more towards the South. But they had only proceeded as far as Frankfort when they were again stopped by illness. This time Mrs. Forbes was the sufferer, and they were unable to attempt any further progress until the end of October. At length, however, by slow, and often weary stages, they reached Naples.

To E. C. BATTEN, ESQ.

'NAPLES, *November 24th.*

'. . . My former letters from Frankfort and Geneva have explained the course of our tedious journey. My dear wife has been not a little delicate ever since; and even when we had arrived at Aix, and had arranged to go by Nice to Genoa, I was reluctantly obliged to give up the idea, and take ship to Marseilles. At Geneva we spent some pleasant days, but after leaving it we had a very cold journey of three days by *voiturier* to Lyons, and down the Rhone to Avignon. At Marseilles it froze in the night, and there was a bitter wind—indeed it was not until we were within sight of Genoa that we met with an Italian climate, which, however, has since followed us. The Bay of Genoa was finer, and the city far more splendid and interesting than I had expected: we had nearly two days to examine it, and I was quite delighted with it. . . .

'The extraordinary beauty of the weather has kept us much on the move in the way of sight-seeing, and my numerous geological excursions—generally three a week—have filled my notebook and my evenings. I have also had a heavy correspondence with my friends at home, both scientific and otherwise, whose kindness at the present juncture I cannot sufficiently acknowledge. My first and main anxiety when I last wrote, was, that my absence should be rightly construed, and my place efficiently supplied. All my interest in Italy was subservient to seeing it with a clear conscience. . . .'

The 'numerous geological excursions' in the environs of Naples, which occupied him until the end of the year, were chiefly devoted to revisiting and re-examining the scenes of his 'Physical Notices of the Bay of Naples:' Astroni and Solfaterra, Pozzuoli, with its ruined temple of Serapis, whose honeycombed pillars had recorded the oscillations of level between the bay and the coast, and lastly Vesuvius. This he visited several times, and accumulated a mass of fresh observations on the structure of its various lava streams, and the mechanical conditions which had affected their course. In condensing the following account from his notes, these scientific details are necessarily omitted:—

Journal, November 30th.

'I ascended Vesuvius under the guidance of Vincenzo Gozzolino, and everything seemed to me just as I remembered it, only that an observatory was rising near the Hermitage, and that the cone was all streaked with the ravaging lavas of 1836 and 1839, up the latter of which we scrambled to the top. . . . The crater seemed very like that of 1826, but filled up on the N.E. side, whence the lavas of 1832, 1833, 1834, and 1839 flowed, and it was also less deep. . . . We were surprised and delighted to see a cone in the centre resembling a blast furnace, which continually ejected stones and fragments of *flagro*, while another opening was emitting a steady liquid current of lava 15 or 20 feet wide. I went down the crater to examine it more closely. Its velocity I estimated at one foot per second, and it was consequently in an extremely liquid state. I carefully compared the phenomena presented by its surface with those presented by the surface of a glacier, but not altogether with effect, for the resemblance failed: firstly, on account of its great liquidity; secondly, on account of a crust which forms as the surface cools, and which, being then broken up by the liquid pressure below, produces peculiarly irregular and fantastic appearances.

'Nevertheless, I observed: 1. That the cracks which form in the dark-coloured slag as the stream spreads itself abroad, resemble the radiating fissures of a glacier under similar circumstances, and are well marked by the liquid fire, shining red through them.

'2. That the slag, where solidified, presented furrows along the surface, parallel to the direction of the structure lines in glaciers, *i.e.* inclining slightly from the sides towards the centre of the current.

'3. That where the lava, having become viscid, forced itself through the obstacles of its slag, it fell like a paste, streaky and drawn out in the direction just mentioned.

'The lava filled the bottom of the crater, with the exception of a large cone in the centre, whose fantastic form could be fitfully seen through the smoke. It from time to time ejected red-hot tears of lava, and cinders, which fell to windward, after rising often as high as the exterior walls of the great crater. The sound which accompanied each explosion was not unlike the roar of a blast furnace, to which, indeed, the whole phenomenon had the most striking resemblance: and when the internal commotion was unusually violent, the lowest mouth from which the lava welled out became disturbed, and belched it forth with violence, scattering the semi-fluid fragments. The lava which had recently flowed over that some days older, and which was still hot, had the exact form of delicate woollen stuff fallen in natural folds on the floor.

'When I had remounted the sides of the crater, we walked towards the eastern side, along a sort of terrace formed within the Punta del Palo, collecting some good crystals of augite among the disintegrated lava. We thence saw very well the course of the lava of 1834 towards Ottajano, and then returned to the point we had first left. As the night fell, the effect of the volcanic fires became most splendid, and we would willingly have remained, but the guides advised our descent, on account of threatening weather. They were right, for before we reached Naples the rain fell in torrents.'

Journal, January 1st, 1844.

'I went in the afternoon to Vesuvius with Vincenzo Gozzolino. . . . He assures me that he and many others saw wide fissures open in the Atrio del Cavallo in 1822, and again in 1834. I ascended the crater by moonlight. There was no lava flowing, but fine explosions, perhaps 400 feet high; and from the principal cone I saw, many times, and without any ambiguity, flames issue. The whole phenomenon bore a wonderful resemblance to an iron furnace, with the exception of the ejected matter, which consists of drops of liquid lava, spurted out by rising steam or gas bubbles, exactly as happens in boiling porridge or polenta. . . . Next day I returned with Vincenzo, his son, and a gendarme to the Vallone Grande, above Ottajano, where there are pillars of Breccia, seemingly composed of lapilli and limestones cemented by calcareous matter. We then went to the crater of 1834, and entered the great crack which has divided the cones after they were formed, proceeding from Vesuvius, and passing right through them all—five or six in number.

At one spot the fissure may be entered by the help of a rope. It was exactly three feet wide, and in one part probably twenty feet high, but very rugged and narrow. I entered it, preceded by Vincenzo's son, and at first had to crawl on my very face, though it afterwards became higher, and we rose up again. The cavern at the end, glistening in the light of the candle, was beautiful: it was covered with crystals of specular iron, producing a splendid effect, while greenish crystals clothed the general mass of the lava, which assumed rugged and singular forms. The chasm closed here rather abruptly, but a kind of dome, like a furnace, rose above our heads, which had doubtless once formed a cone, and into which I managed to climb. . . . Returning to light and air—for we had latterly been in the dark, owing to the accidental extinction of our candle—we had some difficulty in clambering up the precipice again, although helped by the rope which Vincenzo held. . . . We returned in heavy rain to Naples.'

Two days after, they started for Rome. A week before, Forbes had found in the travellers' book at the hermitage on Monte Epomeo the following entry :—

'John
Charles } Forbes, March 28th, 1827.'
James

'And all this time,' he said, 'has passed as a watch in the night. I seem to stand where I had been standing a few days before!'

To E. C. BATTEN, ESQ.

'ROME, *January 7th.*

'. . . At last we are safely at Rome. The climate feels damp after Naples, but oh! what a place Rome is after all! . . . I must explain that we heard nothing of the eruption of Etna until it was entirely past, as I believe the Neapolitans took care not to make any sensation about these things, for fear of disturbing strangers; and the recurrence of an eruption of Etna after what took place last year had never entered my imagination, while, to make matters worse, I have multitudes of letters congratulating me on my good fortune!'

To the Same.

'*January 29th.*

'We have seen the ceremony of making three new Cardinals, and I have made the acquaintance of some of the Jesuit body—their astronomer, their antiquary, and their preacher. The first, Padre Vico, has determined the rotation of Venus, under favour of the skies of Rome; the second, Padre Marchi, has made a wonderful collection of coins of the Etruscans and others of the ancient peoples of Italy; while Padre Grassi, who is considered the most eloquent of the most able Jesuits, accompanied us—my uncle and myself—through the Jesuit monastery, and gave us a discourse on Romanism which was, to say the least of it, most interesting. We passed many hours with him, and in the course of that time there were few points of doctrine on which he did not touch. My uncle—as I rather maliciously alleged—drew him out by hesitations and concessions, so that I fancy he spent rather more eloquence and learning upon us than he might otherwise have done: he is most amiable as well as able, and well fitted to make proselytes, which, it is said, he very frequently does. The impressions left on my mind by the conversation were:

'1. How unsatisfactory was discussion, as a means of arriving at truth!

'2. That the Roman argument lends itself to discussion on account of the definiteness of the things to be proved, and the convenient manner in which many difficulties can be thus solved. But definiteness and conveniency, so far from being tests of a true faith, seem usually to be denied to us, as a trial of our faith.

'3. That an English Tractarian could hardly have escaped from our Jesuit's conclusions.'

To the Same.

'SAN GERMANO, *April* 15*th.*

'. . . We left Rome two days ago by the Ceprano road, with a *voiturier*, and have been greatly repaid for this deviation from the common route, as the country is exquisitely beautiful, and the inns by no means intolerable. Yesterday I went to Alatri, which quite exceeded my expectations, so stupendous are its Pelasgic remains—unequalled, I believe, in Italy. . . . The views are most delightful, and such is the appearance of civilization and comfort, that I own myself to have been long labouring under great delusions respecting this part of Italy, and indeed the Italians generally. I feel that I should have no hesitation in trusting myself alone with them, and there is nothing I should like better than a riding tour in the Apennines. . . .'

They arrived at Naples on the 17th, and a week after, Forbes started alone for Sicily, chiefly for the purpose of examining and comparing the volcanic phenomena of Etna and Stromboli with those of Vesuvius.

To MRS. FORBES.

'ON BOARD THE "DUCA DI CALABRIA," *April* 23*rd.*

'Not yet at Messina, but I have had a very pleasant voyage and the best bed I ever found on board a steamer. We were however disturbed in the night by stopping at a place called Paolo, and these touchings on the Calabrian coast have greatly lengthened the voyage. Stromboli is in sight, and so is Etna, but the latter, from its great distance, does not produce the effect one would expect. . . . I forgot to tell you that when we passed Sorrento yesterday, I had all eyes and my little telescope out, to see you. I could see the Villa Pisani very well, and I thought I saw something white and waving on the terrace—at least I pleased myself with thinking so—as the spot on Venus was seen in Delbrio's glass! My tedious voyage has at least had the good effect of completely calming my nerves, which were a little excited by the bustle and heat of Naples, and otherwise putting me into good condition. We are at this moment passing the hills of Calabria, which are covered with snow, down even among the woods—a bad prospect for Etna.'

To the Same.

'GIARDINI, NEAR TAORMINA, *April* 26*th*.

'I am so far safely on my journey to Catania, having spent more time than I wished at Messina, especially as I may have to return and wait there again; but I could meet with no conveyance until this morning, and I have at all events happily arranged my main point, of seeing Taormina well. I travelled with a chance party of Sicilians, for strangers are scarce, and found the country rather monotonous, with now and then a peep of Etna—but then there was such a fine fresh breeze off the sea! The vegetation in this neighbourhood is far richer and more striking than near Messina, while the abundance everywhere of the Indian fig (cactus) is astounding, and forms one main feature of the landscape. At Taormina we found the loveliest wild flowers already in bloom; wild oleanders, and even palm-trees, growing in the open fields. . . .

'The ruined Græco-Roman theatre at Taormina is a fine thing in itself, but it would be little worth mentioning were it not for its magnificent situation—on a promontory of rock, overhanging a very picturesque town, with other villages around perched on the wildest crags, a sea-shore broken into curved bays on one hand, and the whole "*massif*" of Etna rising in front. I never saw a mountain whose height deceived me nearly so much: it rises more gradually than you can imagine, and in some respects Vesuvius might be called a more striking object. . . . On this side there is much snow, but, as I am told, far less on the south side, from which the ascent is made; so I still hope. . . .'

To the Same.

'ZAFARANA, VAL DEL BUE, *April* 30*th*.

'. . . I have just returned from a most delightful and successful excursion to the Val del Bue. Although the sirocco clouds were hanging about the lower part of Etna, the upper portion was quite clear; and as for the Val del Bue itself, with its enormous and complicated lava fields, I have thoroughly investigated it—in fact "done" it well, with the assistance of an excellent guide recommended to me by Baron Bruca, to whom I brought letters from Monsignor Spada. I found the Baron a most kind-hearted man, and his wife a really nice person. They regret, I believe sincerely, that you did not accompany me so far, and remain at their country-house during my absence, and indeed this would have been a delightful arrangement for all parties—especially for me, who would give anything to have my dear wife within reach.'

Although Forbes still hovered about Etna, hoping against hope for an opportunity of ascending it, a change of weather finally removed this plan to the category of impossibilities. And after a pleasant visit to Syracuse, he rejoined Mrs. Forbes at Salerno, and towards the end of May they left Naples and travelled by easy stages to the Italian lakes.

To E. C. BATTEN, ESQ.

'BELLAGIO, *July 6th.*

'. . . Here we are at this delightful place, which I have long and long thought of revisiting; but our stay must be short, for the advancing summer warns us into Switzerland, and we must be in England about the middle of September or soon after.

'I enjoyed Venice as much as you could possibly desire, and left it with great regret. We stayed a day at Verona, and from Brescia went to Iseo, which we found so charming that we spent a couple of days there in a nice country inn. . . . We then took to the little steamer on the lake, and paddling up to its head, joined the fine road leading from the Camonica to Bergamo. . . . I have become a lazy traveller, and a lazier correspondent: indeed I have almost ceased to write, except to you. All this I excuse to myself on the plea that I am here for my health, and that a year of torpidity will, I trust, renovate my constitution for fresh exertions.'

To the Same.

'SIERRE, *July 29th.*

'. . . I believe I last wrote to you from Bellagio. We went to Lugano and Bellinzona, thence to Baveno and the Simplon, and you may judge of our rate of travelling when I tell you that we took nine days to get across the Simplon! Certainly, four of those were spent at the village of Simplon, and the remaining five with the monks at the hospice. . . .

'I feel more and more that I owe my life to the Alps. I had a threatening of illness in the north of Italy, which left me, not ill indeed, but in a state which admitted of no exertion with pleasure, as you may judge when I tell you that I went up the Monte St. Salvatore at Lugano one morning, and was so exhausted that I could hardly reach the top, and had to rest a whole day after it. The Simplon quite altered matters, and from the hospice I ascended a mountain 10,500 feet high with the utmost satisfaction, and enjoyed from the summit one of the most glorious panoramas I have ever beheld. The mountains in that part of the Alps are, I may say, almost unknown, and it is difficult to obtain even their names. Their heights, too, are in general much underrated.'

Journal, July 22nd.

'I started from the hospice, accompanied by Chamoine Alt and a guide, to explore the Kaltwasser Gletscher. After passing the glacier gallery, we attacked a steep slope on our right, which brought us to the foot of the apparently inaccessible precipices to the left of the glacier, and the top of these we reached by means of a sloping shelf below the Mäderhorn, finding ourselves perched upon a steep but fertile pasture, so inaccessible, however, that not even sheep can be brought thither. We continued to climb until the slope became a moraine, and this was succeeded by the smooth surface of the glacier, which here forms the col between Switzerland and

Italy. . . . The weather was magnificent, and the view towards Italy opened upon us superbly—so superbly, that we determined to ascend the Wäsenhorn, which rose on our left. The summit of this mountain forms a sharp ridge running from west to east, although when seen from either of these directions it presents the appearance of an abrupt peak. It was along the edge of this ridge that we now worked our way up, finding it excessively rugged, without, however, being at all dangerous; the northern face is extremely steep, and covered with ice, but our guide affirmed that once, when a young man, he had slid down this slope.

'The view was one of the finest I ever beheld, and withal, remarkable in many of its combinations, while in all directions it was perfectly clear, without cloud or haze.

'The Italian lakes were not visible, for the great ridge of the Monte Leone and the Breithorn filled up a large space towards the south, and then came the heights beyond the village of Simplon, having the Fletschorn for their culminating point: he towered above us, his height seemingly increased by our proximity to him—a magnificent mountain! . . . Through a narrow gap on one side of him, I thought I saw part of the chain of Monte Rosa, and further to the right appeared the magnificent mountains of Saas, admirably grouped—the Weishorn, the Gabelhorn, and still further behind, the Dent Blanche.'

The remainder of Forbes's stay at the hospice was taken up with an examination of one of those small isolated glaciers reposing in the cavities of high mountains, called by De Saussure 'glaciers of the second order,' which hangs from the slope of the Schönhorn immediately behind the hospice. He established the fact of its motion indeed, but motion of a very slow character, averaging no more than one inch in twenty-four hours. 'This small result,' he observed, 'is quite conformable with the dry and powdery condition of such elevated glaciers, yielding little water, and capable of exerting on their under parts a very trifling hydrostatic pressure.'

Having thus measured the rate of motion of one of the smallest glaciers in Switzerland, he next proceeded to measure that of the largest, and descending to Brieg, where he left Mrs. Forbes, he established himself at the chalets of Marjelen, on the brink of the great Aletsch Gletscher. His work was greatly interrupted by bad weather, but he was able to recognize in this glacier a remarkable instance of the manner in which the centre of an ice-stream can descend towards the valley with double or triple the velocity of its lateral parts. 'I found,' he writes, 'that while the velocity of the ice at 1,300 feet, or about a quarter of a mile from the side, was fourteen inches in twenty-four hours, at 300 feet distant from the side it was but three inches in the same time, and close to the side it had nearly vanished. Facts like these seem to show, with evidence, what intelligent men, such as Bishop Rendu, had only supposed previously to the first exact measures in 1842—that the ice of glaciers, rigid as it appears, has in fact a certain "ductility" or "viscosity" which permits it to model itself to the ground over which it is forced by gravity: still retaining its compact and apparently solid texture, unless, indeed, the inequalities of the ground be so abrupt as to force a separation of the mass into dislocated fragments. This, it is well known, occurs to every glacier when the strain upon its parts reaches a certain amount, as for instance when it has to turn a sharp angle or to descend a rapid or convex slope.'[1]

A few days afterwards, on arriving at Geneva, Forbes learnt from Mr. Hopkins's published papers, which he then saw for the first time, that that gentleman accounted for the differential motion above mentioned, by supposing a glacier to consist of strips or plates of ice, parallel to its course. And that these, according to the pressure severally exercised upon them, slipped past one another: the central ones faster, and so on through the succeeding strips towards the sides of the glacier, the comparatively small motion of which was rendered mechanically

[1] Eighth Letter on Glaciers.

possible by the increased number of these longitudinal dislocations.

Forbes held the contrary opinion, namely, that the forces by which this differential motion was produced, found in the molecular structure of the ice, conditions which enabled them to do their work without such tearing or 'shearing.' He started for Chamounix, and proceeded to ascertain the relative change of position produced by the motion of the glacier, in a considerable number of points fixed across an uncrevassed portion of the Mer de Glace, at right angles to its axis. Such a spot he succeeded, after some trouble, in finding between 'l'Angle' and the Trélaporte, below the little glacier of Charmoz. On a space of extremely flat and compact ice about seventy yards in width, and several hundred feet in length, he fixed a series of marks, thirty feet apart, between three of which he subsequently placed pegs dividing that space into forty-five parts, measuring two feet each.

'You will probably be surprised,' he writes to Professor Jameson, 'when I state, that in seventeen days, that part of the glacier ninety feet nearer the centre than the theodolite, had *moved past* the theodolite by a space of twenty-six inches, and the intermediate spaces in proportion. When I was reluctantly compelled to cease my observations on the forty-five marks, they had, in the course of six days, formed a beautiful curve slightly convex toward the valley; and as the vertical wire of the theodolite ranged over them, their deviations from a perfect curve were slight and irregular, nor was there any great dislocation to be observed in their whole extent; proving the general continuity of the yielding by which each was pushed in advance of its neighbour. . . . All this, viewed in prospective with the theodolite, left no remaining doubt as to the plasticity of the glacier on a large scale.'[1]

But this was not the only result of three weeks which

[1] Eighth Letter on Glaciers, August 30, 1844; 'Philosophical Transactions,' 1846, p. 137.

he passed between Chamounix and the Montanvert. He had resumed work in his old style, and every day found him on the glacier in company with Balmat and his instruments. He took a variety of observations for the improvement and extension of his survey, measured the general daily motion of the glacier, examined the 'glaciers of the second order' which overhang the Mer de Glace, and made a vigorous attempt to fix a series of marks in the side of a *Moulin* of immense depth, in order to test the relation which the motion of the surface of a glacier bears to that of its lower portions. The experiment, however, was a failure; for while engaged in that chilly operation, a storm of wind and sleet burst upon them with such fury that instant flight was their only resource.

During all this time he was not alone, for Mr. and Mrs. Wauchope, who had joined them at Geneva, formed with his wife a pleasant family circle, a bright spot of sunshine at Chamounix when the storms drove him from the glacier. In spite of hard work and exposure which must have tried his scarcely recovered health, the broken autumn weather, and the necessity of observing for hours together under a burning sun, while standing ankle deep in wet snow, Forbes returned to England early in September, with strength so far restored as to enable him at once to enter upon the winter's work of his Professorship.

Forbes spent the summer months of 1845 in the western highlands of Scotland, but the following year we again find him on his way to Chamounix.

To Mrs. Forbes.

'Annecy, *July* 11*th*, 1846.

'. . . Thank God with me that I have arrived in health and safety thus far—I might say, at the end of my journey. I am once more among green fields and bright streams, with white craggy cliffs, glorious to behold, and even some patches of snow. It is indeed a glorious country.

'I arrived here yesterday from Lyons by a real "slow-coach," which took sixteen hours to travel seventy miles. During the whole day the heat was most oppressive, with a stifling sirocco breeze, the thermometer standing above 80°. In the afternoon there was thunder, and to-day wild deluges of rain are falling, with low, brooding clouds—a bad look-out for Chamounix! But at least, having passed through such an ordeal of hot weather, I feel no anxiety whatever as to the remainder of my journey, trusting to the wonted goodness of God to preserve me from any unusual accident.

'My first business here was to see Monseigneur Rendu, and nothing could be more gratifying than my reception, or more pleasing and engaging than the Bishop himself. He was so cordial, so unselfish as to his own claims on the plasticity theory, so much interested in my present undertaking, that I was quite delighted with him. . . . I am to dine with him to-morrow, and expect much pleasure from conversing with him again. I shall not make any stay here, and hope to be at Chamounix on Tuesday evening. . . .'

To the Same.

'MONTANVERT, *July 20th.*

'. . . I sent off my last letter to you this morning, in consequence of my rather sudden move up here, and very glad I am to find myself here once more, surrounded by the memories of so many pleasant days, amongst them those spent here with you. The air is much fresher and more bracing than that of Chamounix. To-day I got as far as the *Moulins*, and feel quite freshened.

'Madame Couttet is really a fine body, and asked most heartily after "ma femme" and "mes enfants," while old Ferdinand at the "Union" looks upon me quite as a "*chamois vivant*" in the house, and does not like my coming up here. But although consideration and respect are undeniably agreeable things—and I get enough of both to spoil me for any other place—yet it is pleasant to exchange them for the quiet and comfort of the Montanvert, where, as usual, everything is better than at Chamounix: bread (most decidedly), butter (most incomparably), mutton, honey, tea, cream, wine. In short, I am in such good humour with my quarters that I shall find it very difficult to descend again. Auguste's eyes continue well, and he finds the railway "goggles" of the greatest service to him.'

Forbes's first care was directed to the examination of the spaces passed over by the blocks which he had marked on the surface of the glacier; and he soon found that one of them (Block R, above 'l'Angle') had in two years moved something less than a quarter of a mile, while the 'Pierre Platte' [c], in the neighbourhood of the Couvercle, had changed its position by about half that distance. He then entered upon a course of investigations on the ablation or melting of the ice at its surface, compared with the actual subsidence of the glacier in its bed. The former was easily ascertained by boring a horizontal hole in the wall of a crevasse, and observing from day to day the gradual diminution in thickness of the ice above it. The amount of the subsidence of the glacier, or the difference between the geometrical depression of its surface and the ablation, was obtained by directing the level telescope of a theodolite on a measuring tape, the extremity of which was pegged into the hole above mentioned, and then held vertically. When the height of the eye above the hole in question was thus ascertained, the level was then moved in the direction of a fixed object (a cross which had been cut on a stone), as a point of departure for the vertical height. The difference in level of the eye above or below this fixed point having been ascertained in a similar manner, the sum, or the difference (as the case might be) of this measure and the last, gave the difference of level between the spot on which the theodolite stood and the mark on the moraine. The mean results of observations at two stations are given in the following table:—

	Slope of surface.	Daily progression.	Daily ablation.	Daily subsidence.	Geometrical depression.	Proportion due to	
						Ablation	Subsidence.
		Inches.	Inches.	Inches.	Inches.		
Station U, opposite Montanvert . .	—	18·7	3·62	1·63	5·25	·69	·31
Station Q, between l'Angle and Trélaporte	2°¾	21·2	2·73	0·97	3·70	·74	·26

'The last two columns,' he writes to Professor Jameson, 'show the effects of the ablation and subsidence in hundredth parts of the whole depression. As we do not know correctly the slope of the bottom or bed of the glacier, it is impossible to estimate how much of the subsidence is owing to the declivity. . . . It is probable, however, that the greater part of it may be thus accounted for.'[1]

While engaged in these operations a remarkable circumstance occurred, the results of which enabled Forbes to calculate the mean annual velocity of the steep and rugged icefall which forms the outflow of the glacier basin of the Taléfre. He was working on the Mer de Glace with his instruments, when he was accosted by a guide of Chamounix, named David Couttet, who informed him that, while searching for crystals on the moraines of the Glacier de Taléfre, he had that very day discovered, at the bottom of its icefall, the fragments of a knapsack which he at once recognized as one lost ten years before on the glacier above the icefall. The next day, accordingly, Forbes went to the spot, accompanied by Couttet and by his invariable companion Auguste Balmat, who also, curiously enough, was able to assist in identifying the lost knapsack.

'Now, to explain how Couttet and Balmat were in a position to speak so positively to the identity of the fragments,' he writes, 'I must observe that David Couttet was then, and has been ever since, lessee of the Pavilion on the Montanvert, and that the knapsack in question

[1] Eleventh Letter on Glaciers.

was his property, and was left at the Montanvert for the purpose of carrying provisions for the small number of travellers who then visited the Jardin. Auguste Balmat was at that time servant to Couttet, and had very often carried that identical knapsack on his back. The figured stripe of green and purple on the shoulder-straps was very marked, and could not easily be mistaken. . . . I also verified this statement by questioning Jullien Michel Dévouassoud himself—the man by whom the knapsack had been lost.

'The accident occurred thus. On the 29th of July, 1836, Dévouassoud accompanied a stranger to the Jardin, taking, as usual, the knapsack from the Montanvert with a supply of bread, cheese, and wine. Having gained the rocks at the top of the Couvercle, the guide, to shorten the way, attempted to take an oblique course to the Jardin, instead of following the usual track on firm ground round the foot of the Aiguille de Moine, so as to make the passage of the glacier as short as possible.

'The ice was covered with snow, and also full of concealed crevasses: into one of these the guide suddenly dropped, leaving the astonished traveller alone in this wilderness of rocks and ice. After vainly calling to his guide, and obtaining no answer, he left the place in despair, and returned to the Montanvert by the way he had come. Dévouassoud, however, having reached the bottom of the crevasse but little hurt, managed, by the aid of his pocket-knife, to cut steps in the walls of ice, and finally, with great exertion and suffering, to make his escape, leaving behind him his knapsack, of which, of course, he had at first disembarrassed himself. Astonishing fact! that the yet undecayed vestiges, together with a part of the bottle which formed his burden, should be brought to light on the surface of the ice after ten years' friction and onward movement!'

The knapsack had been carried in the interior of the glacier over a distance of 4,300 feet, and down a declivity, the angle of which was 14° 55′; the mean annual motion of the ice at this point was therefore about 430 feet.

After having finished this interesting investigation, as it was only half-past 9 A.M., Forbes proceeded to carry out an idea he had long entertained, of attempting the ascent of the Aiguille du Moine.

'I had carefully examined it,' he writes, 'from the rocks of the Tacul, and it was on that side that we now proceeded to make the attempt. There is a great bed of snow which, from a distance, seems to present no difficulty, but which, on a nearer approach, we found extremely steep, and its lower part to be in a glacial state: had the upper portion been thus, we should have had great difficulty in ascending. Once there, however, the passage from the snow to the rock appeared to be the most serious part of the undertaking, and even Couttet thought that if we could accomplish this we should succeed. At length we found a passage from the snow to the rocks, which were very rough and steep indeed, and after a climb of 2½ hours from the Taléfre, we attained the ridge of the Aiguille, and this we hoped to follow to the summit; but an abrupt turret of granite occupying the whole of this ridge, which is perfectly precipitous towards the Glacier de Taléfre, put a complete stop to our progress at about midday. . . .

'I then observed the barometer, which gave to our position a height of 10,360 feet above the sea, or 740 feet below the summit of the Aiguille.'

Shortly afterwards he paid another visit to the Jardin, in company with Professor Schutz of Aarau, whose acquaintance he had made at Chamounix, and on this occasion was able to record some important observations on the mysterious process by which snow is converted into glacier ice.

Journal, July 31st.

'Near the side and medial moraine of the Glacier de Taléfre,' he writes, 'the structure of the glacier is icy and vertically veined, but elsewhere it is decidedly snowy, presenting hardly any bands, either vertical or horizontal; and this is true even so low down as my line of stations across the glacier, not far above the icefall. From this I gather:—

'1. That the vertical structure comes too near the *névé* proper, to admit of its phenomena being explained by a shifting of the strata of the *névé*, through an angle of 90°.

'2. That, when the vertical veined structure is not developed, the horizontal layers have vanished, and are not yet replaced by another structure.

'3. That the conversion of *névé* into ice, is simultaneous and identical with the formation of blue bands.

'4. That these bands are formed where the difference of motion is at its maximum, that is, near the walls of the glacier; but being once formed, they still continue to show themselves under the medial moraine, and this may be traced in the centre of the icefall of the Taléfre.

'5. Its generally imperfect development accords with the slowness of the absolute motion.

'Whilst on this subject I may mention that the appearance of the crevasses between the Montanvert and Trélaporte confirm exactly my observations of previous years at this season. The falling together of the glacier, the undulating forms of the angles of the crevasses, and the choking of crevasses by falling together, all betoken plasticity in the plainest manner. In particular, I had an opportunity of watching individual crevasses opposite the Montanvert, and what I saw incontestably showed that the crevasses were then widening without tearing at bottom, whilst the surface was losing at once by ablation and subsidence of the masses forming the walls.'

After a few days occupied in various experiments, Forbes, accompanied by Balmat, passed round by the Col du Bonhomme to the other side of the chain for the purpose of re-examining the glaciers of Brenva and Miage, in order to see whether the freshly discovered phenomena of the Mer de Glace were presented by them also.

To MRS. FORBES.

'MONTANVERT, *July* 23*rd*.

'. . . The weather continues magnificent, the scenery as glorious as ever, and the air as refreshing. Such a crowd of people as come here every day now, you cannot imagine—people of all kinds, of all nations, and of all characters; every language you can think of at the breakfast table—a farago of tongues. . . . Last night we had a little adventure. I had strolled up the Charmoz with Balmat to my favourite point for seeing the glacier at sunset, and while there, we saw two men crossing the Mer de Glace from the Jardin—a very tired traveller and a very young guide. They had evidently lost their way, and as the sun was setting we saw them plunging into a mass of inextricably intricate crevasses; so there was nothing else for it but to let Auguste go and bring them out. They did not reach the Montanvert until ten. . . .

'Some days ago Auguste and the chief guide dined with me at Chamounix. The curé was also invited, but being a little unwell he preferred to come in to coffee, when he absorbed champagne with great apparent satisfaction, and sat chatting for a long time. The ease and good manners of the guides in company is really surprising.'

To the Same.

'CHAMOUNIX, *July* 28*th*.

'. . . Yesterday I had my theodolite on the Brevent, and to-day on the Flégère—how often I thought of you, and the happy days we spent together there! To-morrow I go to the Glacier de Boissons, which will mostly finish my surveying in the valley, and then back to the Montanvert to look after my various experiments, "*Il y a beaucoup de choses commencées!*" as Balmat says. I have real pleasure in telling you that good Auguste is much the better for his excursions with me, and his sight is now greatly improved. Farewell, my dearest wife, may you receive my letters more regularly than I do yours. I bless you and the dear children. . . .'

'COURMAYEUR, *August 7th.*

'. . . The weather has been threatening, so I could not start for the Glacier de Miage as early as I wished. . . . Wonderful, most wonderful it is in beauty, but for the particular experiments I intended to make on the bending of the ice near the side of the glacier where it touches the rocks without being crevassed, I did not find it suitable. . . . But how shall I describe the wonders and beauties of the Glacier de la Brenva! You must come and see it some day, for it is a most ladylike glacier, and all its beauties can be seen from a mule path, after an exquisite ride amongst old fir woods: with the icy pinnacles and magnificent veined structure, like Pentelic marble, peeping through here and there. . . .

'Last Sunday was such a lonely, lovely, quiet day. Everything was very peaceful, only I feel I cannot spend many more Sundays away from you. Sometimes I wish I had never left you, but it was for the best, and will be so, and I earnestly pray that it will please God to bring me back to you again, and soon. My idea of a route is this: to return to Chamounix, and from that (spending a day on the way with Charpentier at Bex) to go to Sierre, and ascend the Val d'Anniviers to the glacier at its head. Next I must traverse the weary length of the Rhone valley to the glacier of the Rhone and the Grimsel, and having visited the glaciers of the Aar and of Rosenlaui, to cross the Susten Pass to the St. Gothard: next will come Lucerne and Basle, and then—home, home at last!'

Forbes was not alone during his stay at Courmayeur. He was visited there by his old friends Chanoine Carrel of Aosta, and M. L'Eglise, a canon of the Great St. Bernard. M. Guicharda, Vicar of Courmayeur, made a fourth, and they proceeded to examine the results of some experiments made by the latter on the gradual advance of the Glacier de la Brenva since Forbes's last visit in 1842. On that occasion, it will be remembered, he procured evidence that in 1818 the surface of that glacier stood some 300 feet above the level at which he then found it: but since 1842 it had again increased in size—a condition which appears to have been a general one among glaciers in 1846.

'The extremities of all the glaciers of which I have obtained observations,' he writes,[1] 'were advancing towards their valleys during the summer of 1846. . . . The cause is, no doubt, to be sought partly in the great fall of snow of the two winters 1843-4 and 1844-5, and the cold wet summers which followed them. The immediate effect of the snow is to protect the ice, and diminish the annual ablation. Hence the glacier shoots further into the valley before the waste suffices to equalize the supply. . . . In 1846, however, the Mer de Glace, in spite of the intense and continued heat, was much higher opposite to the Angle in the middle of August, than it was in June 1842. . . . Thus, its motion was even accelerated by the great heat of the season. For though it must increase the ablation of the surface, and the melting of the terminal face, thus diminishing the mass of ice, its immediate effect is to fuse the glacier into a state of pliancy, such as to increase its motion in a very perceptible manner (as I have established by direct experiment), and thus discharging its icy burden into the valley faster than even the increased atmospheric heat is capable of dissolving it, it spreads with a velocity which, if it could be supposed to be continual, could not fail to be alarming. Thus it appears from observations made by M. Guicharda, Vicar of Courmayeur, that the extremity of the glacier of La Brenva has protruded into the valley no less than twenty-two metres, or about seventy feet during the two months of summer, being at the rate of a foot a day. . . . The same gentleman has himself made, with considerable labour, observations intended to test the reality of the movement of the glacier during winter, which confirm in every particular those which I have already published regarding the glaciers of Chamounix.'

[1] Twelfth Letter on Glaciers.

The remainder of this letter is taken up by the description of a remarkable analogy between the condition of a portion of the glacier which was embayed behind a promontory of rock, and that of a plastic body acted upon by hydrostatic pressure.

Forbes had measured the diurnal advance of the Glacier de Miage, and found it to be 9·9 inches—the very same quantity as he had obtained in the case of the small hanging Glacier du Nant Blanc. The identity of these results was curious, as proving that two glaciers of incalculably different bulk might have the same rate of motion, their conditions being different. In this case, the motion of the smaller glacier was accelerated by its great steepness, while that of the larger was retarded by the slight declivity of its bed, and the enormous frontal resistance to its exit into the Allée Blanche, already encumbered with its ponderous masses of *débris.* He also found on that occasion among the westerly moraines of the glacier, specimens of dark blue lias slate, which must have come from the very axis of the granite chain, where its existence had been previously unsuspected by geologists.

Forbes then returned to Chamounix, and began a series of difficult experiments for the purpose of determining the difference of movement between the surface of a glacier, and the lower portions of its ice. On the sloping tongue, or extremity of the Glacier des Bois, where the terminal face of compact ice was inclined at an angle of about 40° to the horizon, three marks were made, one behind (and consequently above) the other, in a direction which was judged to be nearly that of the progressive motion of the ice : any variation, therefore, in the motion of these points, could only be imputed to the friction of the soil, for by the motion of the glacier, each would pass in succession over the same spot; and it might be supposed from the analogy of the lateral friction of glaciers, and from the phenomena of rivers, that the retardation of each lower mark upon the one above it would be sensible, but that this retardation would decrease relatively to the height of the observed marks above the supposed bed of the glacier. The results of Forbes's observations confirmed this. The progress of each point, as well in direction as in amount, was rigorously determined by a trigonometrical process, reference being had to two fixed stations, seventy-five feet apart. This was a work of considerable difficulty, and even danger, as the continual fall of blocks, which bounded with great velocity down the terminal face of the glacier, placed both the observer and his instruments in some jeopardy —indeed, during the course of the observations, Balmat received a severe blow on the head from one of them. In order to plant the wooden pins which marked the points, it was necessary to commence by laboriously removing the blocks and rubbish from the glacier above, the fall of which would, at any instant, have threatened the lives of his assistants, Balmat and Bellin—a work which occupied them for some hours.

The results showed that less than fifty feet of thickness between the lowest mark and the next, corresponded to an apparent acceleration of nearly half the motion of the lower point; the acceleration of the central upon the highest point being less considerable. Their heights were approximately, eight, fifty-four, and 143 feet above the bed or floor of the glacier.[1]

Before quitting Chamounix, Forbes attempted to carry out his project of a more complete exploration of the Glacier du Géant, and the clearing up of some doubtful points in the geography of that glacier. This time, however, he was baffled by the excessive difficulty of the ice-fall, and it was not until 1850 that he succeeded in obtaining the observations he desired.

Journal, August 14th.

'I started at five, with David Couttet and Auguste, for the Glacier du Géant. The morning was charming, and by half-past seven we were well advanced on the Glacier

[1] These observations are given in a more detailed form in Forbes's Eleventh Letter on Glaciers, dated September 16, 1846.

du Géant, which, by keeping well to the right, we managed to mount without any particular difficulty until we were at the foot of the "Petit Rognon," and quite on a level with the upper plateau of the Glacier du Géant. Here, however, the tumultuous confusion of the crevasses was so great that we were unable to proceed further, . . in fact, we were entangled in a maze of vast polygonal blocks, divided from each other by huge crevasses with perpendicular sides, and yet of dangerously soft consistence. For three or four hours we persistently laid siege to these obstacles, but after expending all our ingenuity on them, we were obliged to give it up as hopeless. Without a ladder we could not have proceeded, so I was most reluctantly compelled to mount the most accommodating piece of ice within reach, in order to take the angles I required. . . .

'I had an excellent opportunity of examining a phenomenon, the existence of which I had suspected before, namely, that the wrinkles, and probably the dirt-bands also, of the Glacier du Géant take their rise in the great ice-fall below its upper plateau, and in the following manner :—It is quite certain that the ice falls in festoons of alternate masses and vacancies, in the most striking and unmistakeable manner, as I pointed out to my guide, and tried to sketch. Seen with the afternoon shadow, the effect was beautiful, showing bands of fallen pyramids, and tables of ice, alternating with bands of ice crushed and pulverised. Different causes may be assigned in explanation of this phenomenon, but the most probable, I think, is the effect of the summer velocity, which shoves a great portion of the plateau beyond the limit of its stability, thus producing a succession of rapid ice-falls. (Query, does this exist elsewhere, and can it produce the phenomena of the dirt-bands ?) It accounts, at least, for the fact of the intervals between them corresponding with the annual motion of the glacier, or nearly so. . . . Another fact I observed was the small depth to which the horizontal layers of the *névé* extend, and the complete annihilation of structure throughout the mass generally, just as I found in the Glacier de Taléfre. The first appearance of structure is usually vertical, and the first true ice is the mere result of the development of the blue bands. It is probable, therefore, that the process of glaciation consists in the pressure, and working or kneading of the snow, in consequence of the differential velocity of its parts.'

In pursuance of his plans already stated, Forbes now left Chamounix, and as the weather placed a decided veto on his attempting, as he had wished to do, the then unknown passage of the Glacier de Salena, he went to Martigny by the Col de Balme, and established himself for a few days at Bex, with MM. Charpentier and Studer.

To MRS. FORBES.

'SION, *August 22nd.*

'... Yes, I have got to Sion to-day, and that too by the pass of the Diablerets. It is a precious rule in travelling, not to be too easily discouraged, and to risk somewhat; and accordingly to-day, although we lost the best part of the ascent from Bex to the Col by thick wet fogs, on the other side it was comparatively clear, and the descent most interesting—much more so than I had expected—and we had hardly a drop of rain. So I am well repaid by finding myself well over the pass, and so far on my way as Sion. . . . I have got the home sickness very bad, and dare scarcely think of the sweet pleasure of being with you, lest it should make me quite miserable where I am—especially as I must spend to-morrow (Sunday) at Sierre, to see what becomes of the weather, and once there, how can I help thinking of the happy days we have there spent together. I am sure it will make me very, very sad. . . . But I do really think I shall be home to you soon now, very soon: for if the weather continues indifferent, I shall give up the Val d'Anniviers, and go straight to the Grimsel.'

The state of the weather, in fact, rendered this change of plan imperative. He lingered for a day or two at the entrance of the valley, but at last, 'on the faith,' as he says, 'of the universal report of the neighbourhood that the weather was to be bad,' he passed on to Obergesteln, and proceeded again to explore the glacier of the Rhone.

Journal, August 25th.

'I found the Rhone glacier much enlarged since I last visited it, and it now almost completely fills the basin-shaped valley which contains it. I was much struck by the extreme purity of its ice, and the freedom of its surface from stones, its lateral and terminal moraines being, in consequence, extremely trifling: but to this there is one exception, and a very remarkable one. Stones begin to appear on the surface at a considerable height on the terminal slope of the glacier! How came they there? not a stone the size of a fist can be seen on the surface further up, and on examining a number of crevasses, I could not see any engorged in the ice. The explanation seems to be that these stones are introduced into the ice at the bottom of the glacier by friction, and forced upwards by the action of the frontal resistance, which produces the frontal dip of the veined structure, until they are finally dispersed on the melting of the ice. . . . I then ascended to the Grimsel hospice, where I was once more most kindly received by my old host Zippach.

'Next day I started early for the glacier of the Lauter Aar, and ascended the middle of the glacier as far as the Abschwung. Opposite M. Dolfuss-Ausset's new pavilion (which is very conveniently situated on a grassy promontory called Trift) I saw the apparatus for measuring the advance of the central portion of the glacier. It consisted of a board fixed parallel to the axis of the glacier by means of two stakes which are firmly planted in the moraine. There is a painted cross nailed to the board, and the advance of the glacier is measured by sending a man with a graduated staff, which he moves as directed by signals made by blowing a horn. The instrument used for observing appears to be a good circle, carrying a powerful telescope, which is permanently fixed within the window of the pavilion, and is adjusted upon a mark on the other side of the glacier. . . . Besides the mark on the medial moraine (700 metres distant from the pavilion) are two others, one on each glacier.

'Close under the promontory of Trift, below the principal crevasses, and where the glacier, having subsided considerably in level, has settled against the rock which it closely embraces, is the apparatus for registering the motion near the edge of the glacier. This is extremely simple, consisting of a board fixed to the rock, while at right angles to this, and nearly touching it, a horizontal lath is fixed to two supports planted on the glacier: and this lath is of course slowly carried along the board by the motion of the ice to which it is attached. This apparatus is used for observing not only the forward motion, but also any approach to or recession from the rock by the ice, as well as any change of level. The latter takes place continually, and with tolerable regularity in an upward direction, as I could judge from the series of pencil marks registering advance and ascent made from time to time during the last fortnight by M. Dolfuss-Ausset. This ascent appears paradoxical from the situation, the glacier being at that spot in the act of subsiding into a hollow: but I imagine it may be explained by an outward and upward pressure exerted by the main current of the glacier on the ice which fills a sort of bay on the lee side of the promontory of Trift. This ice, which is exceedingly smooth and plane, would thus be squeezed outwards and upwards against the rock, with but little forward motion.

'The apparatus and mode of observation were kindly shown me by M. Dolfuss when I visited the pavilion on my return from the Abschwung.

'On the whole, I was greatly struck with the uncrevassed condition of the glacier in its central part. The glacier of the Grünberg is decidedly concave, while those of the Lauter and Finster Aar are very convex near the

Abschwung. At that spot their structural bands are correctly divided by the medial moraines, but lower down they seem to unite, and cut the medial moraines. (This is admitted in Desor's new work.) . . .

'The rock of the "Hôtel des Neufchâtelois" has split in two, and half of it now lies flat on the ice where we slept in 1841. Otherwise the environs are absolutely unchanged, and I spent a pleasant hour on our poor old block. M. Dolfuss told me that he had made experiments on the compression of fragments of ice, which he found to mould together at a temperature of 32°, but at a lower temperature they were ground to powder.

'I returned to the Grimsel exceedingly pleased by my excursion. This glacier is eminently fitted for making experiments on a line of stakes, while its cohesion to the sides, and the manner in which it moulds itself to the form of its channel, are admirable. Another point is worthy of notice ; like the Glacier of Aletsch it is eminently crystalline in its structure, and large crystalline surfaces, some inches in length, reflect continually the sun's rays (as they did those of the moon during our descent of the Aletsch Gletscher in 1841). Something of the same is observable on the glacier of the Rhone, but nothing or almost nothing similar on the glaciers of the chain of Mont Blanc. It is undoubtedly a crystalline action, which is prevented when the differential motion develops the veined structure well.'

Forbes returned to the Grimsel, and next day took leave of Balmat. His work was over.

Journal, September 6th.—'Happily arrived at Largs at four P.M., and found all well. . . .'

His home sickness was over too.

Early in the May of 1850, Forbes was on the point of starting for Germany, accompanied by Mrs. Forbes and their two eldest children, when his second daughter was attacked by a sudden illness, which caused her parents the greatest anxiety. Her recovery was tedious, and their projected tour was given up: but at the end of the month Forbes left England with his sister, Miss Jane Forbes, and travelled with her to Kissingen.

The month of June was spent at Kissingen, but on the 4th of July Forbes started alone for Basle and Berne, where he spent some pleasant days with M. Studer and Baron Von Buch. On the 10th Auguste Balmat welcomed him to Chamounix—for the last time.

The chief purpose of his visit was the correction and extension of his survey, and he does not appear to have occupied himself at all with glacier observations. His first expedition was devoted to the Glacier du Géant, and on this occasion he succeeded in reaching the point which he had in vain attempted to gain five years before.

Journal, July 15th.

'I started early for the Glacier du Géant, taking with me Pierre Joseph Simond, in addition to Balmat, and in three hours we had skirted the rocks of the Trélaporte, and gained the foot of the great ice-fall. We here found a great deal of soft snow, which choked the crevasses of the glacier and rendered the passage laborious, although not more difficult; on three occasions, however, we were obliged to creep under masses of soft ice almost in the state of snow, which might have fallen at any moment. We fortunately succeeded in passing the point which had completely stopped us in 1846, where the glacier descending between the Grand and the Petit Rognon is crushed and torn with enormous violence. Half an hour more brought us to the upper plateau, and by keeping well to the right, where there was a great hollow containing a watercourse, we got on without difficulty, and deposited our instruments on a favourable spot. . . . Having finished my observations, we walked gradually upward towards the Col du Géant, and became satisfied that there was really a passage between the Grand Rognon and Mont Blanc du Tacul, and it is by this passage that the ascent of the Aiguille du Midi should probably be made. . . .

'At two o'clock we turned to descend. We had greatly enjoyed our day on the glacier, as the heat of the sun was by no means excessive, and making use of the steps by which we had mounted, we descended the ice-fall, and regained the Montanvert without further adventure.'

To DR. SYMONDS.

'CHAMOUNIX, *July* 16*th.*

'. . . I was more than fourteen hours out, and with the exception of about an hour, never touched earth, or rock, or grass, but was most of the time on deep snow. The day was very favourable, not very hot, but Auguste and I are famously burnt! . . . The effect on me was surprising. I did not feel at all very strong or well the night before, but I returned so vigorous and thoroughly braced, that I feel to-day quite my old self again.

'I am quite delighted with my Chamounix visit. At first I fancied I took to it coldly, but the old spirit is now quite returned. I have come to the Croix de Flégère to sleep, and to-morrow Auguste and I mean to ascend some of the heights forming the chain of the Aiguilles Rouges, which, as you may recollect, forms the eastern part of the Brevent chain. . . .'

Journal, July 17*th.*

'After a thunderstorm in the early morning, the weather cleared sufficiently to enable me to start for the ascent of the Aiguille de la Glière, one of the chain of Aiguilles Rouges. We ascended gradually through a beautiful pasture, filled with violet-coloured pansies and other exquisite flowers. It is traditionally stated that De Saussure was in the habit of eating the aforesaid '*violettes,*' as they are called—and they certainly have a delicious perfume, with a taste resembling vanilla. Keeping to the north-west, we came into the *couloir* up which the road to the Flégère ascends, and after a fatiguing march over the bed of snow which filled its upper part, we gained the crest of the ridge a little before ten. Then, turning sharply to the right, we mounted, generally over snow, and by the back of the Aiguilles, to the summit, which has a height of about 8,800 feet. There were many low clouds which spoiled the view generally, although Mont Blanc and some of the higher summits occasionally showed themselves, and I was able to make a careful sketch of the upper basin of the Glacier du Tour, the only glacier descending into the valley of Chamounix which I had not previously examined, and across which a pass to Switzerland was said to exist. . . . Having made some negus, much to Balmat's satisfaction, we left about three o'clock, and taking advantage of the snow beds in a steep *couloir*, slid down with immense rapidity to the Flégère. . . .'

Forbes then took up his quarters at the little inn on the Col de Balme, and after a couple of days of broken weather, he crossed the pass to which he alludes, from the Glacier du Tour, descending into the Swiss Val Ferret by the Glacier de Salena. This was, perhaps, the most interesting, and certainly the most difficult expedition ever made by him in the Alps.

'Having determined to trace the Glacier du Tour to its source, and, if circumstances allowed, to descend into Switzerland by the Glacier de Salena, with which I understood that it communicated, I slept at the Col de Balme on the 19th of July, in company, as usual, with my tried guide Auguste Balmat. The weather proved so stormy, that I expected nothing but a repetition of my former disappointment; but as it improved the following morning, I started, taking Michel Charlet, the tenant of the chalet on the Col de Balme, as guide (the route being as new to Balmat as to myself), although it was already half-past eight o'clock, with the intention of at least exploring the Glacier du Tour.'[1]

By sleeping at the Col de Balme, the party had the advantage of starting from a height of over 7,200 feet, and although the precipices which bound the Glacier du

[1] The following narrative is condensed from an interesting account of this passage in the Appendix to 'Norway and its Glaciers.'

Tour on its northern side rendered necessary that most tiresome of delays, '*monter—pour redescendre*,' they got fairly upon the glacier at a high level. But instead of skirting its northern rocks, the route now followed for the Col du Tour (which would have landed them on the Glacier de Trient, without further trouble) they wandered towards the centre of its basin, and only extricated themselves from a labyrinth of crevasses by bearing to their left: a course which eventually brought them to the foot of a rounded, snow-covered eminence at the north-east corner of the glacier, where they found space to breakfast on a small islet of rock. 'The chief part of the ascent,' he writes, 'was now accomplished, and we stood face to face with the Aiguille d'Argentière,[1] which had a splendid appearance, being curtained with steep glaciers on its northern side. We continued to advance steadily, but with labour, over the snow fields which still separated us from the rocky ridge of the Alps, but our fatigue was soon forgotten in the pleasure of watching the summits which gradually displayed themselves.'

In five hours from the Col de Balme, they reached a point at the head of the Glacier du Tour, a little to the south of its north-east corner, from which they looked down on the Glacier de Salena lying some hundred feet below.

'The weather had still an unsettled appearance. Mists concealed many of the summits behind us, and also the more distant chain of the Great St. Bernard before us. The snow had drifted with violence into this ravine, and we took shelter from the force of the wind on a platform of rock, a few feet below the level of the drifted snow.

'Having reconnoitred our position, I proceeded to observe the barometer. By a single direct barometrical comparison with Geneva (the barometers were carefully compared by Professor Plantamour at Geneva, a few days later), I obtained 11,284 feet above the sea, or 140

[1] In reality the Chardonnet. But at that time both the topography and the nomenclature of that part of the chain of Mont Blanc was in a state of hopeless confusion.

feet higher than the Col du Géant, and nearly 1,200 feet higher than the Buet, which lies towards the N. W. exactly in the prolongation of the axis of the Glacier du Tour.

'This unexpected result suggests some interesting considerations. There are few spots of the same elevation so easily accessible, and it is unquestionable that some of the numerous peaks which rise from this lofty platform could be ascended without risk, to a height of some hundred feet more. The rocks hemming in the Glacier du Tour, present shelter against the severity of the terrific gales which blow at these altitudes. An observer might be stationed here for meteorological observations, with a degree of security and ease which Saussure never enjoyed in his perilous encampment on the Col du Géant. Provisions could be regularly obtained from the elevated station of the Col de Balme, which is within a walk of which a mountaineer thinks little, and devoid of danger. Even the extent of surface which the mountains here present at so great a height, is itself very favourable to several kinds of observation.

'I have called the fact of the great elevation of this part of the chain of Mont Blanc unexpected, both because it was entirely so to myself, and because the existing maps and models gave an entirely different idea. Even the admirable model of M. Séné, which I inspected soon afterwards at Geneva, shows a rapid depression in this part of the ridge, which indeed might have been imagined from the rapidity with which it dies out altogether in the space of a few miles in the direction of Martigny.

'The temperature of the air, as we have seen, was three degrees (Fahrenheit) below freezing. As we turned round and, facing the north wind, clambered from under the sheltering snow drift, we first perceived its biting coldness, and at the same moment the strong draught of air setting through the gorge, nearly detached all our hats in a moment, and actually carried Balmat's over the precipice down to the glacier of Salena. We were then struck, whilst looking in each other's faces, at the pinched and ghastly appearance which all presented. Both the guides looked nearly bloodless; but none of us felt unwell. We took some brandy as a precaution (probably a needless one) against the cold, and tied our handkerchiefs over our ears. Charlet now told us that when here twelve years before, he had succeeded in descending on the glacier of Salena by turning round the north side of the peak under which we stood, in the direction of the glacier of Trient. It was by descending upon this first that he had gained the level of the glacier of Salena afterwards.

'The question now was, should we retrace our steps to Chamounix, or push on to Orsières? Charlet feared that our non-appearance at Col de Balme or Le Tour might create uneasiness; but after some discussion it was agreed that the opportunity of proceeding was too tempting to be lost, especially as the weather appeared fine towards Val Ferret. After scarcely a minute's delay, then, we resolved to seek a safe place of descent to the level of the *névé* connected with the glacier of Trient, which we had to effect over an almost precipitous surface of hardened snow (which in some places presented an overhanging edge of alarming appearance), but which admitted of a passage at one point with little difficulty or danger. This snow cliff scarcely existed when Charlet formerly passed—an instance of the great changes undergone by the glacier regions. Being now on the level of the *névé*, we turned towards the right hand, and found a wall of rock cut through by a magnificent gateway, flanked by two pinnacles of highly crystalline protogine not many yards asunder, between which we passed with the greatest ease, and, descending a snow slope of no great height, we found ourselves on the *névé* of the glacier of Salena. The abruptness of the change and the beauty of the portal (like the *ports* of the Pyrenees, but still narrower) rendered this a very striking and peculiar pass. The basin in which we now found ourselves is enclosed by precipitous rocks, everywhere interspersed with glaciers and perpetual snow.

'The *névé* of the glacier of Salena, seen from the point, might well appear to have no issue. The formidable barriers of rock, between which the glacier descends almost precipitously, might seem to bar a passage in the direction of the valley. From Munier's account (the guide of Chamounix, whom I engaged in 1846 for this excursion), it appears certain that he did not attempt to descend (his words to me, I recollect, were "Nous n'avons pas osé de descendre"), but he had preferred crossing the lofty range which separates the Glacier de Salena from that of La Neuva, by which he descended near the Col Ferret. Trusting, however, to Charlet's report of what he had actually done (for the advancing afternoon left us no time for abortive attempts), we resolved to descend as much as possible by the Glacier de Salena. We accordingly secured ourselves once more together by ropes, and soon came amongst newly opening crevasses as we approached the gorge which offered the great obstacle to our passage. After dining, we were ready to start at 3h. 45m. and again used the ropes for a short space, but, soon clearing the snow, we abandoned them, and following for a little way the left bank of the glacier, as it got steeper and steeper, and began to break into wider crevasses, Auguste volunteered to go on and see whether it might be possible to effect the descent over the broken ice. As we more than anticipated, however, he returned to say that it was quite impracticable, and that, therefore, we must submit to clamber over rocks to a great height above the right bank, and to pass beneath two small glaciers which seemed almost to overhang the path we must follow, so steep was the mound of *débris* which stretched from their foot. We could distinctly see stones on their upper fronts, but the guides pronounced them apparently safe, and recommended the precaution merely of mounting the slope of *débris*, and slanting over to the shoulder of rock as rapidly as possible. Its summit was strewed with enormous blocks, tossed in confusion, shattered and bruised by the mutual shocks which they had evidently undergone no further back than last spring, when they had thundered down with the early thaw, from the upper level of the little glaciers. At present, however, there was no danger, and we paused awhile for breath.

'We were now at a great height above the glacier of Salena, not only on account of the ascent which we had made, but also from the steep fall of the glacier in a contrary direction. Having passed the summit of the knoll which had formed the great obstacle, we were now to seek a safe descent to the main glacier once more. This would have been, in all probability, a matter of small difficulty, had not the fogs which all day had been hovering on the summits, suddenly descended at 5 P.M., and enveloped us almost without warning. Our position was not free from anxiety, for it was impossible to see more than thirty or forty yards in any direction. Charlet continued to advance until we found a small steep glacier in front of us, descending from the heights above, and completely barring passage in a forward direction. We then attempted to descend the rocky ridge upon which we found ourselves, which fell steeply towards the main glacier, but a moderate distance brought us to impracticable precipices. In these circumstances, only two courses remained open, either to wait for the rising of the fog, or to descend from the rock upon the moraine and rubbish which bordered the small glacier, and then attempt to scramble down it. We followed the latter course, and our descent was facilitated by long snow inclines, over which we slid rapidly; whilst so engaged, the fog happily cleared for a few minutes, revealing our entire position, and giving us an opportunity of resolving on our ultimate route, for we knew that sooner or later we must cross the main glacier. Fortunately we had selected what appeared to be the only practicable descent. On the one side of us was an impassable glacier, on the other impassable cliffs. Having made very rapidly a great descent, we diverged to the right, passing (at a safe distance) under the termination of the small glacier, and soon after reached the level of the glacier of Salena without difficulty, which we also crossed with no great delay. We had then a tedious descent over rough moraines, here and there diversified by patches of the most superb vegetation, till we came to a torrent which we understood to descend from the glacier of Orny, where we halted for a short time at 6h. 45m. This stream we also crossed without difficulty, and had now reached the limit of trees; we had a stony and laborious descent through woods nearly pathless before we came to a certain track. By this time we had passed the termination of the glacier of Salena, which we saw distinctly below us. Amidst the wood were vast blocks of the granite of Orny, and, looking back, fine views of the glacier we had left; but the mists were again fallen below the level of the place of our perplexity, so that, but for the momentary rise, we must have remained in much anxiety. At 7h. 55m., we reached the village of Praz de Fort, in Val Ferret, close to the remarkable moraine which protrudes into the valley, and which attracts the attention of all travellers. An hour's sharp walking brought us to Orsières, which we entered at 9 P.M., $12\frac{1}{2}$ hours from the Col de Balme.'

* * * * * * *

This was Forbes's last expedition among the higher glaciers of the Alps. He left them, as he thought, to return, but never again was it given to him to see with his bodily eyes those 'everlasting hills' which he loved so well, and among which he had laboured so faithfully.

He loved them to the last. 'My heart,' he has wistfully said, 'remains where my body can never be. . . . My yearnings towards the Colinton banks, and towards the Swiss mountains, are much on a par—both, *home-sickness!*'

One other scientific journey remains to be recorded, the last ever undertaken by James Forbes. A total eclipse of the sun took place in 1851, visible only in northern latitudes, and he had fixed on Bergen in Norway as a suitable spot for its observation: the more so, as the journey would enable him to examine some of the Norwegian glaciers, and ascertain whether their phenomena were similar to those presented by the glaciers of the Alps. He sailed from Hull to Christiania on the night of the 21st of June, and after an eight days' journey overland from Christiania to Throndhjem, he took the steamer for Hammerfest.

Sailing down the Throndhjem-fiord they gained the open sea, and threaded their way among the countless islands which fringe the coast of Norway. 'We passed the Arctic circle three hours ago,' he writes to Mrs. Forbes. 'It wants but a few minutes of midnight, but the sun has not yet set on a glorious scene: wild mountains, rocks, and islands of most fantastic form, with wide snow-fields in the distance glowing with the red of the midnight sun. . . . We have been singularly fortunate in weather; the sea, even in places where it is quite open, as smooth as glass, and the nights so bewitching that it is scarcely possible to drag oneself to bed. I have seen a great many glaciers, and although I have not as yet set foot on one, I have formed a very exact idea of them.'

To M. BERNARD STUDER.

'. . . The earlier part of the voyage was chiefly remarkable for the surprising prevalence of "*roches moutonnés*" all along the coast, but as we approached the polar circle, the fiords and islands assumed a grandeur for which I was not prepared. At Alten (a few hours' sail short of Hammerfest) I left the ship; for as I could not and would not spend fourteen days there, I was obliged to take the same steamer on her return two days after, during which time I examined the country a little. The vegetation astonished me, and the general character of the scenery was so like that of the Scotch Highlands, that I could hardly persuade myself that I was not in Inverness-shire. I was greatly tempted to explore the fiords between Alten and Tromsö, which abound in true glaciers, mentioned by scarcely any writer, and evidently of the utmost geological interest. The vast interstratification of limestone and gneiss remind one of the Alps, while the hypersthene formation here and there breaking through, reproduces the very forms of the Cuchullin hills in Skye. The necessity of returning to Bergen in time for the eclipse, has prevented me from doing more than noting the phenomena as I saw them from the deck of the steamer, and I made careful drawings (some at midnight!) of the glaciers, which evidently differ in no essential particular from those of Switzerland. After a return voyage of 1,500 miles I reached Bergen, and soon found that its proverbial bad character for weather was richly deserved.'

[1] 'The climate of Bergen is notoriously the most rainy in Norway. Its humidity is a standing joke against the "Bergenser," of whose wardrobe a greatcoat and umbrella are alleged to be, even in the height of Summer, the most important part. . . . It may be thought indeed unwise that I should have staked my second chance of seeing this great phenomenon (having been already disappointed at Turin in 1842), by fixing upon a place of such ill fame in point of climate. . . . But I was unable to resist the opportunity of obtaining a sight of the Arctic regions and perpetual day, to which may be added the desirableness of observers being distributed along all parts of the space of total darkness, as well for the chances of weather as for other reasons.

'No good fortune attended me at Bergen. I was consoled for the disappointment, as far as possible, by the unaffected kindness and sympathy of my friends there, who, with truly Norwegian courtesy, seemed to feel much more on account of a stranger who had travelled

[1] The following descriptions are condensed from Forbes's 'Norway and its Glaciers,' published in 1853.

so far, partly for the express purpose, than they did upon their own. Some notice of my intention had, it seems, been circulated in a Christiania newspaper some time before, so that I found my coming fully expected, and all parties anxious to accommodate me to the utmost. The commandant of the troops most politely reserved for me a clear space on one of the old bastions, and caused a tent to be erected for the protection of my instruments. Thither I repaired shortly before the time of total darkness, accompanied by my friend Captain Lous of the *Prinds Gustav* steamer, whose courtesies had by no means ceased when his comfortable vessel was no longer my home, by several Bergen friends, and by an English gentleman in the same hotel, the only other countryman, I believe, in the place. The morning had been rather brighter than some previous ones—before twelve o'clock the sun had even shone gaily at intervals—but the clouds were throughout so menacing that no one derived thence much confidence for the afternoon. The body of the sun was visible, however, for a little after the commencement of the eclipse, but it soon became more and more overcast, and a portentous sort of calm commenced, exactly as occurred in 1842, the clouds being almost motionless.

'As the hour of commencing total darkness approached, the whole sky became uniformly bespread with dense clouds. The rapidity of the consummation was not so great as I had observed in 1842, which was easily accounted for by the diffusing effect of the thick curtain of vapour which already absorbed so much of the sun's direct light. There was a tendency on the part of those near me to feel disappointed at the degree of darkness, but this was because they were altogether unprepared for the nearly sudden transition from day to night which takes place at the instant that the moon's edge conceals the very last portion of the sun's disk. When that really happened, it was impossible to have any doubt to within perhaps a half, or even a quarter of a minute, from its mere effect on surrounding objects. Our position commanded not only a general view of the country round, but particularly of the town of Bergen itself. The houses which compose it are almost all of wood, and painted white; and the particular hue of the scattered light during the middle of the eclipse was remarkably brought out by this circumstance. It appeared to me of decidedly a bluish tinge, remarkably cold and unnatural. A fire which happened to be lighted in a ship-builder's yard, and which before had been imperceptible, now threw out a striking red glare. Our countenances appeared wan and colourless; a chilly feeling caused an involuntary shiver. One bat, escaped apparently from the rents of the fortress, flew about us very energetically; but some sheep grazing near were remarked not to be sensibly disturbed. There was a considerable concourse of people of all classes, but I did not observe any signs of strong emotion. It is not, indeed, the character of the people to express it. The approach of the eclipse had been denoted by the appearance of a great black cloud in the N.W., which gradually rose above the horizon like an approaching storm; but its boundary (for it was merely the shadow in the sky) was too vague to produce the appalling sense of the onward movement of a real substance, as I had observed on the plains of Piedmont on occasion of the total eclipse of 1842. But the restoration of the light—the new dawn, when the shadow of the darkness had passed by—was perhaps quite as grand: a *copper-coloured* aurora rose in the N.W., shading off the ill-defined limits of total obscuration, and in a few seconds more we were left in the dull dusky atmosphere of Bergen, which soon resolved itself into its accustomed elements of rain-drops.

'Before leaving the subject, I will mention a circumstance which added considerably to the local interest felt in this total eclipse. When a total eclipse was last visible in Norway I am now unable to state, but the popular mind, with singular fidelity to its time-honoured traditions, at once recurred to one which occurred more than eight centuries ago, and which, owing to the peculiar circumstances of the time, was recorded in the Sagas, and has been traditionally recollected ever since. It happened, as has been placed beyond a doubt by the careful and ingenious researches of Professor Harsteen of Christiania, on the afternoon of the 31st August, 1030. King Olaf the saint (canonized for his efforts to introduce at the point of the sword the doctrines of Christianity among the heathens of Scandinavia) engaged in battle on that day with his rebellious subjects, who were urged on by Knut, king of Denmark and England, who desired also to acquire Norway. Olaf was returning from Sweden with the troops he had collected, and entered his own dominions not far from Throndhjem. Meeting the revolters, led by three powerful chiefs, at Stiklastad, in Værdal, about sixty English miles north-east of the capital, he gave them fight, was defeated and slain. In the chronicle of Snorre Sturlason, it is related that "the weather was fine and the sun shone clear, but after the fight began, a red hue overspread the sky and the sun, and before the battle ended, it was dark as night." One of the *skalds* or poets thus expressed it, "the unclouded sun refused to warm the Northmen. A great wonder happened that day. The day was deprived of its fair light." The body of Olaf was secretly conveyed to Throndhjem (then called Nidaros) and interred. A chapel was afterwards built over it, which is now included within the east end of the cathedral.'

The day after the eclipse, Forbes left Bergen to visit the glaciers of the Hardanger-field, on his way to Christiania. He was fortunate in finding a companion for this wild journey in a gentleman of Bergen, already acquainted with the country, who was making a tour in the same direction. The rain fell with hopeless intensity, and the affection of the people for their ponies, which rendered them unwilling to expose their beloved animals to the smallest fatigue, interfered greatly with their progress. At Oos, however, they quitted their *carriole* for a large four-oared boat, in which they passed through the splendid scenery of the Hardanger-fiord to Bondhuus, from whence they visited one of the glaciers which descend from the snowy range of the Folgefond.

Forbes found the Bondhuus Bræ (or glacier) in all respects similar to those of Switzerland and Savoy, showing, like most of the latter, unmistakeable signs of former extension, in the form of ancient terminal moraines. After examining it, they proceeded to cross the lofty snow-fields of the Folgefond to the head of the Sör-fiord.

'I was naturally curious,' he writes, 'to examine what I had seen so often described, as these Norwegian plateaux. The snow, fortunately for us, was of very good consistence. Probably new snow does not frequently fall in summer, for the general level is but little above the snow line. It is, for the most part, in the state of *névé*, a term applied to the stratified slightly compressed snow of the higher Alps, before it is condensed into the crystalline ice of glaciers. The stratification here, however, is not *particularly* well marked. This *névé* moulds itself to the greater or less inequalities of the plateau, forming large crevasses here and there; and the general form of the ground is trough-shaped—the two edges of the *fond* (east and west) being commonly higher than the centre, and the centre or trough inclining gently to the north. One of the first objects I saw was a small but true glacier of the second order, reposing on a rock having apparently a very moderate slope near the middle of the *fond*, and connected with one of the higher domes of snow to the N.E. It appeared perfectly normal, with intersecting crevasses (owing to the convexity of the surface on which it moved), somewhat like the dwarf glaciers of the Trélaporte at Chamounix. I think that bare rock, or at least ground where snow melts, may be considered as almost a *sine quâ non* for a true glacier, whilst a *névé* may or may not be so accompanied. Wherever we have this, with a good *feeder* or snow valley, and not too great an elevation, and even a very moderate slope, *there* a glacier forms as a matter of necessity. I afterwards saw many such in connection with the Folgefond.'

Descending the Sör-fiord to the Hardanger-fiord, of which it forms a branch, they proceeded northwards through moorland and forests, to the Sogne-fiord, which cuts deeply into the rock-bound coast about fifty miles above Bergen.

'The arrival at Gudvangen takes one by surprise. The walls of the ravine are uninterrupted; only the alluvial flat gives place to the unruffled and nearly fresh waters of this arm of the sea, which reaches the door of the inn. After dining and procuring a boat and three excellent rowers, we proceeded to the navigation of the extensive Sogne-fiord, of which the Näroe-fiord, on which we now were, is one of the many intricate ramifications. The weather, which had fortunately cleared up for a time, was now again menacing, and a slight rain had set in when we embarked. The clouds continued to descend, and settled at length on the summits of the unscaleable precipices which for many miles bound this most desolate and even terrific scene. I do not know what accidental circumstances may have contributed to the impression, but I have seldom felt the sense of solitude and isolation so overwhelming. My companion had fallen into a deep sleep; the air was still damp and calm: the oars plashed, with a slow measure, into the deep black fathomless abyss of water below, which was bounded on either side by absolute walls of rock, without, in general, the smallest slope of *débris* at the foot, or space enough anywhere for a goat to stand; and whose tops, high as they indeed are, seemed higher by being lost in clouds which formed, as it were, a level roof over us, corresponding to the watery floor beneath. Thus shut in above, below, and on either hand, we rowed on amidst the increasing gloom and thickening rain, till it was a relief when we entered on the wider, though still gloomy, Aurlands-fiord, in which the sea had a more natural and agitated appearance.'

Before resuming his journey to Christiania, Forbes visited the glaciers of Fjærland and Justedal, the former lying on an unfrequented fiord branching off from the Sogne. Of these the most remarkable was the

Suphelle Bræ. 'I sketched it,' he writes, 'from the hamlet of Suphellen, from whence, as from the sea, it *appears* as if in complete continuity with the snow-fields above. But when I had mounted some way over the ice itself, I discovered, to my surprise, that a very lofty cliff of rock entirely separates the upper from the lower glacier, the latter being in fact what is termed by the later Swiss writers a *glacier rêmanié,* formed altogether of icy fragments precipitated by avalanches from the steep and pinnacled glacier above. It is an exceedingly remarkable arrangement, which has no parallel in the greatness of its scale in the Alps; but it recalls the *glacier rêmanié* which is imposed upon the glacier of La Brenva. Indeed, we have only to suppose the declivity down which that glacier descends from Mont Blanc to be somewhat greater than it is, in order to cause a complete instead of a partial separation (for there is an island of rock in the centre of it, traversed by avalanches) of the lower from the upper ice. Like the small superimposed glacier of La Brenva, the Suphelle Bræ presents a distinct, though far from delicate veined structure everywhere near the contact with the surface on which it rests, after the type of ordinary glaciers. The upper part of compact whitish ice (like the glacier of Bossons) is nearly amorphous. It is dangerous to ascend too high, on account of the frequent ice-falls from above. I was unable to ascertain clearly whether the upper and lower glacier unite in winter. I heard a rumour to that effect, but I scarcely think it likely.'

Having exhausted the glaciers of Fjærlands-fiord, they started on horseback for Justedal, a charming ride through one of the few Norwegian valleys really comparable to those of Switzerland. 'The first glacier we visited was the Krondal Bræ, which descends from the snow plains of Justedals Bræen in a magnificent sheet, remarkably uniform, yet very steep. It is of course much crevassed; yet I have seldom seen so abrupt an ice-fall so unbroken in its character. When it reaches the valley it compacts itself, and then commence a series of *waves* and *dirt bands* in the ice, near twenty in number, which I saw beautifully by the evening light, and at a distance of several miles. It may be proper to mention that by *waves* or *wrinkles* we mean alternate ridges and furrows in the ice, on a very large scale, and approximately transverse to the glacier, or running from side to side, but more forward in the centre than at the side, so that their ground plan is concave to the origin of the glacier: by *dirt bands* we denote bands of cellular or friable ice, in which mud and stony fragments find a lodging, and thus faintly discolour the surface of the glacier in the same wave-like forms as the ridges and furrows just mentioned, with which they are so far identical that they are found constantly together; so that the "wrinkles" are visible at a distance mainly by the discoloration which the "dirt bands" occasion. Accordingly, the latter were first observed by me at Chamounix in 1842, the former in the following year. I am not prepared to affirm that the explanation of this curious phenomenon is clearly made out; but I have elsewhere endeavoured to show that it certainly depends upon the laws of motion of the glacier, and on the peculiar consistency of the ice of which glaciers are composed. It is pleasing to find features long overlooked, yet apparently essential characters of a true glacier, recognized in regions so remote as Switzerland and Norway, and even in the gigantic mountain chains of northern India.

'The Nygaard Glacier, which is of great length, descends the valley which contains it by angular zigzags, resembling a carefully constructed but gigantic highway, embanked at the turnings by its own moraines, and there are three such turnings quite distinct. It is in all probability the most regularly developed glacier in Norway. Comparing it with the Swiss glaciers, it somewhat resembles the Mer de Glace above Montanvert, but in picturesque effect falls far short of it, owing to the want of a fine background; the view here terminating as usual in the flat-topped snow-fields of the "Fond," a poor contrast to the noble buttresses of the Jorasses and Tacul, and the pinnacles of Le Géant and Mont Mallet. The lower termination is also quite unlike, for the glacier spends itself on nearly level ground at the expansion or *embouchure* of the valley which it occupies. And here we have evidence of the immense fluctuations in dimension to which glaciers are subject within periods by no means remote, probably more striking than is anywhere to be found in Switzerland, not even excepting the case of the glacier of La Brenva, on the south side of Mont Blanc, which I have very fully illustrated elsewhere.

"I made the best of my way on horseback across the stony desert which separates the roadway from the foot of the Nygaard Glacier, slanting towards its southern edge. Dismounting, I scrambled along the moraine bordering the ice, in company with the man whom we brought from Rönneid, and who said he had been once on the glacier before. I proposed to cross it from south to north at a convenient place. This is always a matter of some difficulty and uncertainty, and but for my geological hammer, which I used to cut steps, we must have abandoned it. The character of the ice was *very highly* crystalline, such as we find characteristically in such glaciers as that of Aletsch which have run long courses. The ice was very hard, rough, and sharp, presenting many angular prominences; and the sun glanced from the plates of crystalline texture in a way which I do not recollect to have seen so strikingly except in the Swiss glacier just named. Farther, the general surface of the glacier showed the gradual obliteration of the more salient outlines, which I have particularly noticed late in autumn on the Mer de Glace, and elsewhere, as one of the most familiar and evident proofs of the "plasticity" of the ice of glaciers on the great scale—the ridges between crevasses sink, the crevasses themselves are gradually cemented by the cohering of the material of the sides bulging under their own weight—all the forms pass gradually from *serrated* into *undulating*. This was so well marked on the Nygaard Glacier, that I have no hesitation in concluding that, notwithstanding the high latitude and the long winter, the Norwegian glaciers (I speak now of those not within the Arctic circle) are in all respect as "viscous" or "plastic" as those of Switzerland—in other words, will, equality of size being supposed, move quite as rapidly upon a given slope.

'Owing to the peculiar condition of the ice which I have described, it was not to be expected that the *veined structure* would be strongly marked, especially as the glacier, where I traversed it, was in the act of spreading itself abroad, ceasing to be confined by the walls of the valley; and, as I have shown by many instances collected in the Alps, the veined or ribboned appearance of the ice is unquestionably connected with intense lateral constraint. It was, however, by no means wanting, and was well developed near the contact of the ice with the soil, on its northern bank.

'On our return from Justedal to Rönneid, I noticed a fact which appeared to me to be incontestable, namely, that the level to which the *roches moutonnées* extend on either side of the valley progressively lowers as we approach the sea. This fact, which is also observed in the Swiss valleys, is held to denote, on the glacial theory, the highest level which the glacier at any time attained. From Rönneid we took boat to Solvorn. I spent a day of welcome rest with my former friendly entertainer there, and proceeded on my now solitary journey to Christiania across the Fille-field—solitary, because Mr. D.'s engagements at Bergen required him to take the steamer from Lærdal to Bergen, which sailed this season but once a week, ... leaving me to perform the remainder of the journey to Christiania alone, by *carriole* (except a small space by water) without any prospect of communicating with a single individual by the way, except through the few words of Danish which I had acquired. . . . At Christiania I was fortunate in meeting many friends, both Norwegian and English, which proved a welcome change after a long comparative solitude. I returned to England with little delay, except a day or two spent at Copenhagen, crossing the North Sea from Hamburgh to Hull.'

The tour had been a fatiguing one. The constant changes from excessive heat to intense cold, and from active exercise on foot or on horseback to hours of exposure in an open boat, severely tried Forbes's strength. Still worse, the realized wish of an indefatigable observer for 'a day four-and-twenty hours long' had constantly tempted the mind to cheat the body of its rest, and, as he often said, he found it impossible to sleep among the glories revealed by the midnight sun. All this sent him back but ill prepared for his work, and soon after he resumed it, his overtasked energies broke down. The tale must now be told of an illness, which put an end to his active life, and compelled him to break off, at once and for ever, his glacier observations: leaving unsolved more than one problem which still awaits solution.

CHAPTER XI.

FAILURE OF HEALTH.

THE end of September 1851 found Forbes in London purchasing models at the Exhibition, which he had seen with Mrs. Forbes soon after its opening in May. When early in October he joined his family in their summer home at Craigieburn, he complained of a bad cough. November brought a new session, and though the cough had by that time left him, he confessed that he entered on his work with a sense of weariness never felt before.

Of this to him eventful month his journal contains the following notice :—'The cold appeared gradually to subside. Lectured for four weeks without inconvenience. 28th, much exhausted by class and a Royal Society Council Meeting. 29th, gave an extra lecture. 30th, seized with hæmorrhage, the commencement of a long and severe illness.'

Here are the thoughts he wrote down at the time :—

'*November* 30*th*, 1851.

'This morning I spat blood. It was neither caused nor accompanied by any cough, so that I incline to take a favourable view, and to believe that it does not come from the lungs. Nevertheless I felt on the 28th great exhaustion after lecturing. There is enough in this to make me serious, and when I consider that I have lately found it an effort to fix my thoughts on anything beyond and above this world, and that the regulation of my thoughts generally has been far from strict and effectual,

I ought to thank God for this reminder of feebleness and mortality. This I hope I do sincerely, still praying, however, that this visitation be not unto death. I read the Bible and other books, with an interest and application to myself which I have not felt for long, and my neglect of my greatest good appears very strange and wicked. I do pray earnestly my God to bless to me this reminder of mortality; the more striking as it recalls my illness now of eight years ago, which, alas! has left so little impression. The chief good thoughts which have passed through my mind in the last six months are connected with the memories of John Mackintosh. May the same Saviour, to whom he commended me from his dying bed, be my support too, when my hour shall come. Supposing this symptom, for it is a mere symptom, were really to harbinger fatal disease, were it not best to act so now, as to prepare myself for such being the case? The symptoms of fatal disease will and must come some time. Will they be less unwelcome then than now? Probably not. More unwelcome as their interpretation is less equivocal, unless God will please to change my heart, so as to wean it from the love of worldly things, and make me truly fitter than I now am for the great change. I will earnestly pray that this day's reflection may be a step to that change.'

In two days his illness assumed a still more serious aspect. On the 2nd December he writes in pencil to his most intimate friend :—

'. . . I am sorry to grieve you as I fear this note will. I have had a smart attack of blood-spitting. The cause was obscure, but the immediate effect I much fear will be to incapacitate me for duty. I rely much on your affection and prayers. In all probability I shall be able to write you a better letter in a day or two.'

The crisis had at length come. The wear and tear of more than twenty years—winters devoted to his class, summers still more trying to the strength, travels far and near in search of knowledge, close application at home, or bivouacs on the glaciers—these had all done their work on a frame which if more than usually muscular and agile, had in it probably some inherent weakness. Since the illness in the summer of 1843 he had never maintained the same unbroken health he had enjoyed before; ailments from time to time had told him that he 'was not now that strength' he once had been, yet much hard work both bodily and mental had been undergone since. He had passed through the trying Alpine summers of 1844 and 1846; with the exception of one session, he had carried on his winter work as vigorously as ever; and his experiments on heat had been, if intermittent, never abandoned. But that rapid and severe Norwegian summer had strained his strength to the last fibre, and his health, perhaps already undermined, suddenly gave way. On the free use of his energies a hand of stern arrest was laid at once and for ever. As far as scientific discovery was concerned, it was as though he now heard a voice that said, 'Thus far and no farther.' Hitherto side by side with his scientific compeers he had scaled the hard steep of discovery, intent mainly on the truths to be attained, but not insensible to the reputation that waits on their attainment. His foot was already on all but the highest ridge: one step more had placed him on that topmost eminence. But that step was not allowed. It was a severe trial for flesh and blood. We are now to see how he bore it.

All December 1851 Forbes lay hanging between life and death. About the middle of January 1852, he was so far recovered as to be able to move, attended by Mrs. Forbes and his family, from Edinburgh to the milder air of Clifton. This became his head-quarters for the next two years, and there under the kind and skilful care of the late Dr. Symonds his condition gradually improved. The physician to whom he owed so much soon became his firm friend, and continued so to the last.

The summer of 1852 was relieved by a sojourn of nearly two months at Ambleside and Grasmere, and the summer of 1853 permitted him to make a short tour for health's sake on the Continent, and Basle, Zürich, St. Gothard are names that occur in his journal.

During his enforced leisure, when not prevented by illness, he was constantly engaged either in writing a dissertation on physical science for the 'Encyclopædia Britannica,' or in preparing for the press a work on 'Norway and its Glaciers.' In the summer of 1853 he was called from Clifton to Oxford to receive the honorary degree of D.C.L. at the Commemoration at which the late Lord Derby was installed as Chancellor of the University.

During these two years and a half of absence from Edinburgh, his health, though on the whole improving underwent several vicissitudes, and there were anxieties besides about the health of Mrs. Forbes and other members of his family. Allusions to these occur in the letters which follow.

To E. C. BATTEN, ESQ.

'CLIFTON, *April 23rd*, 1852.

'. . . All this makes me feel that I am in a shaky state, and the completion of my 43rd year at the same time makes me regard things and events as far more transitory than they used to appear. At the same time I have a sad deficiency in the power of realizing, and thence taking hold on the unquestionable realities which are unseen. As for ambition, I feel as if I had almost worked out that powerful motive to action. Do you sometimes withdraw your mind from the exertions of business and domestic anxieties to try to realize these things?

'. . . I do indeed concur in your regrets, and melancholy, pleasing recollections of "dear Wilson." His influence as a man of high principle will be long missed in Edinburgh. Ferrier is a candidate for his Chair.'

To the Same.

'GRASMERE, *August 19th*, 1852.

'. . . Our life passes in a quiet, and not unpleasing uniformity. With the aid of Dr. Davy's library, I find my time anything but heavy on my hands. Our situation on the east slope of Grasmere is very delightful. It is perhaps the prettiest of the smaller lakes, and the points of view and rides very varied. We have had some very stormy weather, but also some delightful days. To-day we were hours on the lake. I fear I am not sufficiently Wordsworthian to be worthy of all my privileges—*e.g.* to have Wordsworth's original cottage,[1] where he wrote the "Excursion" within about 200 yards of where I now write this piece of prose. We are happy in having our dear children all together, and so healthy and happy, enjoying the complete country.'

To the Same.

'CLIFTON, *April 3rd*, 1853.

'I have been wishing to write to you often since you were here, and always hoped to have a more cheering account to give of Alicia. But we have been making slow progress, and with some serious drawbacks. She has only been twice out of her room this year, and her weakness is extreme. May God give us grace to trust in Him, and hope to the end. Dr. Symonds is kinder than ever. . . . God has visited us with very great trials in the way of sickness, but has hitherto carried us through without serious loss. I feel sure that the lesson was needed, and trust it may work a due effect, though the chastening is grievous. My own health continues wonderfully uniform. I was much obliged by your sending the *Guardian*. I like the Duke of Argyll much, and we correspond occasionally.

'. . . The former recommendations about *ad eundem* degrees at Cambridge, proposed a year ago, were much more limited, and expressly excluded Scotch Universities, whereon I remonstrated to Dr. Whewell, and sent him a copy of our examination papers. The gross ignorance and studied neglect of what goes on in Scot-

[1] There must be some mistake here. Wordsworth wrote the greater part of the 'Excursion,' not in the 'original cottage,' but at Allan Bank, a large house to the west of the Lake of Grasmere.

land is highly irritating. For instance in the *Times*, &c. the professorial system is described and discussed as if it existed nowhere but in Germany.

'We had a most agreeable visit from Auguste Balmat.'

To the Same.

'CLIFTON, *February*, 1854.

'. . . In all our trials, be they greater or less, wrenching the memories of the past or embittering the enjoyment of the present, let us strive to realize the great blessings and comforts we have; the conviction—surely an increasing one—that a life of mere worldly enjoyment does not conduce, as we imagine, to even temporary happiness, and that the time is surely coming when all these things will sink immeasurably in importance in our eyes.'

To the Same.

(Written in pencil.)

'CLIFTON, *April.*

'We have indeed been in much trouble, but are now relieved and better. My disorder was pleurisy, quite distinct from my old complaint. George has had severe gastric fever. I could not ask you to come and see me for more than half an hour. We sleep in the drawing-room. These continued afflictions are heavy to bear. May we learn to read their intention aright! We have received much kindness from persons we knew but little: but to Dr. Symonds and the G. Fowlers, we are unspeakably indebted. We are brightening now, and may yet see better days if such be good for us.'

In June 1854 Forbes' health was so far restored that he was able to leave Clifton and return to Edinburgh. 'A few days after,' his journal records, 'Dr. Symonds joined us, and we spent a happy week in his company and that of other friends. His letters to me at that time and after showed a depth and sincerity of attachment on his part, of which I had previously no idea; while acting simply as a physician he concealed it.'

The summer was spent in visiting relations and friends in Scotland, until once more he returned with all his family to the home he had quitted in so precarious a state nearly three years before. 'On the 20th September we reached our own comfortable house in Park Place, for which,' says his journal, 'I thank God humbly and sincerely.'

'O God, who has visited us with many trials, and led us like the Israelites of old from place to place without any certain abode, bless, we beseech Thee, our return home, and mercifully grant that the afflictions and anxieties of that long probation may bear fruit in a more self-denying and godly life; and that we may have our hearts fixed on a yet more abiding resting-place, eternal in the heavens, for Jesus Christ's sake.'

The two following short letters, written soon after his return home to that friend with whom he shared so many of his confidences, refer, the first to the impression made by the perusal of the beautiful memoir of the late John Mackintosh, written by the late Dr. Norman Macleod; the second to the friendship he had formed with his physician at Clifton, one of the last and firmest he ever formed.

To E. C. BATTEN, ESQ.

'3, PARK PLACE, *October* 12*th*, 1854.

'. . . I am now reading Mackintosh's life very slowly, having devoured it at first with a rapidity quite unusual with me, and I am enjoying it afresh. His buoyancy in his first tours awakens my own early recollections, when I felt the charms of nature and solitude with a rapture not a whit inferior to his, though I never recorded them. They were perhaps the most thrilling hours and days of my life; and to read his vivid pages makes me young again. I can indeed thank God for having disclosed to me the hidden charms of His beautiful world; and I could live content to any age in the joyful recollection of them.'

'3, PARK PLACE, *October* 12*th*, 1854.

'. . . He—Dr. Symonds—is a man such as few are. You can have no idea, for it astonished myself when I learnt it, of the warmth and depth of his feeling for me and for Alicia and the children, knowing how doctors are circumstanced, and how used-up their feelings are. I never credited it till he left me no room for doubt, which was not till his professional relation to me was drawing to a close. I feel to Dr. Symonds as to very few indeed besides, and I scarcely expected to form such another friendship in this world.'

During the two sessions of his absence, the work of his class had been conducted by his faithful colleague and friend, Professor Kelland, with the aid of a younger assistant. When November 1854 brought the opening of another session, Forbes was once more in his place. He was generally able to lecture three days a week, being assisted on the other days by Mr. James Sime, and he 'found the fatigue much less than he expected, and no injury from it, verifying Dr. Symonds' opinion and justifying his excellent advice.'

To E. C. BATTEN, ESQ.

'3, PARK PLACE, *December* 10*th*, 1854.

'. . . I am very thankful to tell you that things are going on well with us. . . . I am surprised how well I get through my lectures. But the difficulty I at present have is with people coming to see me after, and being kept conversing an hour longer. . . . I am sometimes exhausted for that day, and even the next. . . . Alicia is so much better than she has been for long, that it is a matter of great ease and thankfulness to me; and the dear children are all so well. So our minds, by God's mercy, rest calm after a long time of tossing. How long it may last we feel to be in His fatherly hands.'

To DR. SYMONDS.

'EDINBURGH, *December* 18*th*, 1854.

'. . . I have finished, or all but finished, the "Dissertation." In fact it is out of hands, and will begin to be printed immediately: but I shall carry it slowly through press for the purpose of revisions. As you kindly take an interest in it, and indeed have had a considerable share in the fact of its production, you will I know be glad of this. It extends to about 600 pages such as you have seen me writing. It has been a very useful as well as pleasant labour—useful, I mean, as an employment of interest; and the very handsome honorarium I am to receive—400 guineas—is by no means a matter of indifference. What a profoundly striking event this death of the Czar! I hope it will be the dawn of better days for Europe, and this country in particular. I trust indeed we have learned a lesson in the humiliation to which England has been and is reduced, and which will not be forgotten for one generation at least.

'I do not follow the *Times* in their recent outcry about England being governed by a few families, ruined by its aristocracy, and plebeian talent kept down. Show us the talent! Who has been named to meet the crisis, who cannot get into power? The House of Commons is not elected by the aristocracy; yet where is the commanding talent it possesses? In fact the *Times* only exposes its own inconsistency. A few weeks ago the single man it could name to be the saviour of his country was Lord Dalhousie. I cannot help thinking Layard, who is the present *Times* hero, rather a light horseman. It is a rare event when I enter upon politics, so pray forgive me; and believe me, with Mrs. Forbes' kind and cordial remembrances.'

About the middle of the session he writes thus: 'I can give you a good report of myself. With trifling oscillations my health is good, and I feel really up to my work, and the better for it. I take, indeed, extreme care to avoid cold, especially since east wind set in, which it did some days ago. As I am much in the house at this season, I get in the *Times* for two hours, which is an amusement, though at present a melancholy one; for no one can read about the Crimea and Scutari hospitals without shrinking from the future. The last *Quarterly* is not comforting, though very interesting: I mean the Crimean article. Is it by your friend Kinglake, whom Dr. Symonds wrote me he had been telegraphed to attend some time ago? We fancy so. I have not yet seen the *Edinburgh*, which defends Government.'

As soon as the work of the session was over, Forbes and Mrs. Forbes repaired to Clifton, to pay a visit to their much valued physician and friend there. After a stay of more than a month they returned, stopping at the Observatory at Greenwich, and at Cambridge, on their way to Scotland. About the middle of July the whole family set out for Braemar, where they had taken for their summer quarters the Free Church manse. With its owner, the late Rev. Hugh Cobban, a man of more than common cultivation, an acquaintance began, which quickly ripened into friendship. 'Tis thus that Forbes writes to Dr. Symonds during his sojourn in Braemar:—

To DR. SYMONDS.

'BRAEMAR, ABERDEENSHIRE, *July* 29*th*, 1855.

'. . . Our friend the Free Church minister has honourably acquitted himself of his undertaking. . . Our manse can hardly be said to be in the village. It is at least the last house in it, and open on all sides: a stream, a tributary of the Dee, in a deep rocky channel a little way in front, and a wood of pine and birch—almost the exclusive trees, except the alder, of this country—at the back: both of which are the delight of the children,—I mean the wood and the river. We hear much of Edmund's and George's fishing excursions, though we have not as yet seen any results. The village is small, and irregularly built; contains two general shops and a good inn, a Free and Established church, and a very pretty Roman Catholic chapel, built under the auspices of the Duchess of Leeds, half the population being Romanists. . . . St. Swithin has been a great deal more gracious than we ventured to expect. Both Alicia and I have been trying to revive our dormant sketching powers, though as yet without any great success. I should have mentioned that Braemar is said to be 1,100 feet above the sea, and within a short distance of the house we can see Ben-na-muic-dhui, Cairngorm, and Lochnagar, the first, as you know, nearly the highest mountain in Britain ; and many of the hills have still patches of snow, which, though hardly picturesque, have a certain charm for people as crazy about the Alps as myself.'

After that sojourn was over, it was thus he looked back on it :—

To the REV. H. COBBAN.

'EDINBURGH, *September 22nd*, 1855.

' . . . In looking back to the time I spent at Braemar, notwithstanding some anxieties, I do so with a keen sense of pleasure. Since I have been incapacitated from active exercise, I do not know that I ever felt the exquisite enjoyment of natural scenery, and the fresh air of the mountains more than during our drives about Braemar. I do not except my last visit to Switzerland in 1853. I have also formed a very favourable opinion of the climate of Braemar, which is surely one of the best in this country during summer.'

The close of the year 1855 found him doing, with little interruption from illness, the same amount of work, and with the same assistance, as in the previous year. He thus expressed the feelings that rose within him on the last day of that year :—

To E. C. BATTEN, ESQ.

'3, PARK PLACE, *Dec.* 31, 1855, 9 P.M.

'I shall write you my last letter of 1855, and wish you most cordially many returns of the new year—many blessings for yourself and your family, and much health to enjoy them. When I think of my state in these very rooms four years ago, when you were with us, I wonder to find myself still here, and truly thank God for having given me the measure of health and enjoyment which I still possess. How solemnly and imperceptibly the shade of life draws round us! how gradually we learn to think not indispensable all that we once most coveted! How the ranks of friends and relatives are thinned, and how even people that we have no personal regard for, when we meet them casually on the streets, tell us by the singular change in their appearance how the world is getting older, and how we, too, must pass through change to the unchangeable.'

With Dr. Symonds Forbes henceforth kept up an affectionate intercourse by visits and by letters. In him he found not only a skilful physician and a devoted friend, but one who could enter into all his scientific interests with the sympathy of an adept.

In May 1856 he visited Dr. Symonds for a fortnight at Clifton, and this visit Dr. Symonds returned by a visit to Forbes and his family in the autumn of the same year at a pleasant spot in the Perthshire highlands, where they had this year chosen their summer abode. The village of Pitlochrie, then much less frequented than now, combined for Forbes many attractions not easily to be found elsewhere in Scotland. Lying on the great Highland road, just at the entrance to the Killiecrankie Pass, screened, by the high mountain wall which flanks the Tummel valley, from east and north winds, and surrounded on all sides by romantic scenery, it had the especial advantage of a mild and salubrious climate during the summer months. To that stretch of Athole might with slight change be applied those words of Shakspere:—

> 'The climate's delicate, the air most sweet,
> Fertile the vale, the mountains much surpassing
> The common praise they bear.'

Another great advantage which this neighbourhood offered was the presence of a skilful and wise physician,

DYSART COTTAGE, PITLOCHRIE.

Dr. Irvine, in whose advice he could trust. Here, too, as at Clifton, the confidence of physician and patient soon passed into sincere friendship. The first summer spent in a hired house passed so pleasantly, and suited Forbes and his family so well, that he determined if possible to find there a permanent abode. On the approach of the next summer he obtained from Mr. Butter of Faskally a lease of Dysart Cottage, beautifully situated on the Faskally estate, on a bank, scattered over with birch trees, and looking down on the deep vale of the Tummel. The village of Pitlochrie and the vale of Athole lie immediately to the south. On the north the cottage commands a romantic view of the wooded braes of Faskally, and of the gorge whence comes the river formed by the combined Garry and Tummel. This became the summer home of Forbes and his family for the rest of his life.

The following letters to Dr. Symonds show how largely Forbes confided in him :—

To Dr. Symonds.

'Edinburgh, *May 9th,* 1856.

'. . . It was a touching proof of your kind fellow-feeling, the purchase of those octavo volumes of De Saussure! I am afraid you will distrust Mr. Kerslake's use of my name again. Yet I do own a strong feeling towards the respectable old Genevese. So steady and persevering, so acutely observing, so precisely yet often picturesquely describing; and amidst all his lithology and topography, such a glow of honest enthusiasm at times reveals itself through his occasional, but innocent pedantry. The ascent of Mont Blanc and of the Col de Géant in the latter portion of the book are the most readable. I have lately had to vindicate De Saussure and the Swiss geologists, as well as myself, from the impertinences of Mr. Daniel Sharpe, an English geologist of some little notoriety—likely, I am told, to be the next president of the Geological Society, who made a scampering tour through Savoy, and denounces all his predecessors. It will scarcely amuse you much, but I will send it to you when published. I was preparing for it when at Clifton, but never found time to write it down till the few days' holiday at Christmas.'

To the Same.

'BRIDGE OF ALLAN, STIRLING, 12*th May*, 1856.

'. . . You will have received the commencement of the chapter on electricity in my work. My wish has been to give you a specimen of each of the divisions of it without boring you with too much. Galvani, Volta, Davy, which I am now in the course of sending you, were all rather favourite subjects when I was doing them, and I found reason to rate Galvani higher, and Volta, perhaps, not quite so high as is commonly done.

'The work is now all in type, except a few pages; but there is still a good deal to revise. I do not know whether I should send you any more—Faraday, for instance? I shall try to get him to revise himself. It seems a fatality that I must not only write but revise this work at a distance from books. In other respects, Bridge of Allan suits me well enough.'

To the Same.

'CHESTER, 17*th June*, 1856.

'. . . It is hardly necessary for me to repeat how very greatly we have enjoyed your hospitality and that of your amiable family. For myself I shall ever feel grateful to God who has given me such friends at a time of life when new friendships are made with difficulty. You must suffer me to add that I hope I have profited a little by the lessons of benevolence, unselfishness, and singleness of purpose which your daily life conveys. That you may be rewarded for all your disinterested kindness to us is the wish and prayer of your faithful and attached friend.'

When at the close of the year his Dissertation in the 'Encyclopædia Britannica,' which had been the main occupation of his available working hours from October 1852 till August 1856, was at last fairly published, he sent the first copy of it to Dr. Symonds, with the following letter :—

'EDINBURGH, *November* 30*th*, 1856.

'My impatience to send you a copy of my Opus Magnum, of which I received a few from my publisher last night, will not allow me to wait to have its exterior made a little more comely by the binder. Your first impression will be, I am sure, its small and insignificant appearance, though it represents the best part of three years' work of "these degenerate days."

'I owe you triple thanks connected with it; first, for originally encouraging me to undertake it; secondly, for your valuable revision of the language; and thirdly, for your kind sympathy and encouragement during its progress, when a little mild approbation was much needed to get up my flagging spirits.

'It is now finished and embalmed in that dreadful stereotype, and sent out 1,600-fold into the world, for such is its circulation in the Encyclopædia alone! I calmly await the award of the public so far as it may be expressed, knowing that I have done my best; but so well aware of its deficiencies, that no critic could write a sharper review of it than I could. In short, I am well up to its weak points.

'It is not only published but paid for! and Black said yesterday as much as it is in his nature to do, to gratify me.

'. . . My Pedestrian article has been at an absolute standstill for some time, but I have good hopes of resuming it.'

In his journal he writes, 'I received £420 for the copyright.' During the session of '56-'57 Forbes' health seems to have been better than it had been during any winter since his severe illness. For this year and the two following years he was assisted in his class work by Dr. Balfour Stewart, who lectured twice a week, and was present daily at the class for a great part of the session.

The following letters give some of the subjects which were engaging his thoughts at this time:—

To DR. SYMONDS.

'EDINBURGH, *January 1st,* 1857.

'New Year's Day I was resolved should not pass without conveying to you from myself more particularly—Alicia having already written—my sincerest and warmest good wishes for health and prosperity for you and yours, both for the present and many future years; and may you long enjoy in their fullest extent those domestic pleasures and consolations for which you are so pre-eminently fitted, and which, I may add, the arduous nature of your professional duties render a necessity not a luxury. Sympathy and support at home, so needful to all of us, are more peculiarly earned by those who spend their days in the immediate service of their fellows, and more than this, who relieve by actually sharing with them their sufferings, anxieties, and cares. All this I had in my mind to say at this time. Your truly kind remembrance of us all in your letter to Alicia, received this morning, brings it still more readily before me.'

To the Same.

'EDINBURGH, *January 3rd,* 1857.

'. . . I was very much obliged to you indeed for mentioning the article in the *Saturday Review*, on Glaciers, which might otherwise have escaped me. I am inclined to think that probably it was written by Mr. Alfred Wills, a young barrister in London, whose very pleasant narrative of his Swiss adventures forms one text of my article for the *Quarterly*. If it be so, our mutually speaking well of one another has at least been without collusion. The title of his very pleasant little book is "Wanderings in the High Alps."

'Regarding Mr. Tyndall and his theory, I have received no such precise information as to enable me to speak confidently about it; but I do not, as yet, expect that it will be found materially at variance with anything that I have advanced. Almost all that is mentioned in the

Saturday Review, had been said by me, or by others years ago; and I suspect that it is to be, in part at least, a dressing up of old ideas in new phrases. No one can fairly say that I have not acknowledged the difficulties and mysteries which attend processes which I have ascertained to take place, though I shall gladly acknowledge any advance in this direction on the part of Faraday, Tyndall, or any other.

'My first knowledge of the London world being stirred on the glacier theory came in not the most pleasant manner. Two letters arrived one after the other from a London celebrity, whom I shall not at present name, to my colleague, Balfour, evidently glorying in my coming defeat in terms sufficiently provoking, but which showed that it was to be made a regular party question in scientific circles there. I wrote to Sir Charles Lyell for some information about it, which he kindly gave me in a letter with which I also trouble you. . . . All I can do is to sit still till the indictment is made out, and I cordially wish my enemy to write a book and print it speedily, as anything is better than inuendo and suspense.

'On the whole I can thank God that I take these things more coolly than I once could do.'

To the Same.

'EDINBURGH, 10 & 11 *March*, 1857.

'. . . Now I will say something of our plans or prospects, still rather indefinite.

'Since I last wrote on the subject, the then scale of "home" and "abroad" has certainly been inclining rather towards the latter; and this not as a vague impression merely, but in consequence of circumstances not under our control. . . .

'Alicia thinks, perhaps, more strongly than I do, that the invigorating air and scenery of Switzerland might restore the power of my limbs, and, of course, I cannot help having some hope of it too. If you consider it otherwise a safe experiment, if cautiously made, as, of course, it would be, I shall think about it. . . . What I feel to be the great drawback to the intended journey is leaving the three youngest children behind us, and I cannot even now make up my mind to say that we will do it. We talk of taking Eliza and Minna with us, if we go, by way of making the journey less lonely, and relieving the responsibility of leaving them behind; though of course it will add heavily to the expense. And this anxiety about leaving the children bears on another point, which I must not pass over—your truly kind invitation to Clifton, backed as it is by the friendly letters of your excellent ladies. I can only say now that our wishes are all on the same side, and that the only obstacle would be the long journey, and the long absence from our dear little ones. Of course we have always speculated on the possibility or probability of meeting you—and why not part, at least, of your family—in Switzerland. And although your last letter seems to throw a sad bucket of cold water upon this admirable scheme, we are by no means willing to despair of its ultimate consummation at one part or other of the summer, if our journey ever comes off. But so many things may happen one way or other between this and June, that I hope we need not consider the die must as yet be cast of Forbes at Clifton, or Symonds at Chamouni. . . . With the somewhat presumptuous hope of whetting your appetite for accomplishing the impossible, I send by this post—partly, I must own, under female inspiration—a rough copy of my article on pedestrianism in Switzerland.'

The hope, here alluded to, of once more looking on the Alps was not fulfilled. In June 1857, Forbes and Mrs. Forbes with their two eldest children left Edinburgh, and got as far as Folkestone; next day they were to have sailed to Boulogne, but that was not to be. The friend who was waiting to welcome them again in Switzerland was Mr. Wills, an Alpine explorer, to whom Forbes had been drawn by strong mountaineering sympathy. How their acquaintance first began appears from a letter of the former summer :—

To E. C. BATTEN, ESQ.

'EDINBURGH, *June* 29*th*, 1856.

'. . . I have been enjoying a little book by a co-barrister of yours, Alfred Wills, of the Middle Temple—"Wanderings in the High Alps." He must be a young man of the right stuff; I should like to know him. Try to hear something of him and to see his book, which is modest, sensible, and spirited. . . . The perusal of Wills' book has inflamed my Alpine associations and made me long to taste the air once more. I went yesterday with my children to the paper mill at Colinton, and scrambled with them along the river's banks, where I used to catch minnows, and to build castles in the air, I believe.'

Kindred tastes quickly brought about this acquaintance. They were to have met in the summer of 1857. It is thus that Forbes announces his disappointment:—

To A. WILLS, ESQ.

'CLIFTON, BRISTOL, *July* 19*th*, 1857.

'You will probably be surprised at the place from whence I date. I know that you will be sorry when I tell you that having reached Folkestone with part of my family on my way abroad, I had an illness which though to a person in better ordinary health than myself it might have caused but a slight detention, debilitated me so much that I was forced reluctantly to give up my plan of visiting Switzerland, with, I must own, but feeble hopes that the state of my health will allow me even another year to go abroad with that freedom from anxiety to myself and others which alone can give even to our highest pleasures more than a precarious and doubtful value. All hope of meeting you under the shadow of Mont Blanc is therefore at an end; departed like other pleasant hopes which we have to surrender by slow degrees—let us trust, not without extracting some good from disappointment.'

From Folkestone, taking Clifton on the way, Forbes

returned with his family to Scotland, and the rest of the summer was spent in the cottage at Pitlochrie, which had now become his home.

The following letters will suffice to carry on the narrative down to the close of the year 1857, and to the end of the session 1857–58 :—

To DR. SYMONDS.

'PITLOCHRIE, *August 29th*, 1857.

'Your kind letter reached me when I was laid on my back and unable to revolve through the smallest of the aliquot parts of a right angle without a tremendous twitch in the ribs. But during the few days which have since elapsed, I have made happy progress—especially yesterday, and am now nearly convalescent, and hope to be able to drive out on Monday. It is a week to-day since my accident, which took place in the most innocent way, of sliding down a grassy bank, while vigorously engaged in making a trigonometrical survey of our estate (!), whereby the oak handle of a heavy hammer was dug into my ribs with a pressure due to the whole weight of my body. . . . I had enjoyed my country life very much, and become engrossed in training rose-trees and other horticultural operations; the extremely fine weather making us live very much in the open air, to the manifest benefit of our health. The acquisition of this place is quite a new kind of interest to me, and whatever may be my regrets at missing Switzerland, it is undeniable that many important advantages have arisen from our being on the spot.'

To the Same.

'EDINBURGH, *October* 22, 1857.

'. . . About the very moment that this torpid or deadly-lively process will be going on in Alma Mater (11 A.M. Nov. 3rd), I hear that a much more interesting and I hope more lively solemnity will be taking place in your house. We sincerely desire happiness to all concerned. To part with a member of your cherished household must under every circumstance create a pang. I hear the happy pair go to Naples. I could envy them if I allowed myself to dwell on it. After we were married we too spent a happy winter there. Oh! the charm of the long mild December evenings with rosy sunsets, when Alicia on a donkey and I on foot rambled for hours among the woody craters of Agnano and Astroni, and the charming shores of Posilipo and Baiæ! The spring is the worst thing everywhere.'

To E. C. BATTEN, ESQ.

'EDINBURGH, *January* 17*th*, 1858.

'. . . I had a charming letter of four sheets from Wills some weeks ago. I feel as if he were an "alter ego" when he describes his feelings and doings in the Alps. He has bought a few acres in Savoy, and intends to build. Auguste Balmat is, I suppose, living with him now. Do you know anyone who knows him? . . . . He was educated at University College, London. Singular to say, he and Balmat picked up last Summer on the Mer de Glace, a hammer which I had dropped into a "moulin" in September 1842.'

To the Same.

'EDINBURGH, *January* 25*th*, 1858.

'. . . The hammer ought to have travelled nearly 5,000 feet in fifteen years. This is consistent enough with its having been "opposite the Taléfre" when lost, and "below the Tacul" when found. I conjecture that Wills must have picked it up near where line G H of my map crosses the centre of the glacier. I find a full entry of the loss in my journal of September 25th, 1842.'

To A. WILLS, ESQ.

'EDINBURGH, *January* 18*th*, 1858.

'I feel much indebted to you for your delightful letter of the 3rd Dec. Though I have personally known very many Alpine travellers, and have corresponded with more besides, I have never met with one who entered so thoroughly and instinctively as yourself into what appear to me to be the peculiar sources of enjoyment of pedestrian travel. I find in the pages of your letters—always so lively and so admirably expressed—an echo of what I have myself thought and felt. I seem to live over again years long past, and to become young and robust in the vivid participation which I have in your adventures and successes. But let me add that it is not in these last alone that I feel a deep sympathy; it is even more in that quiet tone of reverential admiration of Alpine scenery to which your letters so perfectly, but unobtrusively bear witness. Your last arrived at a moment of some anxiety and fatigue; Mrs. Forbes being rather seriously unwell. She is now almost well again, and I myself am in fair health, quite as good, when I avoid all fatigue and excitement, as I have enjoyed since 1851: I am giving my usual course of lectures. Since Christmas I have not been quite so strong. . . . First of Balmat. If he is with you, pray give him my affectionate remembrances and sincere good wishes. . . . It would give me true pleasure to see the worthy man once more, and great pleasure to receive him in my native town; but it would be selfish in me to urge so long an additional journey when I consider how little I can do in my present state of health, and with my present engagements, to make his stay agreeable, or to provide him with society and occupation. Being the middle of our long term, it frequently happens that I spend my whole available strength in lecturing, and can neither walk nor talk at any length afterwards. Balmat, were he here, would necessarily be thrown on his own resources during the day, but could always spend the evenings with us, as I never go out then. This is all that I could fairly propose to him. . . . Once I seriously blamed you. Your going alone to the Petit Rognon was, I think, not justifiable. I have stated my fixed opinion on the subject in the *Quarterly*: though, of course, it gives me a high opinion of your appreciation of solitary grandeur, which I do most fully share: still it may be too dearly bought, as a mere luxury of feeling.

'I am almost ashamed to offer you my old hammer, for

which, alas! I have no further use, as a gift; but pray do not send it back. I shall be gratified by the thought that it is in the keeping of one having kindred feeling with its old master; and many an ice-step may it yet cut for you! I copy out on a separate leaf an extract from my journal regarding it, written in September 1842, which you will see gives perfect identification of the circumstances. . . . Mr. Ruskin, of fine art celebrity, has just sent me drawings of sections artificially made by him of the contact of gneiss and limestone, which amply confirm my assertions in my Travels, and in a later paper in reply to Mr. Sharpe, which I believe I sent you. Professor Studer of Berne writes me that a pupil of his has fully verified my section at Courmayeur, so that point is settled.'

The foundation of the Alpine Club in 1858 was hailed by Forbes with great interest, as promising to continue the work which he had begun; and his name headed the list of its honorary members. But although his endeavours to lead its founders into paths of scientific observation met, as he good-humouredly complains, with great obstacles in their passion for 'unbounded muscular exertion, and unfettered freedom of range,' his sympathy and advice were ever freely given, and gladly accepted.

He thus writes to one of his Alpine correspondents:—

'. . . I congratulate you on having so sincere a taste for mountain scenery; and as your acquaintance with it and with physical geography increases, I trust that you will be able to make further and still more valuable observations, for there is still much to be done with the glaciers, as well as other subjects. The state of my own health prevents my renewing my active explorations; but it is a very great pleasure to me to hear from you and others of the prosecution of Alpine research; and if my advice can be at any time and in any degree useful, it is much at your service. Robust I can never again expect to be. The mountains are for you and your cotemporaries.'

Four letters bearing on Alpine matters may here be inserted, although their dates belong to a somewhat later time.

To F. F. TUCKETT, ESQ.

'. . . I thank you much for your sketch map of the western part of the chain of Mont Blanc, which very much exceeded my expectations in point of completeness and interest. I have spent some pleasant hours in mastering the features of the country, and comparing your map with my own notes and those of others. Indeed, now that I cannot myself explore such scenes, it is one of my greatest enjoyments to survey them in imagination; and had I strength, it would not be long, believe me, before I was on the ground. . . . I was not before aware that you had succeeded in reaching Mont Blanc by the "Bosse du Dromadaire." The interest of the feat as a *tour de force* is far exceeded by the important light which it throws upon the topography of the western declivities of Mont Blanc. I cannot tell you the pleasure with which I saw that glorious glacier valley of the Miage, shown, for the first time, in its true importance. . . . I am impatient to read the details which are to accompany your map, and I hope you will not spare topographical minutiæ while your memory of the localities is fresh. The accounts of expeditions which I have seen are sometimes very deficient in that precision of which De Saussure so well estimated the value.'

To the Same.

'. . . Thanks to you, the mystery of the Pelroux group seems now thoroughly solved. It was most agreeable to me to find that my old friend the "Montagne d'Oursine," which filled me with so much admiration when I sketched it from Les Etiaches, proves to be the veritable culminating point, the Pic des Arcines. . . . I allowed for the fact of coming upon such scenery unprepared, and with almost the zest of a first discovery; but your testimony seems to show that I had not overrated the sublimity of these Alps.

'I must not omit to express my admiration of the beauty and evident fidelity of your mountain outlines. It is surprising how much light drawings like these throw on the topography of an intricate country, and how much pleasure they give to one's self and others. No part of my notes of travel have I consulted oftener than such panoramic sketches, or with a stronger wish that I had multiplied them more.'

To A. ADAMS-REILLY, ESQ.

'I must take the chance of this still finding you at Chamouni—not a great chance I fear—just to acknowledge your letter from Courmayeur, and to say how much I appreciate your indefatigable energy in the investigation of the topography of Mont Blanc, and to give you my warmest congratulations on your success. With what zest you must have booked your mountains one after another—old friends with new faces and in new associations.

'I do not pretend to judge of the precise import of your triangulation from the slight, though attractive, sketch which you gave me of it. You must have measured many angles which you do not mention nor set down, and these I trust will be sufficient to make the triangulation a connected one, for it is not so in your sketch. For you are aware that it is not enough that adjacent triangles touch at one angle: they must have a side in common. There are two breaks in the continuity of your sketch: one at Mont Joli and one at the Col de la Seigne.

'You would be sure, I think, to triangulate the Aig du Géant from the Cramont and the M. de la Saxe: for by means of this you could complete the circuit—so far as to be a check—to Chamouni. Another year it would be very advisable to measure what surveyors call a "base of verification" in the vicinity of Courmayeur, for which the ground is well adapted.

'And now, my dear sir, let me exhort you to take care of yourself. You have gone through immense fatigue;

and if you begin afresh with fresh companions from England, I must say that you run some risk of knocking up, without great self-denial. For my sake and the sake of your survey, pray do not risk your health nor your life on the Matterhorn.'

To A. WILLS, ESQ.

'I think the new series of "Peaks, Passes, and Glaciers," an improvement on the first. The light thrown on Alpine topography is very material indeed; the classification is much improved, and the illustrations leave little to be desired. The frontispiece of Bernina is a glorious rendering of an Alpine scene.

'I regret to see Balmat's name never alluded to in these volumes. Do you think him really too old now for heavy work? No doubt the Alpine clubbists are enormous walkers, and it would try any one above thirty-five to keep up with them. Indeed they don't agree with it themselves, as is plain enough from their own admissions. There is no doubt they would enjoy more, and learn more, by doing half as much in the same time. Yet for all this, I hope Auguste will not sink down yet into a mere lady's man: he would still be invaluable to any man with ever so slight a tincture of science, such as would prevent him from scampering from col to peak with an almost insane restlessness. I know the feeling myself, but I always struggled against it. Have you ever any longings yourself for an "ascension" now that your health is restored?'

These letters will suffice to show Forbes' interest in the Alpine Club and its objects. From this short Alpine digression we now return to the regular sequence of his correspondence.

To DR. SYMONDS.

'EDINBURGH, *January 24th*, 1858.

'. . . At times my electricity takes a good deal out of me, and, to own the truth, I have had a physical mathematical speculation connected with the temperature of the globe in hand from time to time for several weeks, on which I now and then lay out my small available stock of thinking power. The end of this week I have a large examination, in which, however, I receive ample assistance from my efficient coadjutor, Mr. B. Stewart. And now I shall make no more apologies, nor should I have said so much but that it tells so much of my story. I am in fair health; no indisposition, but little strength.'

To the Same.

'EDINBURGH, *February* 28*th*, 1858.

'. . . I expect, however, shortly an electro-magnetic coil from America such has never yet been seen in Europe, to give eight-inch sparks in air from electricity excited by only three galvanic pairs.

'My own speculations have lain in a different line, connected with a more favourite subject of my own, the climatology of the globe. I expect to be able to deduce some interesting laws of a numerical kind.'

To the Same.

'BRIDGE OF ALLAN, *April* 26*th*, 1858.

'. . . I am of course much interested in the Lord Advocate's Scottish University Bill, though the House of Commons rather uncivilly indicated a contrary feeling by talking all the time he was explaining it, so that "not one complete sentence reached the reporters' gallery." I presume the object to be to appoint a commission with large powers, as at Oxford and Cambridge. As I am now entirely passive on such matters, I look on, though with nearly as sincere interest, with none of the painful anxiety which prevails when one feels a responsibility to act to one's utmost, knowing how little the probable effect of acting will be. I probably told you, when I wrote last, that I had great reason to be satisfied with my class last winter. It was one of the most diligent, proficient, and best conducted which I ever taught. It was a disappointment, to be sure, to break down a fortnight before the session was over, and to bring it to a close, give prizes, &c. by deputy; but as far as usefulness went, my work was almost done. I believe I had been only four days absent from the class all the previous part of the session, though lecturing only three days a week.'

Two letters must here find place, though one of them belongs to an earlier date. To Cambridge men they may probably be interesting, as they were written, the one to, the other by, the late Mr. Leslie Ellis.

'My dear Friend, 'Edinburgh, *March* 16*th*, 1856.

'It gave me very great pleasure to hear from you, and that on more accounts than one. . . . I should have written to you ere now had I felt sure that the thought of a letter which you might be expected to answer might not oppress you, however little. My belief is that it is good for us all to keep up our ties in this world, even while we are striving to prepare ourselves for that other to which we are visibly hastening. For this purpose we should not drop society nor correspondence altogether, even if we at times feel quietness our greatest solace. I own that in this I am preaching what I do not always practise.

'Your letter brings me back in spirit to your quiet yet cheerful room, which it pleases me to have seen, because I can picture you there. And often after since we met has that scene of suffering patience arisen in my fancy, and to see you again would be by far the chief pleasure of another visit to Cambridge, of which, however, at present I see not the very slightest prospect. I often wish you could have your bed raised so as to see out of the window without the intervention of a mirror. This is a parenthesis. In that small chamber you are working out, my dear friend, a problem greater than you ever grappled with in the Senate-house; a victory as real as any military hero's. You have found that "tribulation worketh patience, patience experience, and experience hope,"—a happy climax. I, like you, have wandered near the confines of the dark river. I have felt the trial of shrinking humanity whereof you speak. I have striven to pierce with sharpened eyesight the deep abyss. I wonder, has your subtler and more exercised spirit penetrated farther than mine, and can you communicate any share of the insight which I fancy you may have attained? I never wrote thus to any one before: and I feel confident that you will trust, or at least forgive me. There are few whose thoughts on such subjects would be worth possessing. I feel as if there must be given to you, in compensation for your greater trials, more knowledge than others.

'But I shall say no more of this. Let me rather thank you for your kind, and to me always valuable present of a paper of yours, however short. In a work of some labour and anxiety which has absorbed the major part of my small intellectual activity for a long time past—a sketch of the history of science of the last 75 years—I have gratified myself more than once by citing you, (1) in connection with Gregory, (2) on probabilities, (3) about higher order of forces. . . . I have been lecturing regularly all this winter, and on the whole feel in better health than for four winters previously. With care I suppose I may weather some years more, and do some little work yet; but hard thinking is as much beyond me as hard walking. I feel very deeply for Dr. Whewell. He wrote me a manly and touching letter two days after his wife's death.'

From R. Leslie Ellis, Esq.

'My dear Forbes, 'Anstey Hall, *March 26th*, 1859.

'You have heard enough of me since October to be willing to excuse my silence. I wonder often, and perhaps you would if you knew all, that I think as much as I do of the subjects in which we used both to take an interest. But this is part of the, to me, great wonder how my life has lasted so long, with probably almost every organ diseased, and in a state which makes the conditions of health impossible. Fix a man as immoveably as Prometheus for four or five years, keep him in the dark for two or three, and however healthy he was at first, what a wreck you will make of him! Your essay was read to me or nearly so, and gave me great pleasure, mixed with a sense of how much my life has been wasted. I know, and I hope allow, for your constant partiality; but if, writing as an historian, you could feel yourself justified in mentioning my name, surely I might have achieved something, if the earlier years of my life had not been wasted, partly from want of settled purpose, and partly from griefs and cares which came upon me early, and which I felt more than most men, and if the last ten years had not been consumed by disease. Yet after all, what good would it do now? It is hard enough as it is to tear oneself from vain and worldly thoughts. Your preface seems to me admirable in tone, and I am sure right in substance. Did it ever occur to you that there may be a connection between the permeability of rock-salt to heat, and its being equally soluble in water at all temperatures? How is it with other salts? I think this hint might suggest something. With my best regards to Mrs. Forbes,

'Ever yours,

'R. L. E.'

This last letter bears the date 26th March, 1859. The writer died on the 12th May in the same year.

Early in the summer 1858, we find Forbes, when relieved from college work, still reverting to his old pursuits, and making experiments on the freezing-point and individual temperature of blocks of ice. Here are some of his findings thereon:—

To A. WILLS, ESQ.

'BRIDGE OF ALLAN, NEAR STIRLING,
April 26th, 1858.

'. . . This arises partly from the circumstance that I can now, I believe, show that "regelation" is, as I have for some time suspected, a correlative and necessary proposition from that other, that the transition from water to ice in congelation is gradual. In my 16th letter,

I think, printed in January 1851, I attributed the softness of ice to the fact, then recently stated, that water gives out its latent heat during congelation down to a temperature sensibly below 32°. Connecting this statement with certain experiments I have lately made, I am in a condition to show that a film of water ready to be converted into ice—viscid water or soft ice—between two surfaces of ice in a thawing state must solidify. I shall send you by and by a short paper on the subject. If I am correct, the cementation or "regelation" of thawing surfaces of ice is merely another phraseology for the assertion that ice softens, or it absorbs latent heat gradually, about 32°, where all substances do, such as wax, spermaceti, &c., which soften by degrees. Of course the fracture may be molecular as well as finite, and then it is termed properly plasticity.

'I am truly glad to know that your visit to us was not altogether a failure, notwithstanding our disappointment about Balmat. My inactive state makes me a bad entertainer, and it is only after years of discipline that I can bring myself to acquiesce in a mode of life, which as you may fancy is strangely at variance with my whole previous habits down to the age of 42.'

The summer was spent at Pitlochrie, varied only by a visit to Fettercairn and an expedition to Loch Rannoch. From his highland cottage he writes to Dr. Symonds.

To DR. SYMONDS.

'EDINBURGH, *May* 30*th*, 1858.

'. . . You must not forget that we too have a very respectable Sleepy Hollow at Pitlochrie, where our friends can lie on the grass and read poetry and listen to the murmurs of the Tummel. Seriously, we will take no refusal about coming to Dysart Cottage this summer. The whole season is open to you to choose. All times from the middle of June are alike to us.'

During this same summer Dr. Whewell and Lady Affleck, whom he had lately married, visited the

Forbes' at Pitlochrie. It will be observed that Forbes' correspondence with the Master of Trinity slackened after illness put a check to Forbes' scientific labours, but their friendship continued unchanged.

During the latter part of 1859, Forbes was subjected to not a little annoyance by discussions which were going on in London regarding his viscous theory of Glaciers. By some of the disputants not only was his claim to priority in discovery denied, but doubts were thrown on his straightforwardness in making that claim. The question as to originality was insignificant to him compared with the imputation thus cast on his character. The grounds of these charges will be weighed by more competent hands in another chapter. Forbes felt the imputations deeply, but from the thought of them he was called off by a very heavy personal bereavement in the death of his brother Charles, who in the beginning of November died at Canaan Park, near Edinburgh. This was the brother nearest himself in age, from whom his boyhood and youth had been inseparable, and for whom, even after their homes were divided, he retained a quite peculiar attachment. The following letters contain allusions to these and other incidents in this year of his life:—

To Mr. Wills.

'Edinburgh, *April* 12*th*, 1859.

'. . . I could not be indifferent to the injustice and bad taste of the review of my book in the *Athenæum* of last Saturday which I read last evening. I felt how sorely at a disadvantage I am placed with relation to the scientific world of London. Your sympathy and your willingness to assist me in obtaining fair play—all I ask—touched me a good deal, I assure you. It is close on post time, and as Balmat's affair is urgent, let me discuss that now, and I will return to the other subject.

'What you tell me of Balmat's feelings towards me was another pleasant revival of old sympathies. Sorry as I am to interfere in any degree with your well-devised plans for his photographic education, you will, I trust, not dissuade him from accepting the invitation I enclose for him, open for your perusal. After what you say I really thirst to see the good fellow again; and considering that all his other friends are within easier reach, and that yourself and many others have the road to Chamouni open to them in prospect, my position with reference to him is very different indeed. Looking to the uncertainties of the future, it is very doubtful whether he and I shall ever meet more if not now. So if you find his heart as much bent towards Scotland as you suppose, do postpone the photography for a week and let him come.

'Do not think me too selfish in all this. I have not a moment more, but will finish this letter another day, perhaps to-morrow.'

To Dr. Symonds.

'Edinburgh, *April* 1859.

'. . . Now I must take my first leisure time to assure you that we were very grateful for your kind invitation to stay with you, and that it is not without much hesitation that we have decided for the present, at least, to deny ourselves this gratification. The plain state of facts is this: putting economy out of consideration for the present, when our session draws near a close I have such a longing for absolute rest that other inducements seem for a time to have little weight. True, indeed, you have always made Clifton Hill House a place of real repose, and such, through your kindness, I have often enough found it, but, had we gone to Clifton we must have gone still farther, . . . to London especially. Though my presence there would be very favourable to obtaining the attention towards my theory which I wish, even that personal gratification might easily be too dearly bought, if I were agitated in mind and exhausted in body. And thus, as I have said, languor has in the meantime prevailed over even the powerful stimulation of both friendship and ambition.'

To the Same.

'PITLOCHRIE, *September 25th*, 1859.

'. . . We returned on Wednesday from Aberdeen, having hesitated about going even to the very last hour. . . . I spent the morning in Section A, and made no communications except a speech of about half an hour, arising out of another person's paper—Jas. Thomson—on the Properties of Ice, which was well received, and there was no painful discussion whatever. Tyndall was not there. About two o'clock some hungry philosophers were generally willing to attack a cold pie which we had provided in our lodgings close to the Sections, and which enabled us to see some real old friends in a quiet way; such as Airy, Faraday, Vernon Harcourt, Lloyd, William Thomson, &c. &c.'

To A. WILLS, ESQ.

'EDINBURGH, *November 14th*, 1859.

'. . . To-day is the first day, for several weeks, that I have had any remission from a heavy and wearing correspondence, which has been thrust upon me by the coincidence of a variety of affairs of an anxious kind at one and the same moment. The illness and somewhat sudden death of my next elder brother—not him whom you saw at Aberdeen—and to whom I have from infancy been deeply attached, has closed with its solemn warning this period of mixed disappointment and success.

'As regards the Royal Society affair, the mere irritation occasioned by apparent neglect and jealousy subsides under the visitation of mortality; but the deepest sting —a moral imputation, from which, notwithstanding the reiterated disclaimers which I have received, I cannot conceive the recent attacks to have been wholly free—remains behind with something like a leaden weight. The generous support which I receive from yourself and a few old attached friends, diminishes even the residuum considerably.

'I believe that the effect of the struggle—though unsuccessful in its immediate object—will be to render Tyndall and Huxley and their friends more cautious in their further proceedings. For instance, Tyndall's book, again withdrawn from Murray's "immediate" list, will probably be infinitely more carefully worded relative to Rendu than he at first intended.'

To the Same.

'Edinburgh, *December 24th*, 1859.

'. . . Bishop Taylor's "Holy Living and Dying," a book wonderfully adapted to readers and thinkers of every class, and from which I believe I have derived more comfort and instruction than from any book whatever of a similar description and character. There is no book less controversial or more intensely practical. The vigour of the writer compels our attention. And this is an advantage. To have our thoughts drawn off for a time from the cares, the pleasures, the ambitious projects, and even the praiseworthy employments of this life must be to us all a matter of unspeakable importance.'

To the Same.

'Edinburgh, *February 13th*, 1860.

'. . . I am glad you take an interest in Jeremy Taylor. You ought to make great allowance for the style of writers and thinkers—even the greatest—of his time. Redundancy, quaintness, and ingenuity in devising arguments, were inbred in all men of mark of the seventeenth century, who had any imagination at all. If old Jeremy had written like Bishop Butler, all hard arguments, should we have thanked him for it? not I for one. Let us take what our taste and right reason approves, and recollect that in the very many-sidedness of his views he may bring conviction to another though not to us. Thus, for example, about the fear of death, I must say that I differ from you. I feel its universality to be one of its greatest alleviations. A singular lot is full of bitterness. The fellow-feeling mitigates distress, whether it be the fatigue of an exhausted army, the visitation of cholera, or the breaking of a bank.'

In the autumn of 1859, an important event took place, which turned the remainder of Forbes' life into a new and unlooked-for channel. The Principalship of the United College, in the University of St. Andrews, became vacant by the transference of Sir David Brewster from that post, which he had held for more than twenty years, to the Principalship of Edinburgh University.

Forbes was induced to offer himself for the vacancy, and his eminent achievements in science as well as his long and faithful service in his own University determined Sir Cornewall Lewis, then Home Secretary, to recommend him for the appointment. This choice is the more honourable to the justice and impartiality of Sir C. Lewis, that Forbes not only had no personal acquaintance with him, but, in as far as he belonged to any political party, it was to the one which opposed the ministry then in power. Others of the then ministers interested themselves in Forbes' promotion,—his sometime student and faithful friend, the Duke of Argyll, and Mr. Gladstone. Besides these public men, Forbes' colleagues in his own college, as well as his many friends and former pupils, rejoiced, not only because it placed him in the position which he desired, but because in his appointment they saw an act of public justice done.

It is clear that ever since his illness the duties of his Professorial chair must have been a burden. To appear each day at a stated time, to lecture for an hour to a large class, to see students afterwards and answer questions, to prepare experiments or superintend their preparation, and to look over masses of examination papers—these duties, which effectually task a man in the prime of vigour, fall too heavily on one who feels that life within him hangs by but a slender thread. All this, no doubt, Forbes felt from the close of 1851 till the end of his Professorial work. The change from the daily class-room to his own study—for there a large part of his work as Principal could be done—was just the thing that suited him. His duties out of doors lay mainly in attending College and University meetings, and in addressing students now and then, at times which he could generally choose for himself. His weak health seldom unfitted him for the duties of the new office he undertook. And the daily, hourly sense of the frail hold he had on earth, only made him feel that he must labour the more earnestly while his day lasted.

The date of Forbes' commission as Principal of the United College is December 2nd, 1859: his induction took place in the Hall of the College on the 10th.

He did not, however, at once, resign his Professorship and migrate to his new position, but continued, with the assistance of Mr. Keith, to carry on his class-work till the close of the session 1859–60. From time to time during the winter, he visited St. Andrews, to attend meetings of the United College and of the University.

On the 1st of April he sent in his resignation of his Professorial chair in Edinburgh, on the 10th delivered his last lecture to his class, on the 16th received from his own University the degree of LL.D. in company with Mr. Gladstone, Dean Mansel, Dean Ramsay and other distinguished men, and on the 25th delivered an address to the graduates of arts, in which he bade farewell to that University which his own labours had so greatly benefited and adorned.

The attempt, even if I could make it, to sum up what these labours amounted to, and what were their results, is happily not required. Some letters from those who studied under Forbes, at different dates in his Professorial career, will convey their impressions of his characteristics as a teacher, and will fitly close this portion of his life and work.

Mr. Thomas Cleghorn, now Sheriff of Argyllshire, writes thus to Mrs. Forbes:—

'Edinburgh, *August 24th*, 1869.

'My recollections of the Natural Philosophy Class in 1835-6 and 1836-7 are very fresh, and among the most delightful of my life. I have vividly before me the form and features of our then very youthful Professor, as he commanded with his dignified bearing the respect of a large and miscellaneous class of somewhat rough students, while he charmed all by a most eloquent, polished, and skilful exposition of his subjects. It was quite the kindling of academic life for me, and I doubt not for many others ; and it was not only by his lectures and examinations, but by the most constant and varied endeavours to interest and attach his students, that he inspired into very many of them an enthusiastic love of the studies he so delighted in himself. I well remember the evening gatherings at the Dean House, and those delightful breakfasts there, to which his students were by turns invited, and how he was ever stimulating us by exhibiting his apparatus and objects of scientific interest, by walks in the neighbourhood to test philosophical instruments, and by stirring us up to form societies among ourselves for scientific experiment and research. I was deeply impressed with his kindness to myself, as I had no introduction to him but the class-ticket, and ultimately our intercourse ripened into a lifelong friendship.'

Mr. J. T. Harrison, civil engineer, says :—

'LONDON, *July* 31*st*, 1871.

'. . . As Professor of Natural Philosophy, Mr. Forbes not only gained the esteem of all the students by the great interest he evidently took in them, and by the great pains he bestowed in the preparation of his lectures, and on the experiments, illustrations, and mathematical solutions, which enabled all those who were really anxious to do so thoroughly to master the subjects of them, but he endeared himself to them by his kindly and genial manner, and by the assistance he was ever ready to afford those who asked for further explanation of any difficulty they were unable to overcome.

'It was, I believe, his greatest pleasure to see his pupils striving to acquire information, and to help them in doing so. Many of his pupils thus became sincerely attached to him. His breakfast parties, and the delightful excursions with him into the country for the purpose of taking barometrical and other observations, will ever be remembered by me and many others with the greatest pleasure; on these occasions he was the joyous spirit of the party, and he not only instilled a real love of nature, but he was instrumental in forming and cementing many sincere and lasting friendships among his pupils. His interest in old pupils was a striking feature in his character. I never met him without his inquiring about some one respecting whom he thought I could give him information. Although ever anxious to encourage the love of natural philosophy for its own sake, he was fully alive to the great importance of sound theoretical knowledge as the basis of the education of the civil engineer, and he showed his high appreciation of that profession by choosing it for his son. It gives me great pleasure to note down these few remarks in affectionate remembrance of my dear master.'

Professor Balfour Stewart, who, besides being a pupil, was for several sessions assistant-lecturer to Professor Forbes, thus gives his recollections:—

'PENDYFFRYN, NEAR CONWAY, *May* 4*th*.

'. . . I believe it was during the session 1845-6 that I attended Forbes' lectures as a student. Two things struck me: first was the gentleness of his manner, united with something that carried authority, so that he was not only perfectly obeyed by his class, but also venerated and beloved. Had he been a general, I conceive he would have been worshipped and beloved by his men. The next thing that struck me was the noonday clearness with which he explained everything, especially points requiring explanation. The truth was conveyed to his hearers in the best possible words. He was more gifted than any man I have met with in diffusing light. I am certain that in his own mind he was not content with merely apprehending a truth, but he viewed it in all possible lights, and finally selected some one as the best point of view from which to paint it to his class. There was thus engendered in the minds of his pupils, not only the effect directly due to such a description, but the indirect feeling that their teacher had taken great pains and care to produce this result. Perhaps you are aware that I acted as his assistant during the sessions 1856-7, 1857-8, 1858-9, during which time it was my duty to lecture on mechanics and the more mathematical part of his course, while he lectured on the subjects which are generally embraced in treatises on physics.

'I had the benefit of very full notes of his lectures, and I often think that it would be to the advantage of science if his lectures could be published.

'I learned, during the time that I assisted him, several points regarding him that I had not previously known. He was among other things very clear in his pronunciation, and he told me that in order to obtain it he had taken lessons from Mrs. Siddons. During the time I used to assist him in his experiments, I learned to notice the caution combined with sagacity, the union of the centripetal and centrifugal principles, that was an eminent characteristic of his mind. He was not satisfied in his own researches with viewing a thing in one light, but he insisted on verifying his conclusions by corroborative evidence derived by regarding the subject from another light.

'I ought not to forget his great kindness to me in some experiments on radiation which I made when I was his assistant, and for which I have since received the Rumford medal of the Royal Society, London. He very generously gave me many hints, and allowed me to use not only his own apparatus, but to make use of his valuable specimens of rock-salt. Had it not been for these facilities, I should not have succeeded in the investigation to any extent.

'These are my chief remembrances of Principal Forbes.

He combined to an eminent extent the philosopher with the man of science; and so far as I am personally concerned, I know that I owe any success which I have attained in a very great measure to those habits of thought which a man like Forbes was so well qualified to communicate.'

CHAPTER XII.

LIFE IN ST. ANDREWS.

The University over whose oldest and largest College Principal Forbes was now called to preside, is one of the few fragments which survived the wreck of the Scottish mediæval Church. Whatever the shortcomings and corruptions of that Church for two centuries before the Reformation may have been, it ought not to be forgotten that it is to her that we are indebted for our Universities. Three out of the four Universities of Scotland had Catholic Bishops for their founders. This was pre-eminently true of St. Andrews, the most ancient of them all. A Bishop it was—Henry Wardlaw—who, near the opening of the fifteenth century, founded that University, and the accomplished First James smiled upon its infancy. Each of the three Colleges which were successively incorporated into it, owed their origin to a separate prelate. The oldest of the three Colleges, that of St. Salvator, was founded and endowed by the successor of Wardlaw, Bishop James Kennedy, kinsman of the king, and the wisest man of his time, both in Church and State; a prelate of such pure and beneficent character that even George Buchanan, prelate-hater though he was, has no word but praise to speak of him. To him, in the old sea-tower at St. Andrews, his cousin, the Second James, turned for counsel when the violence of the three banded earls, each almost a king, had all but driven him from his throne. The next College in order of time was that of St. Leonard, founded by the youthful Archbishop of St. Andrews, Alexander Stewart, and by John Hepburn, prior of the monastery. One of the charters of the foundation was signed by the young archbishop, and confirmed by his father James IV., the year before they two, father and son, fell together on the field of Flodden. The foundation of St. Salvator's College by Bishop Kennedy, was one of the many efforts made by that wise prelate to counterwork the corruptions of his Church, and to reform those abuses which he saw were eating out its life. In the original charter he states the aim of his foundation to be 'the strengthening of the orthodox faith, the increase of the Christian religion, and the removing the pestiferous schisms of heretiks.' And in conclusion he requires and exhorts 'in the bowels of Jesus Christ our Saviour, who is the Patron of the College,' that there be appointed pastors and defenders, who shall conserve the said College for the purpose to which it was devoted. Again, Archbishop Stewart in his charter states that his object in founding St. Leonard's was his desire to preserve 'the tempest-tossed bark of St. Peter, and to uphold the declining state of the Church.'

This last College was scarcely founded when it became the nursing-mother of many of those ardent spirits who bore a chief part in working that Church's overthrow. To have drunk of St. Leonard's Well, was another expression for having adopted the principles of the Reformation. When the Reformation had got itself established, George Buchanan became Principal of St. Leonard's, which he adorned by his scholarship more than by his character. He received one pension from Queen Mary, and a second from Queen Elizabeth, for slandering his first benefactress; so that, as has been said, though he did not serve two queens, he at least took wages from two.

With the Reformation these two Colleges, which had been founded mainly for the rearing of clergy and the teaching of theology, were so far secularised that they were devoted exclusively to instructing students in classical literature, science, and philosophy. Instruction in theology was handed over by the Reformers exclusively to the younger College of St. Mary's, which, having been founded and endowed chiefly for this purpose by the last three Roman Catholic archbishops, James Beatoun, David Beatoun, and John Hamilton, was soon after the Reformation presided over by those two stout anti-Prelatists Andrew Melville and Samuel Rutherford.

The two older Colleges, restricted to the more peaceful pursuits of classics, mathematics, and philosophy, were less heard of in the turbulent conflicts of the seventeenth century than their younger theological sister.

About the middle of the eighteenth century, the finances of St. Salvator and the tenements of St. Leonard's having fallen equally into disrepair, the more flourishing finances of the one were transferred to the better buildings of the other, and the two Colleges were by Act of Parliament conjoined, under the prosaic name of the United College. From that time, 1747, there have continued to be two instead of three Colleges in the University; and at this day St. Andrews remains the only place in Scotland where native Scots have an opportunity of learning the distinction between a College and a University.

In the later years of the last, and the three first decades of the present century, the names of certain families recur so frequently in the roll of Professors, that one is almost tempted to imagine that the spirit of the St. Andrews Culdees of the eleventh century still survived in the University of the eighteenth—so faithfully had the Professors copied from these old secularized churchmen their practice of handing on their benefices from father to son. Yet among these frequently recurring names, some there were who served their University so well, as to suggest a doubt whether their disappearance in modern times has been pure gain.

At the time of the Reformation, in his 'First Booke of Discipline,' John Knox, when laying the foundation of

Scotland's system of education, and naming the three then existing Universities—for that of Edinburgh had not yet appeared—speaks of 'the First University and Principal, viz. S. Androes.' The primacy which Knox attributed to it, St. Andrews then undoubtedly held, not only by right of antiquity, but also of mental and theological eminence. In the interval between Knox's time and our own, the Universities of Scotland, like all other things, have greatly changed. The three younger Universities, owing to their situation in the heart of large and rapidly increasing populations, have far outgrown in size and number of students their elder sister, placed as she is apart from throngs of men, and amid a stationary community. But during these three centuries St. Andrews has not failed to contribute to the service of her country an amount of ability and trained intellect out of all proportion to her comparative numbers. In the generation just gone, three distinguished contemporaries,—one, the greatest preacher and divine which Scotland has, during this century, produced; another, the Lord Chancellor of England; and a third, the Lord Justice-General of Scotland,—were all alumni of the United College.

These historic details are not out of place in this account of the life of Forbes. For him the ancient records, monuments, and traditions of his newly adopted University possessed a peculiar charm, and called out a faculty and taste which had hitherto lain dormant within him, only because it had nothing to feed on. Among the predecessors of Principal Forbes for more than a century, no distinguished name is to be found till we reach that of the venerable Dr. Hunter, famous in his day as a scholar and philologist, who, after filling with great success the Humanity chair for nearly sixty years, was, towards the close of his long life, raised to the Principalship of his College. Him followed, after a brief interval, Sir David Brewster, who, during his twenty years' tenure of the office, if any remains of the family system still lingered, scattered them, somewhat turbulently, to the winds. The reputation of Sir David's name was, of course, an honour to St. Andrews, as it would have been to any University; but he laboured under a delusion, of which he could not dispossess himself, that it was the peculiar calling of St. Andrews to train practical men of science, especially engineers, for the whole nation. So illusory was this idea, that it may be doubted whether so much as one student ever came to St. Andrews in quest of training for this profession. Principal Forbes, devoted though he was to his own subjects, did not share this delusion. He saw clearly enough that in St. Andrews these could have no special prominence; that it must continue, as in the past, to give a general education to young men meant for any one of the professions, and that if it had a specially professional calling at all, it was to prepare ministers for the several churches, and teachers for the borough and parish schools. But to Sir David's unflinching opposition to all jobbery in choosing Professors, and his determination to elect the best men that could be found, Principal Forbes owed it that he found his College equipped with a staff of Professors not then surpassed by the staff of any other Scottish University. Prominent among these was the late Professor Ferrier, who, by his subtle philosophic genius, expressing itself in a perfect style, not only adorned his own chair, but maintained for another generation his country's ancient fame for metaphysical genius. Two others may be named who now adorn other Scottish Universities. Professor Sellar, now Professor of Humanity in Edinburgh, then taught with great power from the Greek chair in St. Andrews, and Professor Veitch had just then begun to teach in St. Andrews that vigorous course of logic and metaphysics which he has since transferred to the corresponding chair in Glasgow University. Others might be named who were the colleagues of Forbes in St. Andrews and still occupy the chairs they then held.

Just before leaving Edinburgh to reside at St. Andrews, Principal Forbes had wound up a letter to a friend with this remark: 'All things become solemn when the past perspective of life is the predominating object.'

This was the prevailing tone of feeling with which he entered on his new duties. He felt strongly the frail hold he had on life, and whatever his hand found to do he set himself to do it with his might.

The time at which he entered on his new sphere was a momentous one for St. Andrews as for all the other Universities. The Scottish University Commissioners were in full session, busily framing ordinances which should control the course of study, the University finances, the library privileges, and the Professors' salaries for a long time to come.

To supply the Commissioners with the information they required, and to offer his own suggestions for their guidance, was one of Forbes' earliest tasks.

The finances of his own College he found in a confused and dilapidated state; and to understand these, and devise measures for their restoration, he first addressed himself. At whose door the largest share of blame for this confusion lay need not now be inquired. Suffice it that it mainly arose from a long habit of dividing among the Professors the annual rents of the College lands, without laying by a reserve fund adequate to meet the necessary outlays for repairing farm steadings or other such contingencies. To unravel the tangled mesh Forbes applied himself with characteristic diligence, method, and business faculty; and it was mainly owing to his exertions that the Commissioners were enabled to place the finances of the United College on a footing which, if somewhat burdensome to the present generation of Professors, promises to provide for their successors ampler incomes than those now living are likely to enjoy.

The next subject he had to tackle with was the University finances, or the contents of the University chest. These he found in a much more flourishing condition than those of his own College. But this prosperity was mainly owing to a source of revenue which the Universities'

Commissioners were understood to regard with no friendly eye. This source was the granting of medical degrees—a function which St. Andrews, though it possessed no thoroughly equipped medical school, had yet, in virtue of its original charter, been accustomed to exercise time out of mind. It is said that there had been a time, extending down to the early years of this century, when these degrees had been granted, without sufficient examination, to persons but poorly qualified. This practice, indefensible if it ever existed, had, however, long ceased, and under the able management of the late Dr. Reid and the late Dr. Day—successive occupants of the Chair of Medicine in the United College—a system of examination had been instituted in which the candidates were thoroughly tested, and the granting of degrees to none but duly qualified persons was adequately secured. With the St. Andrews Medical Professor as Chairman, a Board of able examiners, sanctioned by the University, had been got together, consisting of the best of the extra-academical medical lecturers in Edinburgh and Glasgow. To the half-yearly meetings of this Board flocked, in ever increasing numbers, medical men from England, who, having begun practice without possessing the degree of M.D., found by experience the advantage they would derive from such an addition to their name. In later years the candidates were numbered by hundreds, of whom, while a sufficient percentage were rejected, not a few out of the majority who passed now stand high in the medical world of London and elsewhere.

At the date of the Commission, and for a long time before it, the examination was such that no fault could be found with it. Still it was so far an anomaly in the Scottish system, that in all other cases candidates for degrees of any kind were required to have studied at the University in which they graduated. Traditions, too, of a bygone time and a laxer state of things were freely urged by its opponents to discredit the improved system. To Edinburgh medical Professors especially, the existence of the St. Andrews privilege had long been a standing offence.

These urged, it may well be believed, with no measured vehemence on the Commissioners, the former abuses and the present anomaly: and the Commissioners were known to sympathize with the views of the Edinburgh men. Principal Forbes, too, while he saw and felt the great benefit that accrued to the University revenues from medical graduation, shared, though from no interested motives, the dislike of his former colleagues to the St. Andrews system, so that he would not defend the retention of the old privilege unimpaired. The whole subject cost him much anxious thought. The result in the end was that the Commissioners sanctioned a compromise limiting greatly the exercise of the right in future, but allowing it to continue under certain very definite restrictions, which, while they meet an acknowledged need in the medical profession, are still a source of some revenue to the University.

A third project which early engaged the attention of Principal Forbes, was the founding of a College Hall. In St. Andrews, as in the other Scottish Universities, it had long been customary for students to live where they chose in lodgings in the town. All that the University requires of students is regular attendance at the Professors' lectures, good conduct within the College walls, and without them to keep the peace. In old times, St. Andrews had been resorted to as a place of education by the sons of many persons in the higher ranks. Indeed the shields attached to the silver arrows in the old College, attest how largely it was frequented by the sons of the oldest and most honourable families in Scotland. This had, however, almost entirely ceased more than thirty years before Principal Forbes' advent to St. Andrews. The Professors who had once been in habit of taking boarders had ceased to do so, and the general set of the educational tide southward had borne from St. Andrews to England almost all who could afford to go thither. It seemed to Principal Forbes and others, that the idea of a University, as originally held in Scotland, was not fulfilled, unless it contained students of all ranks; and it occurred to them whether by providing a fitting place of residence under proper superintendence, some of those who had left it, to the loss of the University and of themselves, might not be lured back. St. Andrews, with its noble historic memories, its academic aspect, its healthy climate, and its fine Links, which have been for ages the Elysium of Golfers, seemed to offer peculiar outward advantages for the trial of such an experiment. But for the energy and business talent of Principal Forbes, this idea of a College Hall might have continued till now only a dream. As soon as it was mentioned to him, he adopted it with all his energy and less than his usual caution, and straightway set himself to realise it. The result of his exertions was the formation of a company, whose members subscribed for a sufficient number of shares to set the institution on foot. Within two years from Forbes' appointment, the College Hall was opened, with twelve students in the first, and an increasing number in the following session. These lived in a hired house, one of those which occupied the old site of St. Leonard's College. Over them was a warden, an Oxford graduate, who superintended the discipline and management, presided at the common meals, and assisted the students in preparing their College work. During the first four or five years, the College Hall prospered so well, and attracted so many desirable students, that Principal Forbes conceived the more ambitious project of building a large Hall, which should be specially fitted for its purpose, and should accommodate a greater number of students. As the institution itself possessed no funds except those necessary to carry it on from year to year, and as no University revenues could be used for this purpose, the venture was a bold one. There were some who thought that it was too bold—that the institution had not yet struck its roots deep enough to warrant so large an experiment. But Principal Forbes was not to be turned from his purpose. By his almost unaided advocacy the old shareholders were induced to take more shares, and new shareholders were added, and by the joint contributions of these, a sum was raised which proved nearly sufficient to erect a large and commodious Hall within what was the ancient Garden of St. Leonard's. The completion of this structure will be noted in due time. Whether the venture was altogether prudent or not, is a question which time has not yet finally answered.

A fourth project which deeply interested Principal Forbes, was the restoration of the College Chapel of St. Salvator's. This chapel is not only the oldest unruined fragment of ancient St. Andrews, but along with the noble tower of St. Salvator's, which rises above it, forms the earliest piece of University building still extant in Scotland. Tower and chapel had both been built by the good Bishop Kennedy, and are the only remnants of his workmanship. The original roof of the chapel is said to have been of a peculiar and rare construction,—massive blue stone, deeply engroined. Within the chapel is the tomb of the founder, a Gothic structure wrought in Paris, of blue stone, in the middle of the fifteenth century, which must originally have been of wonderful beauty, since even in its cruel defacement it still shows so fair. As the old stone roof is said to have been nearly flat, the Professors, about a hundred years since, either themselves conceived, or were persuaded by some architect, that it would one day fall in and crush them. They therefore resolved to have it removed, and a common lath and plaster ceiling placed in its stead. So solidly, however, was the old roof compacted, that the workmen, in order to remove it, had to detach it from walls and buttresses and let it fall *en masse*. The fall is said to have shaken the whole city. But however this may be, it is only too certain that it shattered the richly wrought columns, canopies, and pinnacles of the founder's tomb. A maimed and mutilated fragment that tomb now stands, beautiful still in its decay, proving that Professors of the eighteenth century could be more ruthless and insensible to beauty than were the ruder Reformers of the sixteenth or seventeenth. But besides the mutilated tomb, of which no restoration was possible, parsimony and Philistinism had combined to make the rest of the church hideous. High bare fir pews, an unsightly gallery at one end, lath, plaster, and whitewash, floods of harsh light from many windows, —ugliness could no farther go. To the removal of these deformities and the restoration of the church, not to its ancient beauty— that was not possible—but to somewhat greater seemliness—Forbes gave his undivided attention for one whole winter. At his earnest advocacy, a more liberal government than we have of late years enjoyed undertook to restore the body of the church, and what with seemly oaken seats, a raftered roof, and more appropriate mullions and tracery in the old pointed windows, the work has been so done as to render the College church, if not as perfect as might be desired, at any rate a great improvement on what it was. By his solicitations, the College and private persons were stirred up to substitute for the former common glass, with its untempered garishness, new painted windows, which, by toning down the light, give at least solemnity to the church. As the result of all these exertions, the College church, if it has not re-attained its pristine beauty, has certainly lost its former repulsiveness, and been rendered one of the most soothing and attractive places of worship in which Presbyterians at this day meet.

But though the church was restored, Principal Forbes' troubles with it were not yet over.

The chapel of St. Salvator's had been used for a generation or two as a parish church also by the congregation of St. Leonard's. This had come about in the following way. When the old parish church of St. Leonard's fell into disrepair, the Professors of the United College, who were the chief heritors in the parish, being loth to incur the expense of renewing it, offered to accommodate St. Leonard's congregation in their own College chapel. And so it came to pass that the congregation, after they had worshipped in it for some time along with the Professors and the students, came to believe that the College had

CHAPEL AND SPIRE OF ST. SALVATOR'S COLLEGE, ST. ANDREWS.

denuded itself of its original 'propriety' in its own chapel, and made it entirely over to the parishioners of St. Leonard's. This belief, which had been growing insensibly, was startled into distinct consciousness, when Principal Forbes challenged the right of the congregation to anything more than mere accommodation, and proceeded to readjust the sittings, on the assumption that the chapel belonged of right to the College alone. A controversy between the congregation and the College arose, accompanied by some not painless collisions. But Forbes was resolute to maintain the rights of his College, which were at last confirmed by a decision of the Court of Session entirely in their favour.

Such were some of the enterprises in which Forbes engaged for the good of his College and of his University. Besides these, he lectured as Principal to the students in the Hall of the College from time to time on subjects which had been long familiar to him, such as 'Glaciers,' 'Climate,' 'The History of Discovery.' These lectures were numerously attended by residents in the town: not so numerously by the students as might have been expected from the eminence of the lecturer and the excellence of the lectures. Young men in their ignorance often undervalue privileges which in after years they would give much to recall.

One of the latest labours in which Forbes engaged was an examination and arrangement of all the ancient charters and other documents preserved in the charter chest of the United College. To this his attention was turned by a disputed claim of feudal superiority which the College was called on to defend in court. This led Forbes to plunge deep into the charter chest, in which he found materials of daily growing interest. Gradually he taught himself to decipher and read the crabbed characters and strange contractions of the mediæval documents. Had he lived to complete this work, he would no doubt have extracted from these a large amount of valuable history and curious antiquarian lore. As it was, he completed an inventory of all the College records, and left them arranged and ordered as they never had been before.

All these various subjects of interest, as well as others which engaged his time and thoughts, will best be illustrated by the following selections from his St. Andrews correspondence :—

To DR. SYMONDS.

'ST. ANDREWS, *January* 6*th*, 1861.

'. . . You heard, I believe, of our spending a fortnight in Edinburgh among our old friends—at least those of them who still remain there—for death and change have now materially diminished their number. My wife and I and the two boys stayed with my sister, and the three girls with Mrs. Wauchope. Notwithstanding the excessive severity of the frost, we spent a pleasant time, and the change to Edinburgh was very much in favour of the children, especially Eliza and George, who have returned to St. Andrews quite well. My chief and almost daily business was sitting for my picture, for the Royal Society, to Sir J. Watson Gordon. Fortunately he is a pleasant informed old man, and I did not grudge him the time I spent, as he was well satisfied with his work, and he took the greatest pains with it. It is considered a good work of art, and some good judges think the likeness excellent, but Alicia is not quite pleased in this respect. I believe myself it is too favourable. However, it is an honour of which I am very sensible, to be handed down to posterity in such good company as that of the worthies of the Royal Society. My duties and occupations here employ me sufficiently without being burdensome. I believe that I enjoy the confidence of my colleagues on most points. A very searching examination of the financial state of the College, which has landed property to the extent of nearly £3,000 a year, has occupied a great deal of my serious attention : and I have of course a great deal to learn on such matters ; but we have taken the highest professional advice, instead of working on with our own crude notions. The sitting of the Commission, and the influx of younger men into our body, have been favourable circumstances for a remodelling of our finance, and I have reason to hope some of the main difficulties are now past.

'Before Christmas I delivered four lectures on the Art of Study—homely, practical advice to students. To-morrow I commence a short course on Climate, a subject with which I am familiar. They will be altogether oral, delivered twice a week, and they impose no feeling of burden on me, since even if from bad health I should have to interrupt so short a course—which I hope is not likely—no one would be academically or professionally a loser by it.'

'You will not expect any general news from our nearly peninsular and out-of-the-way abode. A curious circumstance occurred last night. In the course of an incredibly short time a heavy sea began to beat on the coast, it being all the time a dead calm. The fishing-boats which had gone to sea a few hours before, ran in wherever they could get shelter, with much peril and some loss.

'The thermometer is at 20°, or lower, every night. It is now three weeks since this severe snow-storm and cold commenced.'

To the LORD JUSTICE CLERK (INGLIS).

'EDINBURGH, *April* 10*th*, 1861.

'On considering the conversation which by your Lordship's kindness I had with you on Saturday, I feel desirous of adding some explanatory statements. Though they will probably not affect your Lordship's views on the merits of the questions discussed, they may possibly place in a juster point of view than I fear I succeeded in doing my own position with reference to the affairs of the University and Colleges at St. Andrews.

'From the time of my first acquaintance with the duties and responsibilities of my new office, the state of the finances and debt of the United College attracted my anxious and I may say alarmed attention. Any services which I could hope to render to the College appeared insignificant for the time, compared to the duty which I felt devolved upon me of first ascertaining the extent of our liabilities, then the cause of the annually increasing debt, and finally the means of diminishing or extinguishing this, and of at least putting a bar to its seemingly indefinite increase. My own experience in, or aptitude for, such financial arrangements were very inconsiderable. It was in every respect an unwelcome task, especially as it might seem to reflect more or less upon the conduct of those with whom I for the first time became intimately associated. To the honour of those gentlemen, I must say that I received no discouragement, but the contrary, in making the investigation; but it is not the less true that the inquiry, except in so far as it was conducted by the accountants, fell chiefly upon myself. The scheme for reducing the debt by a grant from the University chest and by means of a terminable annuity, to be defrayed by all the Professors of the United College alike, was concerted between Mr. Jamieson and myself. It was ultimately acceded to by my colleagues, and also by the Professors of St. Mary's, in so far as their interest in the University was concerned. I had a very sanguine hope that it would be favourably regarded by the Commissioners also, and that my tenure of the office of Principal would have been inaugurated by a complete relief of this noble and ancient foundation from the burden of debt under which it laboured, and the professors from the painful conviction that they were every year burdening their successors with a yet more insupportable load.

'You will grant, I hope, my Lord, that while aiming at such results, I was not actuated by personal or selfish motives in feeling and expressing an anxiety about the pecuniary arrangements of the University and Colleges, such as I never either felt or expressed in any matter affecting my personal pecuniary interest. It is easy to see that if through the miscarriage of the system for granting medical degrees or from any other cause the

£4,000 from the University chest should not be forthcoming, it is not to be expected that the terminable annuity—insufficient as it would then be—can be proceeded with, and that the scheme which I have been now for twelve months promoting, and which had been adjusted apparently to the satisfaction of the parties more immediately concerned, must fall to the ground, to the infinite discouragement of all who had part in it, and who felt perhaps that they were at some small personal sacrifice lending their aid to effect a public benefit.

'It will be proportionably still more painful and disheartening to me, should I have to announce to my colleagues, who have reposed so much confidence in me already, that the hopes of adjustment which I held out as the result of their proposed relinquishment of a proportion of their annual profits are at an end, and that the debt is immediately to be increased to a larger amount than it has ever stood at since the foundation of the College.'

To A. Wills, Esq.

'Near Crieff, N.B., *September* 1861.

'. . . I am very glad that the "Eagle's Nest" has so completely answered to all your hopes and expectations. The picture which you draw of your complete enjoyment there, presents itself somewhat in contrast to the more prosaic and less Elysian character of my own summer experience. We have had a summer and autumn of unusually broken weather. Since the middle of July we can count but few days without rain, and none of genial warmth. Our plans for residence and occupation have been very much upset. Proposing to return to St. Andrews early in August, we let our cottage at Pitlochrie for three months. But scarlet fever and small-pox having become somewhat rife at St. Andrews, we did not like to return thither with our children, and after an anxious search we secured lodgings in a farmhouse near Crieff, where we still are: but other circumstances as well as weather have rendered our stay less pleasant than we hoped—though the country itself is delightful—and we are now anxious to get away. St. Andrews, however, has not yet a perfectly clean bill of health; perhaps it is vain to wait for such, but we shall probably go first to Edinburgh for a time. Thus without any of the heavier ills of life, the amenity of our summer has been much interfered with, and the "Eagle's Nest" is like a glimpse of fairyland, or a return of one of those happy Alpine summers which I used to know some twenty or twenty-five years since. But life does not always pass so smoothly, and the chariot-wheels become clogged as we advance; of which we ought not to complain.'

To the DUKE OF ARGYLL.

'*December* 31*st*, 1861.

'. . . Your Grace asks what is thought of the Ordinance. Always excepting Dr. Tulloch's peculiar case, it has, I think, been generally accepted as a fair solution of a difficult case. No doubt our Professors complain a little that the entire burden of extinguishing a debt which few of them shared in contracting, should be thrown entirely on existing lives, and their successors gain on a sudden both principal and interest, I mean by the rent-charge for the principal ceasing at the same time with the interest on the debt. Still it is a relief to all, I should think, to see this discreditable debt question fairly buried, and the repetition of such transactions expressly prevented. In my own case I am slightly a loser by the Ordinance, the extra rent-charge being deducted from my salary as recently augmented by the Treasury. But your Grace will give me credit, I hope, for not being a grumbler.

'I am surprised, and a little amused, to learn that my old colleagues in Edinburgh University, who very justly and rightly get the lion's share of the £10,000, are much more discontented than the poorer brethren at St. Andrews.

'I hope your Grace will kindly take an opportunity of lending your powerful aid with Mr. Cowper in favour of completing the restoration of our beautiful College chapel.

Now that the mullions are partly in the windows, and the roof is visible through the scaffolding, the grandeur of the whole is really conspicuous. The wretched fir-boxes of pews cannot, we hope, be allowed to remain, but will be replaced by the open seating, for which Mr. Matheson has sent an estimate to the Board of Works.

'It is also an encouragement for the Board of Works that there is every probability that several of the windows will be very soon filled with painted glass through private exertions. The Lord Justice-General has undertaken to fill one; another has been nearly subscribed for to the memory of Dr. Chalmers; and two more have been undertaken. Considering that it is the last relic of old St. Andrews which admits of thorough restoration, I think it is to be desired on public grounds that it should be well done.

'Your Grace may perhaps recollect that you are president of a literary and philosophical society here. We have recently memorialized the Board of Works to put up a bulwark against the encroachments of the sea on the eastern side of Beaton's Castle, which are annually becoming more desperate. Lumps of the masonry on the eastern side of the Castle are annually falling on the beach below, owing to the excavation of beds of soft shale by the waves. In a few years the fourth side of the castle court will be no more, unless promptly protected. Will you be so good as to plead this too? I sent a photograph to Mr. Cowper, clearly showing the state of matters.

'Will you have the goodness to say to the Duchess that the St. Leonard's Hall continues to prosper, and that we like Mr. Rhoades more and more. We have now fortunately secured a thorough housekeeper, who has put the whole establishment on an excellent footing. We have also got a five years' lease of the premises, subject to the life rent of Lady Playfair.'

The destruction by the waves of the fine old sea-tower, the palace of the Primates of Scotland, goes on apace. The ruin belongs to the Crown, and is perhaps more rich in historical associations than any other castle in Scotland. Forbes' pleadings had some effect at the time; but to all antiquarian and historical appeals the present heads of the Government Department turn a much deafer, more defiant ear than their predecessors in office, to whom Principal Forbes eleven years ago addressed his remonstrance.

To the REV. DR. WHEWELL.

'ST. ANDREWS, *January 12th*, 1862.

'In answer to your inquiries, I can state with confidence that I have found every reason to be grateful for the change of position which I made two years ago. The burden of a duty felt to be imperfectly, and in part vicariously performed—with the almost certainty that year by year it will be more severely felt, until it becomes simply impossible—is one which was to me crushing both to body and spirit. In coming to St. Andrews I left many pleasant and lifelong associations. But I found a practicable sphere of not unwelcome duties. The climate is not at all worse than that of Edinburgh, and at this season, I should think, has a considerable resemblance to that of the very country where you now are; the winter isothermals running as they do nearly due north and south. Moreover, as a general rule, I am not obliged to go out in bad weather; and having a tolerably comfortable old-fashioned house in a sheltered situation, with my books about me, I feel that Providence has in every way been very kind to me. My health is certainly better on the whole than even when we last met. I have not attempted experiments. This place is singularly unadapted for them. There is the want of assistance, locality, apparatus, and artists, which I have always been almost luxuriantly supplied with, and I have not had the courage to do much without them. I am about to commence a short course of lectures on Glaciers—voluntary—twice a week, of which I enclose a syllabus.

'You kindly ask concerning the success of our College Hall. I rejoice to say that it has succeeded beyond our expectations. A commodious house, fortunately vacant, was fitted up for twelve young men and a Warden, with every comfort—some people say too comfortably. The Warden, Mr. Rhoades, of Rugby and Oxford, proved to be all that we could wish, and the institution promises to be permanent.

'I can warmly enter into your feelings on the loss of the Prince Consort. Even in this remote place, the loss was indeed felt as you describe it—a family affliction coming home to everyone. I never remember an impression at all approaching to it. Your closer relations to the Prince, and also to the Prince of Wales, must have made the shock still more severely felt. The poor Queen must find a melancholy joy in the perfect sympathy with which her truly hard case is regarded by her subjects at large. I hope and trust that the Prince of Wales has the character requisite to rebound to the blow and not sink under it.

'When your letter arrived the second volume of your Plato was lying on my table. I have since begun it. I had read the first when it originally came out "from cover to cover." It was entirely new to me, and very interesting. I apprehend that your treatment of the author must be very skilful. I shall be greatly obliged if you will present me, as you kindly proposed, with the third volume, which I think is the one you mention.

'I think the appointment of the Duke of Devonshire a very fit one. How well I recollect seeing him write for a College examination in the gallery above the Hall of Trinity, shaking back his long hair from his eyes!

'With many kind remembrances to Lady Affleck, believe me. . . . .

'P.S. The Duke of Argyll, who is Chancellor of this University, was here lately with the Duchess, inspecting our new College Hall, and intends to send his two eldest sons to reside next winter.'

To E. C. Batten, Esq.

'St. Andrews, *February* 16*th*.

'. . . I should say that four hours in the day fatigues me more than nine or ten used to do; and as I have been having meetings about five times a week, and as a meeting at St. Andrews nominally lasts two hours as a minimum, you see at once how much of my brain-juice is expended. In fact, to make up at all, I have at present to breakfast in bed, that I may rise the fresher. All after dinner is mere play-time, of necessity. If I happen to dine out it counts for work done. I have had my lectures once or twice interrupted from simple incapacity to deliver them. To a large mixed audience, an hour's lecture even on a familiar subject must be carefully prepared. I think that they have been pretty successful. Then I have, secondly, my own College work, meetings, committees, and correspondences. This is not at present particularly heavy. Thirdly, I have University work, which is and has been very heavy. The Commissioners' Ordinances have turned our finances topsy-turvy, and I have been the chief actor in making a scheme of expenditure and income, and wrote with my own hand a finance report of eighteen quarto pages—now printing. Much of it lies on debatable ground, and will have to be fought inch by inch. Fourthly, I have the church restorations chiefly on my hands, which divide into two parts: the architectural, on which I have to deal with Government people; and painted windows, in which many private parties and my own colleagues are concerned. We are making progress in both departments. Fifthly, there is the College Hall, which cannot be expected to progress always with perfect smoothness. When any hitch occurs—and it is sure to be when I am otherwise most busy—as Chairman of the Council, I have to take an active part, and last week we had three long meetings about it.'

To the Same.

'Pitlochrie, *July* 20*th*, 1862.

'. . . Mr. Jerram has sent me two copies of his "Thoughts on Revelation," and his "Thoughts on Miracles." I admit to you that it is an effort to me to sit down to read carefully books on religion professedly controversial. But in this case, I had pleasure in making the effort, and like Mr. Jerram all the better since I did so. I like his style, which is very plain, earnest, unaffected, and self-convinced. Though he can hardly hope to silence his opponents—whose writings I have purposely not read—he may confirm his friends, and by feeling the stability of his own convictions may be thereby much more fitted for a minister of the Gospel in these troubled times.

'I think I agree with him as to the reasonableness of his argument on every point. On some he expresses convictions which have long been peculiarly my own:—as at p. 12, &c., that a revelation must, to be of use, be kept within the range of what it is possible for us to know and in terms which mankind can understand; and at page 38, that as we accept the opinion of eminent lawyers on points of law and astronomers on astronomy, so the convictions of theologians and of the Christian world ought to have the greatest weight with every man not inflated with vanity. In the "Thoughts on Miracles," p. 21, I find a very striking analogical argument from human interference with the ordinary course of laws of nature to the impossibility at least of disproving Divine interference in ordinary providence. I was the more struck with this, because there is a chapter of "Reflections" most inappropriately thrown into Dr. Tyndall's book called "Mountaineering in 1861" on the folly of prayers for fine weather, &c., which even from a purely scientific point of view seems to me far from convincing; and which, if carried out, would suspend prayer in almost every human contingency; such as the extreme illness of a child or a parent, when, mercifully, the heart over-rules the judgment of the head and—as

Mr. Jerram justly says—the common voice of mankind protests against the logic of the metaphysicians.'

To DR. SYMONDS.

'ST. ANDREWS, *October* 12*th*, 1862.

'. . . We reached our St. Andrews home yesterday afternoon, and I am most thankful to say that we are all quite well, and have been so for the greater part of the summer. We had, as everywhere else, rainy months of June and July and a good part of August; but since that time, and especially latterly, we have had as fine an autumn as Scotland ever produces—some truly delightful days, in which the country looked far more beautiful than in the height of summer, not merely from the tints of the foliage, but from the slanting lights which contribute so much to the effect of country in high relief. The last day the boys were with us before returning to school—the 8th September—we had a very charming excursion to Tummel Bridge, between Loch Tummel and Rannoch. One day we went a large party with Mr. and Mrs. Airy and their two sons to Glen Telt, and lunched almost on the spot where we did a like act six years—yes, it is six years!—before. The day was not so positively hot as that day—there has been no such hot day in Perthshire since—but it was equally pleasant. We had not visited the place in the interval. Alicia has told you, I believe, that we took a good deal of interest in the navvies, whom we found an interesting and grateful set of people. Though they quarrel occasionally among themselves, no instance of dishonesty or rudeness has occurred at Pitlochrie, which is saying a great deal! A temporary hospital was erected just above the Pass, which I visited regularly; but the cases were not numerous, and when I left there was not a single patient. You will be amused, I think, to hear that only three or four days ago a delicate clergyman from the south of England came with his wife to look for winter quarters at Pitlochrie, as a sanitary measure. The gentleman has a bad cough, and our friend —— had told him to pass the winter in a bracing climate, and mentioned two places to choose between, Pitlochrie and some place in Westmoreland. As he has eight children and heaps of servants, they found Westmoreland out of the question, and Pitlochrie not much better. So much for Pitlochrie. You ask where my popular articles on Glaciers appeared. It was in *Good Words*, the editor of which, Dr. Macleod, is an old friend of mine, and asked it with an urgency I could not well refuse, and his publisher remunerated me in a way which seems almost extravagant. But I do not intend to allow myself to be led into the snare of much popular writing, the effects of which on the quality of the product are too lamentably apparent in all the magazines, even those of most name. You have seen, I daresay, in the *Times* that our poor friend Auguste Balmat is dead! It was to me almost like a family bereavement, and has removed one of the lifelong hopes to which I clung of revisiting Chamouni in his company, with his sympathy and help. He was devotedly nursed by his friend and mine, Mr. Wills, in the very chalet at Sixt in Savoy, which he had contributed to construct and embellish for Mr. Wills. I trust that you are now quite relieved about your son's health, though your last mention of him seems not absolutely free from anxiety. I hope you will induce him to let his fellowship alone for a year or two, and to put himself to grass. . . . But I must wind up. Our College Hall is quite a success—overflowing. Though we have added four rooms, at least two pupils, probably more, have been refused already.'

To the RIGHT HON. W. COWPER.

'ST. ANDREWS, *February 9th*, 1863.

'I feel much obliged by the attention you were so good as to give to my suggestions relative to the excavation of the moat of the Castle of St. Andrews, where it is bounded by Mr. Hope Scott's ground. Mr. Scott has appointed his architect to meet Mr. Matheson on the ground, and I hope that Mr. Matheson will soon have time to inspect it and proceed with the work.

'I am going to venture upon another request in behalf of our old College. Of late there has been a plan on foot for providing a common table at a low rate for such of the students as choose to avail themselves of it. The Duke and Duchess of Argyll, when here in November, interested themselves very much in it, and I have corresponded with them about it since.

'The great difficulty is to find a suitable locality. It has only lately occurred to me that an open corridor with arches, which forms the south side of the quadrangle of the College, and immediately adjoins the College chapel, now under repair, could be converted at small expense into a hall very suitable for dining in, and which might easily be warmed by the same hot-water apparatus which has just been erected in the chapel.

'The corridor at present is universally allowed to be of no use whatever, and could be at once converted into a hall suitable for the purpose suggested, by putting windows and doors into the open arches. It would be also useful for other purposes besides dining, as for examinations and the like. I have reason to believe that Mr. Matheson sees no difficulty or objection in the way of the proposed alteration.

'It is much wished to begin the dining scheme next session.'

To E. C. BATTEN, ESQ.

'PITLOCHRIE, *July* 10*th*, 1863.

'. . . I have had a considerable trial in the course of last week. I had a return of hæmorrhage on Thursday, to no great extent, indeed, yet with accompanying indications which leave little or no doubt that it came from a vessel in the lungs—such is my able doctor's opinion. I have been kept very quiet since, but having had no return, nor indeed any uncomfortable further symptoms of any kind, it is likely that the trouble is over for the time. But it leaves me with a depressing conviction of permanent disability, which a complete remission of such symptoms for a good many years had enabled me in a measure to shake off. It is ascribed to a very moderate exertion which I made on one of the late warm days, to which in this country we are strangely unaccustomed. We have not had any heat to speak of here since 1857 until now. I am afraid Dr. Irvine at least will never give his consent to my venturing on the Continent in summer, which I had ever looked forward to.'

To A. WILLS, ESQ.

'PITLOCHRIE, *July* 13*th*, 1863.

'. . . Mrs. Forbes and I spent most of the month of May at St. Andrews, I being engaged in writing an article on "The Antiquity of Man," which, if you are curious about it, you may see in the forthcoming *Edinburgh.* It was not volunteered on my part, and, in fact, cost me a great deal of study. It is rather unfortunate that the singular story—or *cause célèbre* as it has been called—of the human jaw of Abbeville was developed after the article was in great part written, and new facts —or rather assertions—kept pouring in up to the final correction of the last proofs. There is perhaps in consequence some want of homogeneity in the article. My own belief from the first was favourable to the genuineness of the relic, as I confess it is still ; but I am assured that the oral traditions of the Geological Society of London are now decidedly contrary, and the editor, as he was quite entitled to do, made some modifications at the very last moment. I do not know whether you have attended to the question ; but the fact, admitted on all hands, that many genuine flint weapons have been found in the sand formation with elephantine remains seems to make it a matter of comparative indifference whether the human jaw found in similar circumstances was genuine or not, so far as any scientific argument is concerned ; and consequently the excessive anxiety displayed to elude the admission seems to me to show a very needless scepticism. However, probably all this concerns you very little.'

To the LORD ADVOCATE.

'PITLOCHRIE, *August* 10*th*, 1863.

'Dr. Cook has intimated to me his intention of resigning the charge of St. Leonard's parish, St. Andrews, towards the end of next month. I cannot help stating to your Lordship that this appointment, though small in emolument, is one of much importance to the United College and to the students.

'Not only is the United College by far the largest heritor of the parish of St. Leonard's; but what is still more important, the College chapel and the parish church being one and the same, the interests of our body and the religious instruction of many of our students depend materially upon the good sense and high character of the parish minister. A good incumbent may be a real blessing to the institution, whilst one not gifted with discretion as well as zeal would be extremely the reverse.

'I do not know of a single candidate as yet for the expected vacancy, therefore I write to your Lordship on behalf of no individual; but simply to request that the qualities and aptitude of the person to be appointed to this peculiar and interesting, though not wealthy, cure, may receive mature consideration, with reference especially to the academic body with which it is so intimately associated.'

To the DUKE OF ARGYLL.

'ST. ANDREWS, *February* 6*th*, 1864.

'Mr. Cowper had the kindness and courtesy to write to me lately about the proposed conversion of the College cloister into a dining-room for students.

'I have just learned that a private individual is going to start a restaurant in the town for students and others: and our plan being so far forestalled, it will be prudent to allow a fair trial of the experiment thus to be made. I am, however, not a little desirous that the alteration of the College cloister, so far approved of by your Grace and by Mr. Cowper, should not be allowed to drop.

'The dinner scheme may yet be revived at any time; but the benefit of the change would go much farther. The cloister is universally regarded as a useless deformity. It is a cold, shivering, miserable place. By inclosing and heating it at a moderate expense, a common room would be obtained for the resort of students between class hours, and it might be furnished with books and periodicals. It would be available for their debating society, which at present meets in a room far too confined, and urgently required for other purposes. For these and other objects the conversion of the cloisters would be hailed as a great boon to the College, and as it was quite understood by Mr. Matheson that it was to go into this year's estimates, I hope that your Grace will use your always friendly influence in having it retained there.'

To SIR CHARLES LYELL.

'ST. ANDREWS, *July* 19*th*, 1864.

'I was exceedingly gratified by the proposal made to me through you by the office-bearers of the British Association, that I should consent to be nominated as President of the meeting intended to be held next year at Dundee.

'It is with unfeigned regret that I feel it my duty to relinquish the hope of holding so honourable a position. The objection arises solely from the uncertain state of my health. This acts in two ways. In the first place because with every precaution the personal risk would be considerable; and secondly I feel that, even under the most favourable conditions, my powers of endeavour would fall short of what would justly be considered necessary by those most interested to be the President's duties.

'I therefore beg most respectfully to decline the nomination which you so kindly propose, and which under other circumstances would have been so congenial to my wishes.

'Allow me to add that the office-bearers of the Association could not have made their proposal through a channel more gratifying to me than through yourself.'

To the Same.

'ST. ANDREWS, *July 26th*, 1864.

'I received your kind letter of the 20th, and in compliance with your wish I have carefully considered again the question of the Presidency. The natural promptings of ambition did not fail to urge me in the direction which you and Sir R. Murchison are good enough to recommend. But in honesty as well as prudence I feel bound to act on my original decision. To hold out to others or to myself hopes of efficiency, more than likely to prove delusive, would only bring anxiety and disappointment to all. Allow me, then, finally to decline the flattering proposal. I shall ever gratefully remember your friendly offices, not only on this but on many other occasions. When I made, in the spring of 1831, the—to me—memorable journey to London in the character of a scientific aspirant, I made very numerous acquaintances and friends. Of those who survive I look upon you as one of three who have during the long period since elapsed been unswerving in their friendship towards me.'

To PROFESSOR STUDER, BERNE.

'*October 30th*, 1864.

'. . . The cause of my forgetfulness probably was a tedious illness, though not of a severe kind, under which I laboured for many months last winter and spring. I caught cold on Christmas-day, and suffered from the effects of it for many months—so much so, that I thought it advisable to remove in March to my old quarters at Clifton, where I remained till June.

'I have now pretty much recovered my usual health, though perhaps I lose a little as years roll on even in my diminished average state of bodily vigour. This last summer there was at one time a prospect of realizing my cherished plan of revisiting Switzerland. Being already with most of my family so far south as London, and with our house *ménage* broken up, my family were very anxious that I should go to Switzerland: but just at that time—May—my strength was reduced by long confinement and illness to its lowest ebb, and my courage finally failed in the face of a long railway journey with a party of ladies. Besides, the accounts I received of the changes of travelling in Switzerland in the last twelve years, the enormous pressure of tourists, want of lodgings, and general absence of repose, seemed to me to deprive it of most of the advantages which it formerly offered even to a confirmed invalid. So in the middle of June with reluctance we turned our faces homewards. I have no reason to doubt that I acted wisely. But it was sad: the more sad because I fancy I shall never think of it again, unless indeed I were to begin by passing a winter in the south. Excuse so much about myself and my health.'

To PROF. WM. THOMSON.

'ST. ANDREWS, *December 2nd*, 1864.

'I hope very soon now to complete the careful reduction of my experiments on the conductivity of iron, and to publish them. In the meantime I can state confidently that the results will confirm—even numerically—very closely the approximate results I before published. I think that the diminution of conductivity with temperature will also approach nearly to the coefficient for iron given by Mathieson for electricity. For myself I surrender to public clamour any hope of present justice on the glacier theory. I have no longer strength for it, and perhaps I can bear disappointment better than I once could; but I feel a little of the old fire when I see an old acquaintance like Ruskin take up the cudgels for me. . . .

'I am going—health permitting—to hear the Duke of Argyll's address on Monday; but shall be returning here almost immediately, and hope to hear from you when you have time.'

To ALFRED WILLS, ESQ.

'EDINBURGH, *December 7th*, 1864.

'. . . My interest in Mr. Reilly's map and its suitable and prompt publication is not unnaturally a strong one. The best, and possibly happiest, years of my life were spent in work more or less connected with the exploration of the chain of Mont Blanc, and I fondly, perhaps too sanguinely, expect to be remembered not only by the theory of glacier motion, which I still call my own, but by the execution of the first detailed map ever made of a glacier.

'Surveying was with me from early boyhood a passion. To improve and rectify and extend my map when made was a subject of reasonable ambition, and a labour of love. When disabled by hopeless infirmity from doing more in this way, I have tried year by year to incite younger men to take up the work and unravel the remainder of the chain of Mont Blanc. I tried for years in vain. At last I see it done in a masterly manner by an amateur like myself, based on a system of triangulation of my own, which, being unpublished, I furnished to Mr. Reilly, and I also advised in great detail as to the course to be pursued in carrying forward the work. I see all this done, and what is even more difficult and surprising, I see it embodied in a map, which I have no hesitation in saying would do credit to a bred engineer and draughtsman. I see the way paved for its appearance as a British work, the first true delineation of the most interesting ground in Europe; and then some indefinite and shadowy obstacle seems to be interposed in the way of its appearance, and I see that the really later French work, in a mutilated form, embracing but a section of the chain, will, if more energy be not used on this side the Channel, be actually laid on our tables while the Alpine Club are talking about the proposed publication. I ask you, therefore, as a friend of my own, and as one unquestionably deeply interested in the subject, and in the credit due to British enterprise in this matter, to obtain—as you, from your position as President, so easily can—no doubtful decision of the Club on this subject, and I trust that you will have the kindness to give me early information of what is intended. I will answer the other part of your letter, in which I am also much interested, at another and more fitting time.

'P.S.—I intend returning to St. Andrews to-morrow.'

To E. C. Batten, Esq.

'St. Andrews, *Easter Day*, 1865.

'. . . I cannot resist writing you a line to tell you how happy I felt to-day in kneeling at the altar with my dear boy beside me. How vividly it recalled the like anniversary of 1838. To think that nearly a generation of men have passed since that! We are indeed shadows, and have no abiding stay.'

To Dr. Symonds.

'Fettercairn House, *April 23rd*, 1865.

'I am afraid it is a very long time since I have written to you, but I sincerely hope you will never misinterpret my silence, for I shall never cease to value your friendship, and to feel grateful for your great and steady kindness. I have chosen the tranquil opportunity of writing to you from my brother's house and one of my old homes, where, in this quiet country scene, which you recollect, and removed for a time from daily cares and anxieties, one loves to recall the past and the absent. Indeed, as I have not now the power to visit the real home of my childhood, Colinton, near Edinburgh, no place can bring back to me the scenes and the friends of early years so much as this, where I spent two or three months most years, from the age of fourteen to twenty-four. It is a district which has been altered singularly little by violent or obtrusive changes. The houses—all the principal ones—are the same, only their tenants are a new generation, or new comers. The woods and gardens and fences and roads are the same, only the trees and shrubs, which I well remember being planted, have grown to a startling height, and seem to reflect visibly my own age; the rivers and hills are absolutely unaltered, and with the help of an old pony I can get to places where I used to take my solitary walks. Though sad, it is indescribably pleasing, and brings, beyond anything I know, a tranquillizing influence and refreshment. . . .

'Our visit is a very short one, at least Alicia's and mine. Our boys, who are with us, are to spend their fortnight previous to the summer session of the College Hall, and they too are acquiring the same deep love of the localities which I did forty years ago at their age.'

To PROFESSOR STUDER, BERNE.

'PITLOCHRIE, *October* 1865.

'. . . You will believe that the catastrophe of the Matterhorn powerfully affected me. Having friends in the district, particularly Mr. Reilly, I suffered for some moments intense anxiety, and the whole details of the accident were of a kind to occasion the profoundest interest and regret. I have lately been at the Birmingham meeting of the British Association, where I met Reilly and Whymper, and from the latter I learned privately every detail. Whymper's letter to the *Times* was in excellent taste, and he has most prudently abstained from the temptation of being made a hero of.

'Finding myself rather better than usual, I ventured to attend the meeting of the British Association at Birmingham, where I saw some friends, but found a large preponderance of new faces. Sir R. Murchison and M. Von Dechen were staying in the house where I was. There was not much original matter brought forward in any branch of science; but the meetings were interesting. I returned home very tired, but not otherwise the worse for it. Phillips was the President; Sir C. Lyell was there for a day, but I did not see him. The Cambridge men were mostly absent, which was a disappointment to me.

'In about a week I hope to be at St. Andrews for the winter, where I hope you will write to me and tell me the scientific news of central Europe. There is a curious book, full of clever drawings, published, called "Frost and Fire," by Mr. Campbell, not a professed geologist, but a clever fellow. He ascribes much to the action of icebergs carried by ocean currents. Ramsay—who is wilder than ever about ice—praises it highly in an article in the *North British Review.*'

To the Duke of Argyll.

'St. Andrews, *November 6th*, 1865.

'When I visited Birmingham at the meeting of the British Association in September, Sir Roderick Murchison called my attention to the extreme importance of having the science of Geology thoroughly taught in the Edinburgh University, with which is now connected the Geological Survey Museum there.

'Sir Roderick, when he spoke, was aware that I had long lamented the decay of the study and teaching of Geology proper in the Edinburgh School, to which I had an opportunity of specially adverting in an address delivered two years ago to the Royal Society of Edinburgh. Considering the immense impulse given to the science by the Edinburgh School of Geology, the present torpor of the University as a source of geological knowledge cannot be too much regretted. Sir Roderick as a Scotchman, though unconnected with Edinburgh in a direct manner, feels it keenly, as I also do, and many more. It may be in your Grace's recollection that at the time of the death of Edward Forbes, the division of the chair into two departments, of Geology and Natural History proper, was seriously discussed.

'A great part of the magnificent and increasing collections of the museums is rendered unavailing for students. I do, as a Scotchman and as an old Edinburgh man, feel keenly the reproach, and desire to see it removed. Sir Roderick has in view for the post an admirable man—Geikie—who is believed to be most desirous to accept it, and all that is wanted is a moderate pecuniary endowment, which cannot, we think, be withheld, if your Grace will give the proposal your active support. It is a sad thing to see a study dying out in a country where it formerly so conspicuously flourished.

'I trust that your Grace will excuse me for urging

upon you these considerations, and that you will use your deserved influence towards obtaining an endowment.'

To DR. SYMONDS.

'ST. ANDREWS, *December 24th,* 1865.

'The close approach of Christmas is an agreeable call to write to you. We have not heard very directly from your house for some little time. I hope your health has been good, notwithstanding your culpable neglect of a holiday, which I sincerely hope you will not again suffer yourself to omit. You will no doubt be drawing some of your olive branches round you for Christmas time. Our dear Edmund came home from Wimbledon last night, looking, I must say, strong and hearty, though just emerged from a whirl of examination papers. I don't at all know as yet what opinion to form of the result of his going to school: but I expect it will be on the whole favourable.

'. . . George is going on very steadily and well at the College Hall: mending his classics, and going forward with his mathematics. I suppose he will go to Cambridge in October 1867: the question of Trinity or a small College will come very soon.

'Eliza and Alice are at home, Minna with Mrs. Wauchope—all well and sweet and good. . . . We are too large a party to migrate to Edinburgh, and therefore spend our Christmas at home.

'Not to omit myself from the family tableau, I can thankfully say that I have been in stronger health for the last two months—fully—than for a long time previously. I have more elasticity and less liability to colds. For a considerable part of that time, I had indeed a feeling of health unusual to me. Such breaks always keep me from desponding: and my lectures on the History of Science on seven successive Saturdays were only a pleasure to me and no drag. They were well attended considering all circumstances. As you kindly interest yourself in my reputation, and as I always consider myself as in part owing to you the measure of health which I have got, and power of work, you will be pleased to hear that the Royal Society of Edinburgh have given me the Keith prize for my paper on the conduction of heat, which is probably the last tough job I shall do in physics.

'Our students, by a strange freak, have elected John Stuart Mill the Rector for three years—an office more of duty than show, in fact quite prosaic. It must be a great bore to Mill, who has begged off delivering his address till January 1867. But the business of the University cannot stand still till then.'

To E. C. BATTEN, ESQ.

'*December* 31*st*, 1865.

'. . . The return of the anniversary is not needed to make me feel old. I have lost nearly all the elasticity of youth, and much of its hopefulness: and the thoughts of perhaps living on a few years more, are intensely associated with the anxious prospect of my sons' settlement in life. It is an anxiety so great as to absorb almost every other care, and I feel, alas! how little I can—as situated —now do for them.'

To A. WILLS, ESQ.

'EDINBURGH, *January* 2*nd*, 1866.

'I met your brother at Birmingham at the house of Mr. C. E. Mathews. There was a large Alpine party, and I had the melancholy pleasure of hearing from Mr. Whymper's own lips the details of that awful accident. For a long time it quite haunted me. Mr. Whymper's letter to the *Times*, so perfect in taste and tone as well as in narration, raised him immensely in my estimation, and the impression was confirmed by his bearing and behaviour at Birmingham, which was everything that could be wished, though he was subject to the temptation of being violently lionized. But I was going to say that I was concerned to hear from your brother that Mrs. Wills' health had prevented her accompanying you to Sixt as usual. It must have thrown a damp over your autumn excursion, but I hope you will be able to give us good accounts of her now. I shall like to hear whether you extended your walks this year into any new regions.'

To Dr. Joseph Robertson.

'*February* 3*rd*, 1866.

'The references which I have frequently been forced to make of late to old charters and documents which I am unable to read,—or but a few words here and there,—have led me to wish to make an attempt to learn the old hands of the period, in particular from 1450 to 1600. I fear I am too old to learn, and I feel that it may prove to be a waste of time; but I think of making a short trial, and I think you may be not unwilling to give me a hint of how to set about it. I have been told that I might begin by comparing some published facsimile of an old hand with the printed interpretation, as set down in such books as the Bannatyre Club books or some other. If this is your opinion, perhaps you would recommend some specimens about the period I mention.

'Any other hints I shall much value. Remember I do not pledge myself to persevere, for it may prove a loss of time: but I am almost desirous to try. Another point may I ask your opinion upon. Do you know of any silversmith who could safely be entrusted with our ancient mace, which is really in a very shaky state. You will know its value. The artist would need to be not only a man who would do no mischief, but who would not charge exorbitantly.'

To Sir R. Murchison.

'Pitlochrie, *August* 29*th*, 1866.

'Many thanks for your kind letter, and for thinking of me for the vice-Presidency, which I most readily accept. As the meeting is to take place, it would be inexcusable if I held back from doing the little I can, and promoting its success in a small way amongst my colleagues at St. Andrew's. It will not be necessary for me to engage to attend every day, nor at dinners and evening meetings in general, which are what I am afraid of. You are quite right in saying that my health was the sole cause of my declining the Presidency so kindly offered, and I should have been disappointed had I not been allowed at least to show my goodwill by appearing as a vice-President.

'I congratulate you on your generalship in getting the Duke of Buccleugh to act. When I first saw the report I feared it might not be true.'

To E. C. BATTEN, ESQ.

'ST. ANDREWS, *November 4th*, 1866.

'Your kind letter crossed mine. . . . I inclose a fragment of my opening address on Wednesday, but it will appear as a pamphlet. It is quite historical and statistical, and gives an account of the origin of the British Association and of the first meeting at York. I have not for many years been busier than during the last three weeks, and I have a large arrear of letters to write. I have had the whole arranging of the new constitution of our College Hall. The general meeting of shareholders was on Monday; and went off well. We have had the plans also to consider, and our factor having very inopportunely taken to his bed, I have had the whole common business of the College and the Hall to look after. However, I have been very well. You will be glad to hear that the old shareholders of the College Hall are not likely to be called on afresh for perhaps nearly a year. We have about 75 new shares taken. I am much pleased with William Thomson's honours. Cholera is occurring near Dundee and elsewhere. One of my many engagements has been inducing the Provost and baillies into getting up a cholera hospital; in which we have partially succeeded. I must now stop.'

To PROFESSOR STUDER, *Berne.*

'ST. ANDREWS, *November 17th*, 1866.

'. . . I hope your health continues good, and that you are still able to make your geological excursions, to your own pleasure and to the benefit of your pupils. My own health remains exceedingly stationary; has rather improved within two years; but still compels a comparatively inactive life. I was in London for some weeks in May and June. I had not attended a meeting of the Royal Society nor indeed seen anyone almost in London since 1851, the year of the first Exhibition! so you will imagine that I seemed to myself and others a sort of resuscitated antediluvian, or one of the stone weapon people come to partial life.

'The occasion of my visit was sad: my eldest and last remaining brother, Sir John Forbes, whom I think you slightly knew, died in London on the 28th of May, after a painful illness, and I came to be with him.

'The British Association meets next year at Dundee, not far from here. Shall you not take one trip to Scotland and see us all? my family now nearly grown up would alone be a reason against attempting to travel, besides my infirmities. My eldest son is with a civil engineer at Newcastle. My second son I hope to send to Cambridge.'

To E. C. BATTEN, ESQ.

'ST. ANDREWS, *April 6th*, 1867.

'Weeks pass by almost unperceived by me, and anon I find myself reminded by a letter from you that I have been remiss. It is curious how much miscellaneous business falls into my hands, perhaps from having more normal leisure than most people. Setting aside a not inconsiderable correspondence, it is surprising how many reports, drafts, minutes, advertisements and circulars it falls to me to write out with my own hands; the number of these things I do, gives me of course a certain facility in doing it. One of the last things is organizing a subscription for an entertainment of a select party of the British Association here on an excursion from Dundee in September. It must be done, and as well done as we are able. Then our new College Hall and the management of the present one costs me no little anxiety. We are told it is not at all known, and the Englishmen we have had have liked it so much. Then it is the fate of this College never to be out of some legal or financial botheration and slough of despond: and just as I am seeing daylight through some complicated trouble, I have some new set of ideas to master, and claims to meet. Now it is Teinds; dreadful word! with its titulars and tacksman and underpayers and overpayers, its localities and decreets. It is enough to drive one wild, with claims of interest exceeding the whole amount of principal, and accumulating with no regard to even forty years prescription!

'I am not very well, and the long consequent confinement to the house has told upon me rather severely.

'Ever affectionately yours,

'JAMES D. FORBES.'

To PROFESSOR INNES.

'ST. ANDREWS, *April 9th*, 1867.

'Could you favour me with your opinion of the word "Fermoraria:" I think it is of the first declension. I guess it to mean either "farmery" or "infirmary;" it was an appendage to the Priory of St. Andrews, within the Abbey wall, and might well enough be either one or other. I find no such word either in Ducange or Maigne's Lexicons. . . .

'I have now nearly completed my examination and analysis of the old Chartulary from the advocates' library. The notes I have made would materially assist any future searcher.'

To MRS. FORBES.

'LONDON, *May 22nd*, 1867.

'. . . The weather is really beyond precedent. I was awakened in the night, last night, by the cold, and I see by the papers that the parks this morning were as white as in January. It is this partly I suppose which keeps up my cough, which is still troublesome, and I feel less active than I did a week or two ago. . . . George talks of going to Arran for a few days; he is a dear boy, and I miss him much; never a shadow came over him with all the hitches about his plans.'

To SIR RODERICK MURCHISON, BART.

'PITLOCHRIE, *August 5th*, 1867.

'. . . . The burden of my letter must be to tell you of how little avail my co-operation is likely to be at Dundee, if indeed it is not absolutely *nil*. The fact is I have been and still am much out of health. You recollect what the weather was in London the last fortnight of May. I was only beginning to recover from a long winter's cold, when it recurred under the severity of that aggravated weather. I reached Scotland with some difficulty, and was immediately confined by an illness at St. Andrews for some time; my recovery from which has been slow and uncertain. How often I have congratulated myself on having withstood the temptation—not an inconsiderable one—of the Presidency. I must have surrendered it absolutely. Even as it is, I dare not count with any certainty on being able to appear at Dundee, or even—what is still more essential—to take official part in such hospitalities as we can offer at St. Andrews to the excursionists.

'I am happy to say, however, that Mr. Geikie takes in hand the entire scientific conduct of the excursion, which, if the weather be fine, will, I believe, be a pleasant and interesting one. . . . Your efforts on behalf of the Association and of Scotch science are invaluable. May they be long continued! I hope your walking powers have improved since I had the pleasure of dining with you at the R. S. Club, which was almost the last day of tolerable health which I enjoyed.'

To PROFESSOR FISCHER.

'PITLOCHRIE, *August 28th*, 1867.

'I must once again offer you my cordial thanks for your important services to George in the promotion of his studies, and for your and Mrs. Fischer's kindness to him. I am sorry that I must now formally give up any idea of attending the British Association at Dundee, or of receiving our guests at St. Andrews. I have not been so well, and the effort of conversation is plainly injurious. . . .

'I was much obliged to you for sending the "Comptes Rendus." I have read the greater part of the Pascal papers and controversy with attention. It is a question of some magnitude and importance. The mass of contemporary documents in M. Charles' possession is evidently very large. It would be worth no one's while to forge all those. My conclusion would be—provisionally, of course—that some possessor of the mass of genuine contemporary correspondence, to give it piquancy and value, has forged a number of Pascal's notes on gravitation, and probably all of those bearing Newton's signature. I cannot believe that Newton would write in French instead of Latin. Sir David, I see, makes the same remark. The exact masses of the planets, and, above all, the exact flattening of the earth given by Pascal appear to me incredible. . . . It is all very strange and bewildering. Photographic fac-similes of Newton's letters ought to be asked for by competent authority.'

When Forbes entered on his duties as Principal, it was with high hopes of what might be done to make of the United St. Salvator's and St. Leonard's a College which should be a model of its kind: and with this view he took the whole thing more in hand and threw himself more earnestly upon it than perhaps any Principal ever before had done. If the results he achieved fell far short of his first hopes, this was not from any lack of zeal or energy on his part. It is no more than what befalls most men in their ideal aims. Of newly appointed Bishops some one has said that they must spend the first years of their office in learning how little power the law has left in their hands. The same may, perhaps, be said of Principals. Forbes, with his ceaseless activity, found this; he had his own share of that experience which Herodotus expressed and Arnold so often quoted—*ἐχθίστη ὀδύνη πολλὰ φρονέοντα μηδενὸς κρατέειν*—that direst pain, to have many thoughts and no power to realize them.

F F 2

With aims so high, and so unflinchingly pursued, it could not be but that Forbes, in acting publicly with other men, should meet with much to thwart and disappoint him. These things, from his earnest and anxious temperament, he no doubt felt severely. It was not in his nature to take things easily. He could not, like some men, pass off a serious matter with a humorous jest. In his pursuit of public objects, those who worked with him may sometimes have felt as if his demands on them were exacting. But none could ever say that he asked for any labour or sacrifice in others, of which he had not first set the example. And so it came to pass, that however much some may have felt the strain which his rigorous pursuit of duty imposed, and however widely others may have differed from his views of duty, all were compelled to accord him sincerest respect, I had almost said reverence, for the purity of his motives and the elevation of his aims. The example of high moral purpose and energy which he left behind, will survive in all who witnessed it, as a salutary and bracing memory.

But if the course of public life did not always run smooth with Forbes any more than with other men, he found, when College meetings and business were over, abundant comfort and refreshment in the quiet of his home. His children had reached that age at which they could be more or less companions to him. And his necessary abstinence from dining out left his evenings the more free to enjoy the society of his family circle. A high authority at St. Andrews used to announce that it was one of the first duties, whether of Principal or Professor, to receive and reciprocate hospitalities. Even had Principal Forbes agreed with this dictum, his health would not have allowed him to carry it out on so large a scale as the author of the saying might have approved. But in a quiet and informal way he and his amiable family to dinner or in the evening gladly welcomed friends and neighbours who appreciated that refined and unceremonious intercourse.

The correspondence given above forms the best narrative of Principal Forbes' active life at St. Andrews. In the spring of 1867 his health sensibly declined. His old enemy, the cough, got a firm hold of him, which he could not shake off. The last public act he performed at St. Andrews was to preside at the ceremonial of the laying of the foundation stone of the new College Hall, a building which owes its existence entirely to his single exertions.

A great concourse of people from town and country had assembled to witness the ceremony. What with freemasons, professors, town-councillors, volunteers, bands of music, all marching in procession, and flags flying from windows and housetops, it was a gala-day such as the quiet old city does not often see. In painful contrast to these festal attempts was the pale, worn look of him who was the chief inspirer of the whole movement. Yet he braced himself for an effort which he was little able to make. After Mr. Whyte Melville, as Grand Master of the Freemasons, had laid the foundation-stone in due form, Principal Forbes delivered over it an elaborate address, detailing the objects of the building they were met to found, the brief history of the institution, and his hopes regarding its future. This was his last public appearance in St. Andrews or elsewhere.

CHAPTER XIII.

THE CLOSE.

In May 1867, though his cough continued unabated, he undertook a journey to London, and two journeys to and from Cambridge, with the view of choosing a College for his son George, who accompanied him. It was a more than usually cold and ungenial May, little likely to benefit the health of an invalid. On returning to Scotland in June, instead of retreating at once to the shelter of Pitlochrie, he made a detour by St. Andrews for purposes of business, and in doing so met with one of those bleak north-easters which St. Andrews alone can blow in June.

Summer and early autumn passed at Pitlochrie with not only no improvement, but with a gradual decline of strength. A winter abroad was recommended by the physicians on whom he relied. Towards the end of September the Principal left his home at Pitlochrie, passed through Edinburgh, where he halted for a day or two, and thence set out with Mrs. Forbes and his three daughters for Cannes.

To E. C. Batten, Esq.

'Villa Josepha, Cannes, France,
October 29*th*, 1867.

'. . . As I am tired of the labour and expense of moving, it is probable that we shall remain here, unless driven to Mentone or San Remo by the spring winds. I cannot tell you how I feel the difference between wintering here and at Rome or Naples. The charm of antiquity and historical association is here all wanting; and though the scenery is certainly very pleasing, and the climate beautiful, and the vegetation exquisite, I fear that I shall soon feel a stagnation of interest, which reminds me more and more it is an exile on account of health merely.'

There nearly the whole of November was passed, but with no benefit to the invalid. The parching drought of the air irritated his lungs, and brought on one of his attacks. In the last days of November Hyères was resorted to in search of softer air, and though the weather proved unusually severe, yet the change did good.

To DR. SYMONDS, *Clifton.*

'*December* 11*th*, 1867.

'. . . We have now been fully ten days here—at least Alicia and I. The girls we left under the kind care of Mr. and Mrs. Symonds, and they—the girls—joined us on Monday. Since we came to Hyères, the weather has been nearly as unfavourable as possible, snow having fallen over all this district as well as over almost all other parts of Europe. Cold northerly winds have prevailed. Notwithstanding this, I certainly improved under the change; that is to say, I found the air much less irritating than at Cannes, and I have no doubt whatever that this is the case. Indeed, the fact that we are seventy miles from the snowy Alps sufficiently accounts for this. I had no idea until quite recently that the snowy range so completely predominates over Cannes, and by making a permanent current of N.E. wind, set from their cold summits to the shore of the warm Mediterranean Sea, at once produces that preternatural clearness and dryness of atmosphere which belongs to Cannes. The screen of low wooded hills, which shelters Cannes from the immediate force of the Alpine blasts, is not high enough or extensive enough to change the quality of the air. In these respects Hyères is in a totally different position, and has the advantage of being farther from the sea.'

At Hyères, Forbes' health continued to improve till towards the middle of January 1868.

On the 13th of January, feeling stronger, he was tempted to go out for a longer ride than usual. His youngest daughter accompanied him on foot. They went up one of the heights in the neighbourhood, and feeling well, and enjoying lovely views, and delicious air, instead of returning by the usual road, he fixed to make a round. This road turned out to be both very steep and very stony. The animal he was riding had to jump from stone to stone. This shaking was most fatiguing, and he came home about five P.M. much exhausted. The second day after this, the 15th January, a slight spitting of blood commenced. From that day till he died, nearly twelve months later, he never was again able to be dressed, or to be out of bed beyond an hour or an hour and a half at a time. Towards the end of March he had another severe attack, and he never again sat up at Hyères. He generally wrote what he wished for in pencil, as he dared not raise his voice; but latterly at Hyères he could hardly do even that. At this time he was offered the Presidency of the Royal Society of Edinburgh, which he refused. His correspondence now almost ceased. Some letters on business he dictated, and the few that there are to his family and to Mr. Batten were written in pencil by himself.

As May approached, it was felt that to remain at Hyères and meet the summer heats there must in his weakened state prove fatal. To attempt a journey home seemed hardly less perilous. Yet a choice had to be made, and he determined calmly and deliberately, in full view of the risk, to set out for Clifton. If he were allowed to reach that place, he would there be in a good climate, and under the care of his true and tried physician and friend Dr. Symonds. The journey, under all the circumstances, was hazardous in the extreme, and must be described in the words of one who kindly shared in this and in all the other anxieties of that time.

'I shall, as requested, endeavour to give you in this letter a few particulars of the journey home from Hyères, made by the Principal and his family in May 1868. It was throughout a most remarkable one. Of those who witnessed our departure from Hyères, there were, I am sure, few who anticipated that both invalids—for his eldest daughter was also suffering from long-continued and dangerous illness—would reach England in life; and fewer still, who had the slightest expectation that in little more than a week we should have accomplished the journey in safety. But so it was; and I believe that our doing so was in a great degree owing to the intense desire which the Principal had to be once more, and as soon as possible, within reach of Dr. Symonds, of Clifton. The distance from Hyères to the station is about five miles. This part of the journey he had to perform in an omnibus and over a rough road, and this, in his weak state, was perhaps the most anxious part of all. A return of the hæmorrhage was very likely to occur, and had it occurred would in all probability have proved fatal. So fully was he aware of this himself, that the day before we started, he wrote in pencil, and placed in his wife's hands a paper with clear and distinct directions as to how she was to act in such an event; and having done this, and feeling that he was in God's hands, he quietly left himself in them. Nothing during his long-suffering illness proved more completely his entire acquiescence in whatever might be ordered for him. There was no shrinking fear or dread of the future. I have seldom seen anyone who had the same simple and almost childlike confidence that all was right, so long as he submitted and trusted himself entirely to his heavenly Father's care. One great lesson was to be learned from him in this: he knew what a wonderful gift of God life is; he could not but be aware of the great intellect which had been given to him; and therefore he did all in his power to continue that life—the least thing which was ordered for him, being taken with a view of prolonging that life, and yet withal being fully ready to yield it up whenever the end came.

'His weakness was such that it was necessary to have him carried down-stairs on a stretcher. He gave his own directions to the porters as to how this was to be done; and from the stretcher he was placed in the omnibus. It was a lovely afternoon and evening; and when he reached the station he saw, for the first time for two months, his beloved eldest daughter. He recognized her with a gentle smile, for they could not speak at that moment, being in separate conveyances. As soon as the saloon carriage was ready, both invalids were placed in it, with an ample supply of such food and medicine as was thought might be required on the way; and waited for the arrival of the train from Cannes. Several hours elapsed before it came up, some accident having occurred on the line. We were to have continued our journey through the night, but on reaching Marseilles, it was found that the express train for Paris had started without waiting for ours; and therefore it was impossible to proceed further till the following morning. All that could be done was to have the saloon carriage placed under cover for the night. This was explained to the Principal, who as usual quietly and cheerfully submitted. We started for Paris at eleven o'clock next forenoon, but another day and night had to be passed in the train; and it was not until nine o'clock on Saturday morning the 1st May, that we reached the station in Paris; and then came the slow drive to the hotel, through the busy streets, thronged with people even at that early hour. During all these hours the Principal maintained the same calmness and self-possession. Once during the night, when all was quiet, one of the party noticed how faint he was looking, and went to him and administered some brandy: he said himself afterwards, he had felt sinking, but had been too weak to call out. Three days were passed in Paris, namely, from the Saturday till the following Tuesday morning. He seemed to enjoy the comparative rest and comfort of these days; but he felt he must not linger on the way, and was well pleased when the time arrived for starting for Boulogne. We arrived there at five o'clock; and our intention was to remain only one night, and cross the Channel on the following day: but a gale having sprung up, and the sea being so rough and stormy, it was deemed advisable to delay the passage till Thursday, when fortunately the wind had abated, and the voyage was made quietly and easily. Thursday night was passed at Folkestone, and then came the last part of the long, anxious journey to Clifton, which was accomplished successfully, and he reached the house which had been prepared for him there at seven o'clock on that day.

'It is almost impossible for anyone who did not witness the difficulty of this journey, to understand the wonderful presence of mind and resignation of the Principal during the whole of it. There must have been so much resolution to struggle with his increasing weakness—moments when he must have felt almost incapable of any exertion, and ready to wish just to stay where he was. But the indomitable perseverance of the man, and his firm conviction that an overruling Providence ordered all things, were shown throughout. This would never have resulted, had he not during his early and later years, done all for God; and it is a lesson which the young men entering life in the pride of their strength and intellect would do well to follow. Theirs may not be accompanied by the long-continued suffering, so patiently and yet so manfully endured by the Principal, but it must come to all in some form sooner or later. May they take an example from this great leader in the paths of science, and in entire submission be led by their God and Father.'

Arrived at Clifton he thus wrote in pencil to one of his most attached friends:—

'CLIFTON, *May* 13*th*, 1868.

'I am deeply thankful to be on British soil again, and am not solicitous about forming future plans. It is a limited future, I think, and I leave it contentedly, so far as I may, in God's hands.'

When he reached Clifton his inclination was at once to resign the Presidency of the United College, as he felt convinced that he would never again revisit St. Andrews. But from taking this step he was deterred by the advice of Dr. Symonds, who suggested that the Principal should wait quietly for four months and see what change the summer brought: if it were not a change for the better, he might then resign. On this advice he acted implicitly. Meanwhile that his College and its interests were not absent from his thoughts the following letter written in pencil to one of his colleagues then in England, shows:—

To PROFESSOR CAMPBELL.

'CLIFTON, *May* 18*th*, 1868.

'. . . I learn that you are likely to be here before long. Pray give me some notice of your coming. Should I happily be then as free from urgent illness, I could hold cautiously some conversation with you on matters of importance affecting St. Andrews, but which I could not have held sooner without extreme imprudence. I began this letter under the impression that you intended being at Clifton shortly, but Mrs. C. now writes of the 30th June. This is a long time for me to look forward to, and I may either be better or worse by that time.

'I thought that I might perhaps be able to explain to you the drift of the notes which I have made—and which are here—on the College holdings in St. Andrews. . . . In my present state of weakness it is vain to hope that I can put all this in writing.

'It is very doubtful whether I could convey it to you in a few desultory conversations of uncertain length, carried on somewhat in despite of the doctor. But you will easily understand the anxiety I have that the materials I have gathered—and which it may almost certainly be predicted, no one will ever again work out from the beginning, though they might follow up what I have thought and done—should not be utterly lost.'

By the beginning of October the four appointed months had expired, and then Principal Forbes, finding his weakness gradually increasing, with the full consent of his physician, sent in to the University Court his resignation, with the request that it should take effect on the 15th November. On acquainting the senior Professor that such was his intention, several of his colleagues who were then in St. Andrews sent him a joint letter, requesting that he would postpone his resignation. Having, however, acted thus deliberately, he abode by his resolve, and his connection with St. Andrews ceased on the day he himself had named.

With this last act his connection with the outer world may be said to have ended, and he lay calmly and devoutly looking forward to the close.

On the 5th October, to his eldest daughter on her sick bed at Pitlochrie, he wrote in pencil, these his last written words :—

'CLIFTON, *October 5th*, 1868.

'DEAREST ELIZA,

'I am glad to take up my pencil again to write to you. I hardly know why I have not done so sooner. I need not tell you, my dearest girl, how deeply I have been interested in the reports of your changing condition of health, and my sorrow that you have been such a sufferer.

'I rejoice to hear that you have borne it all with so much patience and cheerfulness. We are much indebted to Dr. Irvine and Aunt Jane for their full and frequent letters, and I fondly hope that the latest improvement will be the beginning of better times.

'I myself have had many ups and downs; but at present I am rather better. We have some cold nights. It must be getting cold at Pitlochrie, but I do not know where you would be better at present. We expect J. Mackenzie back from Bournemouth to-day.

'With much love to you and Aunt Jane, I remain, my dearest Eliza, your affectionate father,

'JAMES D. FORBES.'

What remains cannot be given otherwise than in the words of her, who for five-and-twenty years had been the companion of his life, and during the long decline, his unwearied attendant and consoler. 'He saw with calmness the slow but certain progress of his decline. He had no doubts, no anxieties; there was no earthly excitement in his death. Every day to the last he lived, so far as increasing weakness allowed, as he had done for seventeen years, thinking of others and taking the same interest in every subject that he ever had done, yet looking always steadily in the face of death. During these last months at Clifton, each afternoon, when his religious reading was over, he would lie and listen, while his children, to fill up the long intervals of enforced silence, read to him various books of travels, history, biography, and the Waverley novels. But during the last month and more, even such small diversions as this were no longer possible.

'On Christmas-day he had a very bad turn, and on the 27th he sank so low that Dr. Symonds scarcely expected him to live through that day. On Monday he rallied somewhat, but on Tuesday, the 29th, he was so much weaker that he appeared barely conscious, not knowing day from night. About eleven o'clock, A.M., however, he suddenly roused himself and asked, "What day is this?" We said, "Tuesday." "What hour?" "Eleven o'clock," we replied. "Remember," he continued, "that Mr. Mathers comes at twelve o'clock to-day." Mr. Mathers was the clergyman who had attended him, and my husband had settled a week before, when no immediate change was expected, that he should come on Tuesday, the 29th, to administer the Holy Communion. None present will ever forget that solemn hour. His four children and myself all received it with him. He answered to all the responses, and those who heard will remember, while they live, his last Amen.

'After that he called his two sons, and told them he could give them handfuls of advice, but that they knew all his wishes so well he would not. After giving them a few last directions about his watch and other things, he asked them if they would like to consult him about anything, for he was still able to listen and to advise them. When I came into the room, he said, "I leave you two good boys, keep them so, and don't leave me. It won't be long now." A few days before this he told me "he had not a doubt or a fear."

'The 30th of December was a day of sad suffering from breathlessness, but of clear consciousness. He was sensible of myself and his children, and watched with a silent look of love all the efforts to help him.

'On the morning of the 31st he took the hand of his eldest boy, as he stood beside him, and five minutes afterwards, at eight of that Thursday morning, the last day of the year, he passed away, still holding his son's hand. To the last moment he was perfectly conscious, and seemed to prolong his latest breath that he might have us a little longer with him. On January 5th he was buried in the Dean Cemetery, in a lovely spot chosen by himself the last time he was in Edinburgh, before he went on that last journey to Cannes. His grave is shadowed by the yew tree which overhung his window when his home was in the old house of Dean. In little more than four months his eldest daughter was laid beside him in the Dean Cemetery. A year and a few months passed, and his second daughter, who had spent her last strength in nursing the two dear ones gone before her, followed them to the same grave. The sisters, one in life, in death were not divided; and words, used by a friend in writing of the elder, are equally true of both, they "being made perfect in a short time, fulfilled a long time, for their souls pleased the Lord."'

This Memoir may perhaps be fitly closed by the letters of two friends of Principal Forbes, to whom he had probably opened his heart more than to most.

From Dr. Symonds.

'Clifton, *October 17th,* 1869.

'. . . When he was writing, nothing was more remarkable than the perfect ease with which he executed his work. His manuscripts nearly always went to the press just as they were first written; almost without a blot, or an erasure, or a correction. I asked him one day, when I saw his table covered with sheets of the fairest manuscript conceivable, with lines so even and interspaces so regular and orderly, and words so irresistibly legible, I asked him how he had acquired the power of writing off in that singularly finished manner. His answer was that from early studentship he had made it a rule not to put pen to paper till he had mastered his subject, or at all events till he was sure that he had something to say, and that he had a clear notion of what it was that he wished to say.

'With regard to Principal Forbes' powers of conversation, it must be remarked that I only knew him as at best a valetudinarian, so that he was not robust enough, had he been disposed, to take a leading or animated part in the discussion of a dinner-party. In a *tête-à-tête* he was most agreeable and instructive. Perhaps nothing was more remarkable than his liberality or tolerance of differences of opinion. One knew that he held very decided views in politics and religion, and that they were of a conservative character, but they never seemed to tincture his mind so as to prejudice his opinion of individuals in private life, or of scientific workers and writers. However little he might have exerted himself in company, he always left an impression of gentle wisdom and urbanity—nay, of high-bred courtesy. Indeed, nothing could surpass the gentlemanliness of his manners and deportment. His amiability, together with that quietness of sympathy, and penetration into character and motives which constitute what we call tact, both disposed and enabled him so to recognize and adapt himself and his words to the feelings of those about him, that it was impossible for him to neglect the very faintest *nuance* of kindly consideration and forbearance.

'It was a serious loss to me when he left this neighbourhood; but I had, in subsequent years, by visits with which he and his charming wife favoured me, and by visits which I had the privilege of paying to his hospitable home, as well as in frequent correspondence, abundant opportunities of ripening my knowledge of him and of his never-failing virtues and attractions. It would be presumptuous in me to attempt the portraiture of such a mind and character; but were I asked to state in one or two words my sense of them, I should say that intellectually Principal Forbes was distinguished by comprehensiveness of details, accuracy of memory, clearness of insight, and strength of judgment; and that in the emotional part of his nature he was pure, high-minded, gentle, and affectionate.

'But those who knew him as intimately as did the writer of this scanty memorandum, will feel that no words, however multiplied or carefully considered, could afford an adequate idea of that which inspired in their hearts so deep a reverence and so warm an affection, and the remembrance of which will never allow them to cease from deploring that so much genius, knowledge, and wisdom could not have been longer vouchsafed to the world of science, nor an embodiment of so much goodness and benevolence and love to his family and his friends.'

BISHOP FORBES' *Letter*.

'DUNDEE, 1872.

'You ask me to put on paper anything that occurs to me regarding the interior life of my gifted kinsman. I feel the greatest difficulty in doing so, from the obvious reason that the intercourse of a sick man with what the ancient Celtic Church beautifully described as "his soul's friend" is too sacred for publication, and no one would have shrunk more from any manifestation of his religious feeling than he. Still it is due to the cause of truth in this sceptical age, that I should put on record the impressions which a series of years of affectionate intercourse with Principal Forbes have left upon me.

'Of all the characters I have read of in history, no nature seems to me so like that of my friend, as that of Blaise Pascal,—with many important modifications be it said. There was the same sensitive organization; the same intense love of truth for its own sake; the same fearlessness in facing facts, however they might militate against preconceived notions, or established theories; the same bright intelligence which to the end triumphed over the exhaustion of the bodily frame; the same unquestioning submission of will and intellect to the Supreme Being; the same lowly acceptance of the supernatural truths of religion, how incomprehensible soever to man's weakness. On the other hand, there was absent from him that dark and ascetic side which distinguished Pascal—there was nothing of jansenistical rigour about him; his was a religion rather of hope than of fear. I never saw in any man such fearlessness in the path of duty. The one question with him was, Is it right? No dread of consequences, and consequences often bitterly felt by him, and wounding his sensitive nature, ever prevented him from doing that to which conscience prompted. His sense of right amounted to chivalry.

'I cannot be silent with regard to that exquisite tenderness that ran through all his home relations. To see him at the head of his own table, surrounded by his family that adored him, after a day of hard mental work, or of worrying College business, and to observe how, in the sanctities of home, he took interest in the sports and occupations and studies of the children, was what can never be forgotten by me. Neither can I fail to remember the intelligence and eagerness with which he threw himself into lines of thought foreign from his own specialities, always seeking for the good and the true in whatever presented itself to his cognition. If my recollections of the life at St. Andrews are thus edifying, what shall I say of that, to me never to be forgotten, month I spent with him at Hyères, when his illness was so developed that his return to England was as unexpected as it was gratifying? To that interesting spot—the birthplace of Massillon, the port at which St. Louis landed from the Crusade—I was summoned by telegram from Paris on February 11th, 1868. I found the whole family in a state of great distress—the eldest daughter most dangerously ill, lying in the next chamber to her father—and the rest of the party worn out with continued watching. The brilliant sunshine, the blooming almond, the graceful palm-trees, ministered no comfort to the sick man. He was a close prisoner to his bed, and weary nights succeeded to exhausting days with a perfect monotony: and yet all the time the spirits fainted not. His delight was to be read to. The works of Walter Scott were a never-failing resource. He took the keenest interest in the fortunes of the Presidency of the Royal Society of Edinburgh, which he reluctantly declined. Nor was his soul neglected: "Now minister to me," he would say after a conversation on the things of the world, and then he would have prayers read to him, and the Psalms in the beautiful rhythmical Prayer-book version. Night after night it was so; but the most touching scene of all was, when the door was opened between the two sick rooms, a temporary altar fitted up, and the Divine mysteries celebrated. The rites were indeed maimed, the invalids too weak to bear the entire service, but they who knelt at that lowly table will never forget the devotion of those who communicated these blessed days. Three are now withdrawn within the veil, and we who remain, retain within our hearts the never-fading memory of these hours, revived in all its freshness, whenever in holy church we bless God's holy name for all His servants departed this life in His faith and fear.'

The outward work which Principal Forbes did in the world has, it is hoped, been sufficiently set forth in the foregoing pages and those which are to follow. His character, as that appeared to men, if it does not shine out in the narrative and letters already given, would not be made clearer by any condensed estimate of it. None such shall here be attempted. It is probably true of most men, but of Principal Forbes it was more than usually true, that his moral nature and his mental were entirely of a piece. The basis of both was thoroughness: it did not seem possible for him to do anything slightly or carelessly, or by halves. Akin to this was his definiteness and exactness of thought, which, as has been said, is a special form of the love of truth. Springing from this fundamental root were three qualities which especially distinguished him—method, perseverance, and conscientiousness. In the manner in which he thought out any subject, in which he carried on any work, in which he wrote, arranged his materials, and preserved them afterwards—in short, in the orderliness of his whole life—method was so conspicuous that it could no farther go.

Then whatever may have been his natural endowments, their power was doubled by his perseverance. As long as health and strength lasted, his tenacity of purpose was carried to an extent that must have severely taxed his bodily frame; and when health failed, and the body could no longer second the mind's requirements, this power of active perseverance was turned into its passive form of silent endurance.

These same qualities, seen on their moral side, became conscientiousness so rigorous, and carried into such small details, as to seem to some over-anxious and excessive. So straight did he go to what he believed was the right—in the pursuit of this so entire was his disregard of consequences—so little did he shrink from opposition, keenly though he felt it, that, as Bishop Forbes has said, this amounted in him to a very chivalry of duty-doing. To those who met him only in work and business he seemed a man too much on the stretch, too intent on duty ever to relax into a more playful or humorous mood. But on those within the more intimate circle of friendship or family life, he turned another and tenderer side not apparent to casual observers. When the strain of business was removed, the deeper affections had time to unfold; and these in him partook of that intensity which belonged to his whole being. Indeed, during all his later years, when Alpine work was over, it was in these affections that he found a shelter which more than compensated for the loss of the ambitions which stirred his youth. To afford a glimpse into his heart's secrets, a few extracts from his journals have been given, but not more than were necessary to a faithful representation.

From these extracts it will be seen that he was a devout man before he became scientific, and his scientific habits of thought never seemed to have disturbed or cast the shadow of a doubt over his faith. Religion and science in him were so far independent that he never troubled himself to build bridges of reconciliation between them, such as are so common now-a-days. Still less did it ever occur to him to import into religion those physical modes of thought now so much in vogue. I think I can see the look of grave contempt with which he would have regarded any one who proposed to trace the origin of all things to the working of a blind, soulless force, rather than to the ordering of a living and All-wise Mind. With Faraday he believed that no man by reasoning or science can find out God; and if any tried to do so, they would have got little help from Forbes. But in him they would have seen one well acquainted with all that modern science has really made good, and with all the methods by which it works: one who had himself added real contributions to scientific discovery, and who looked forward hopefully to far greater things yet to be discovered, but who yet held by the religion of the Bible and the great Christian verities as the anchor of his hope: and this in no dim struggling way, but with the most practical and ennobling faith. The outward form in which these truths came home to him was that in which they are presented by the Church of England Liturgy. This he was content unhesitatingly to accept. He never sought to rationalize it, and had small sympathy with those who did.

About his religious convictions he never argued, but acted on them silently and consistently. In this way, while his own mind was not disquieted, others took note of him, and received more good from the silent transpiration of his character than they could have got from any arguments. There were those associated with him at different periods of his life, who owed to his strong but unobtrusive faith the deepest and most lasting effects on their own character; and that faith which had guided himself, and influenced others, in the days of his activity, did not fail when the great trials of life fell upon him. During the long shadows of declining life, as earth with its interests receded, the things unseen came closer to him, and continued to be his strength and consolation to the end.

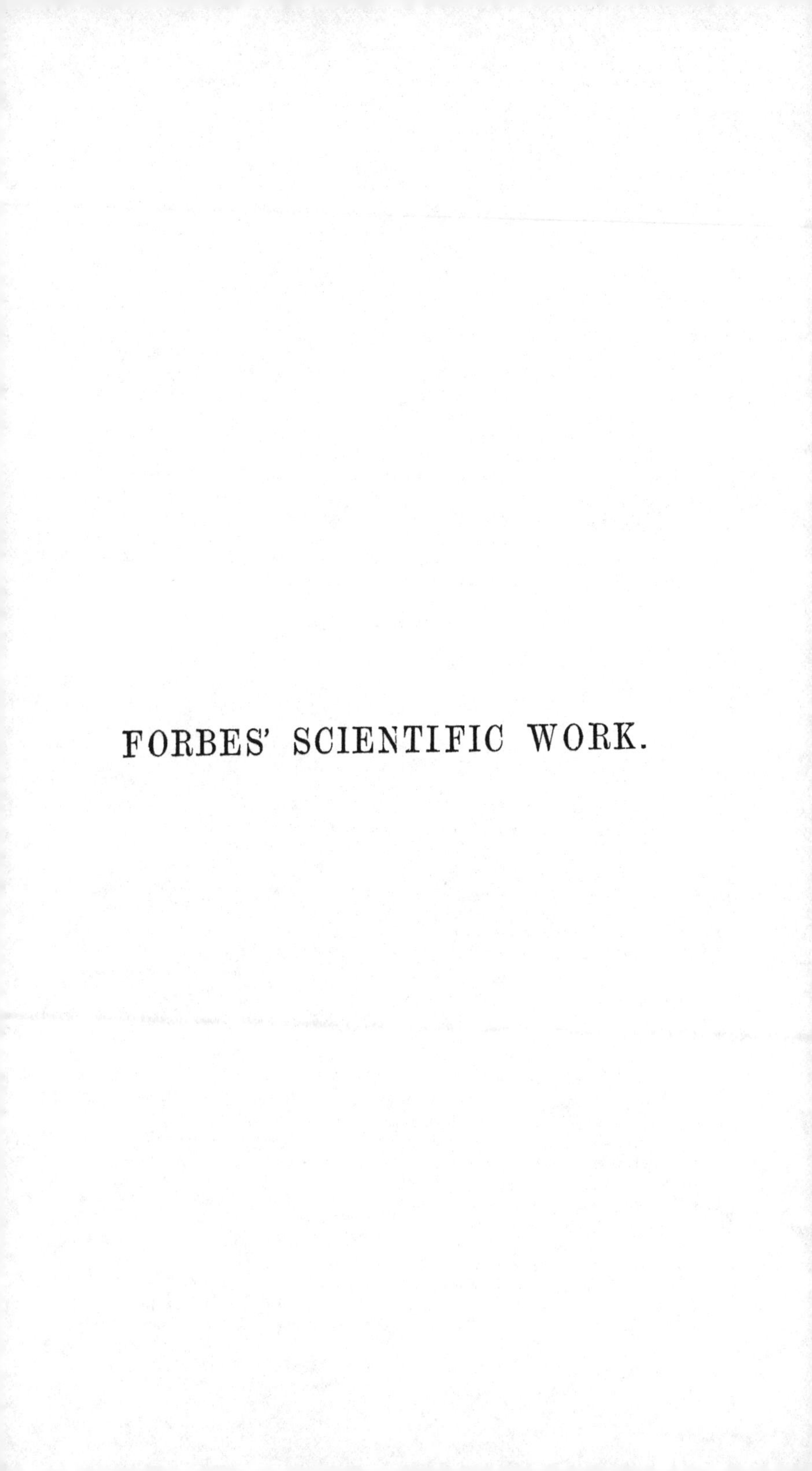

FORBES' SCIENTIFIC WORK.

CHAPTER XIV.[1]

FORBES' SCIENTIFIC WORK.

It is not easy, in a chapter necessarily short, and as little technical as possible (because this work is throughout intended for the general reader), to notice, with sufficient exposition of their object and of the advances made in them, even the more valuable of Forbes' scientific researches. The titles of all his published papers will be found in an Appendix to this volume, and we must refer to the preceding chapters for some general account of them, especially the earlier, and those which treat of questions connected with geology, mineralogy, and meteorology, with which we cannot attempt to deal.

In addition to his M.SS. *Lectures*, in which the whole range of Natural Philosophy was gone over—lectures fully written out, and scrupulously kept up to each advance in science, but which he unfortunately ordered his executors to destroy—Forbes distinguished himself as regards the general subject by writing a very remarkable *Dissertation on the Progress of Mathematical and Physical Science, principally from* 1775 *to* 1850. This was published in the last edition of the *Encyclopædia Britannica*. Such an article is necessarily of a historical character, and could be attempted only by a man of very extensive reading, as well as of great knowledge of the subject. But it also requires for its successful

[1] In part reprinted from the *Scotsman*, January 6th, 1869. The remainder of this, and the whole of the following, chapter, were written in January 1872.—P. G. T.

execution a calm, clear, and perfectly unbiassed judgment. In this respect, no man could have been found more excellently qualified than Forbes. No man was ever more strictly conscientious, or more thoroughly dispassionate; as is amply proved by the Dissertation itself, in which he had to deal with several very delicate matters—such as, for instance, the discovery of Neptune. In this work the author himself is one of the few to whose claims substantial justice has not been done. He detested controversy and, though (as is usually the case with peaceful men) often forced to take part in it, did so solely for truth's sake, never for the exaltation of himself or for the humiliation of his adversary.

Perhaps the most prominent of Forbes' investigations is that long series devoted to the nature and motion of *Glaciers*, which he pursued with intense application, and to the serious, possibly the lasting, injury of his health. The result was an account, of the structure and nature of the motion of these marvellous ice-masses, so full and luminous that very little indeed has since been added to it. While preserving what was true in the mass of statements made by his predecessors, he replaced an immense amount of nonsense by many new facts of the very highest order of importance. Critics have attempted to show that he had been anticipated in some of these; but they have produced only vague statements, probably meant in a totally different sense. Considerable confusion seems still to obscure the common apprehension of what Forbes really did. His careful observations led him to the general proof that a glacier moves like a viscous or plastic mass, though ice is usually regarded as a very brittle solid. Here no physical explanation is involved. In virtue of what property the ice behaves in this way is quite another question, and one the answer to which is even yet in some parts obscure, though there can be no doubt that a very important contribution, at least, is furnished by the beautiful discovery of James Thomson regarding the lowering of the freezing-point by pressure. It is not very easy to discover which theory appeared from time to time to Forbes to be the correct one, and it seems to us that a good deal of the hostile criticism which has been lavished on his works is due to the erroneous supposition that in his earlier papers he attempted to propound a physical explanation ot glacier motion. Another cause seems to lie in a mere verbal quibble about the meanings of viscosity, plasticity, and other similar terms. But the description of the ribboned, or veined, structure, of the shearing motion throughout the mass, and various other important points which must have been seen, though not observed, by thousands before Forbes, forms one most important part of his claim in this matter. If mankind had till now neglected to notice the annual rings of growth in trees, would not the naturalist who first should call attention to their existence and importance deserve immense credit? The veined structure of a glacier is as profoundly and intimately connected with its formation and motions as are these successive layers with the growth of a tree. There are several very singular peculiarities of this veined structure and its relations of position to cracks and crevasses, which appeared, especially as shown by Hopkins, to be inconsistent with the well-known dynamics of stresses. A very recent observation of Sir W. Thomson seems to promise an explanation of this difficulty; for he has shown that, in a viscous solid, continued shearing, parallel to one set of planes, gives rise to rupture along one of these planes. As the subject of glaciers admits of being more easily popularized than any of the other more important scientific work of Forbes, and as it is not difficult to select from his writings enough of simple but comprehensive passages to enable us to let him tell the story of his own doubts, of his determination to solve them by direct measurements, and finally of his unequivocal success, we shall devote to it another chapter, instead of unduly expanding in favour of one particular topic the present general sketch.

The first experimental work of real novelty in which Forbes successfully engaged was with reference to the

Polarization of Heat, the excessively modest account of which in the Dissertation above mentioned forms one grand defect visible to all who are acquainted with the immense value of Forbes' discoveries, and with the extraordinary difficulties under which, although he had the very best instruments of the time, his experiments were necessarily conducted. Yet, even from his own modest statement, the reader may see that something of importance was discovered. He says: 'I have just referred to 'my own early experiments on the subject (which were 'likewise inconclusive), in order to explain that it was 'natural, on hearing of the application of the thermo-'multiplier to measure radiant heat, that I should wish 'to repeat them with the new instrument. This I did 'in 1834. I first succeeded in proving the polarization 'of heat by tourmaline, which Melloni had announced 'did not take place; next, by transmission through a 'bundle of very thin mica plates, inclined to the trans-'mitted ray; and afterwards by reflection from the mul-'tiplied surfaces of a pile of thin mica plates placed at 'the polarizing angle. I next succeeded in showing that 'polarized heat is subject to the same modifications 'which doubly-refracting crystallized bodies impress upon 'light, by suffering a beam of heat, even when quite 'obscure, after being polarized by transmission, to pass 'through a depolarizing plate of mica, the heat traversing 'a second mica bundle before it was received on the pile. 'As the plate of mica used for depolarization was made 'to rotate, in its own plane, the amount of heat shown 'by the galvanometer was found to fluctuate just as the 'amount of light received by the eye under similar cir-'cumstances would have done. This experiment, which, 'with the others just mentioned, was soon repeated and 'confirmed by other observers, still remains the only 'one proving the double refraction of heat unaccompanied 'by light; and though somewhat indirect, it will hardly 'be regarded by competent judges as otherwise than 'conclusive. Iceland spar and other doubly refracting 'substances absorb invisible heat too rapidly to be used 'for effecting *directly* the separation of the rays, which 'requires a very considerable thickness of the crystal. 'I also succeeded in repeating Fresnel's experiment of 'producing circular polarization by two internal reflec-'tions. The substance used was, of course, rock-salt.' Taken in conjunction with the experiments of Melloni and others on the absorption, &c., of radiant heat, this splendid series of researches formed the conclusive proof of the *identity* of thermal and luminous radiations—a fact of the very greatest consequence to the further progress of one of the most fascinating branches of physical science.

The discovery of the polarization of heat will certainly form an epoch in the history of Natural Philosophy.

Most of Forbes' experiments are now easy to repeat, even on a large scale, as class illustrations, so much have galvanometers been improved since dynamical notions of construction have been introduced by Sir W. Thomson. The astatic pair of stout, long, needles common, till very lately, to all the better instruments, has within a few years been rapidly disappearing, and the repetition of Forbes' experiments need now give no one any trouble; but the discoveries he made are not the less meritorious on that account—any more than Faraday's grand discovery of magneto-electricity by a feeble motion of a delicate galvanometer is rendered of less account by the tremendous currents developed of late in the machines of Wilde and others. Rather let us be ashamed that, with the more perfect appliances at our command, we have not added to Forbes' results others of equal or greater value, certainly *now* within our reach, if we knew where to look for them, and destined, perhaps, to be the class experiments of the next generation.

Of great importance also are Forbes' researches on the *Conduction of Heat* by iron bars, and his quantitative measurements of the conductivity of iron at various temperatures. He had called attention, as early as 1833, to the fact that the order of metals as regards thermal conductivity is the same as that for electric conductivity; and one grand result of his new inquiries was to show that, in iron at least, *increase of temperature diminishes the thermal conductivity* as it had long been known to diminish the electric conductivity ; and, curiously enough, in something approaching to the same proportion. [M. Langberg, of Christiania, writing to Forbes in November 1853, says: 'Some days since I got the Report 'of the British Association for 1852, from which I see 'that you have experimentally proved the important fact 'that the conductivity of iron for heat diminishes as the 'temperature increases. I have felt much interest in 'reading this statement, as I think I have myself shown '(*Pogg. Annalen*, v. 66, p. 1, 1845), that the conductivity 'of tin, lead, and steel, is a function of the temperature; 'but the nature of my experiments did not allow me to 'find if it increased or diminished with the tempe'rature.'] The method employed by Forbes, which was entirely devised by himself, was a singular improvement upon that adopted by Biot and others, who, having assumed the conductivity to be unaltered by temperature, and having adopted for all temperatures an empirical law of cooling by radiation and convection, deduced by Fourier's methods the corresponding stationary distribution of temperature ; and then, by the method of least squares, endeavoured to make their observations agree with it. Forbes deduced, by separate experiments, the rate of cooling of his bar for all temperatures afterwards to be observed. He also carefully measured the distribution of temperatures in the stationary experiment, and used Fourier's method only for elementary portions of the bar. His results are of great value, and accord well with those since obtained by Ångström by a totally different, but also very ingenious, process.

Another result of the same investigations, and one of great interest and importance in modern science, is his determination—the earliest of much real value—of the *absolute conductivity* of a metal ; *i.e.*, how much heat passes per second per unit of surface through an iron plate of given thickness, whose faces are maintained at constant given temperatures. As a proof of the value attached by scientific men to these ingenious experiments, it is only necessary to mention that the British Association has given a grant for their repetition with the best attainable instrumental means, and for their extension to other substances than that to which Forbes was obliged by failing health to confine himself.

There is one point of Forbes' observations in this matter which is so exceedingly curious and unlooked-for, that in spite of Forbes' scrupulous accuracy and experimental skill, we must hesitate fully to accept it until it is confirmed by other observers: especially as he may have been led to it by the imperfection of his thermometers, or the exceeding difficulty of properly conducting this part of the experiment at very high temperatures. The result in question is that, while (as has long been known) the rate of cooling of a body by simultaneous radiation and convection increases at first faster and faster as its excess of temperature over that of the air increases, this is true only for a range of 150° C., or so, after which the rate of increase becomes slower and slower as the excess increases. In other words, if a curve be drawn whose abscissæ are the excesses of temperature, and its ordinates the corresponding rates of cooling, this curve will be concave upwards till about 150° C., and will there have a point of contrary flexure.

Another extremely valuable investigation of Forbes is that of *Underground Temperatures.* The principle on which to proceed had, however, been here clearly pointed out by Fourier; and had been made use of by various observers in France and Belgium, and even by Sir John Leslie in this country. The value of Forbes' results is, nevertheless, very great. The five-year series of observations on the sets of thermometers buried in three totally different surface-materials near Edinburgh gave, by comparison, very important information; and, separately, enabled him to give, we believe for the first time, a satisfactory *absolute* determination of the thermal conductivity of three different materials. In his reduction of these experiments, Forbes unfortunately followed an imperfect and inconvenient rule given by Poisson. Sir W. Thomson has, however, applied the correct analysis, and finds results not very different from those obtained by Forbes. Another important feature in this inquiry is that the series of thermometers buried in the trap-rock of the Calton Hill, within the grounds of the Edinburgh Observatory, are still (with the exception of the twelve-foot one, which was broken by the intense cold of the winter of 1860-1) read once a week, and have recently furnished materials for a remarkable paper by the Astronomer Royal for Scotland.[1]

We may next mention Forbes' elaborate paper, *On the Transparency of the Atmosphere, and the Law of Extinction of the Solar Rays in passing through it,* based chiefly on a valuable series of experiments made in the Alps by means of the pyrheliometer ; which had the distinction of being made the Bakerian Lecture for 1842.

In a paper entitled *Hints towards a Classification of Colours,* read before the Royal Society of Edinburgh December 4, 1848, and January 15, 1849, Forbes called attention to the importance of a method of defining colours with precision both for scientific and for artistic purposes. In this paper he adopted from Lambert and Mayer not only their arrangement of colours in a pyramid or a triangle, but their choice of the colours which are to be regarded as primary, namely red, yellow, and blue. He afterwards attempted to form a permanent diagram of colours selected from the collection of artificial enamels employed in the Vatican fabric of mosaic pictures, by comparing these enamels with the tints formed by the mixture of the primaries on a rapidly revolving disk. He found, however, on attempting to form a neutral gray, by the combination of red, blue, and yellow, that the resulting tint could not be rendered neutral by any combination of these colours ; and the reason was found to

[1] Edinb. Observations, 1860–70. *Proc. R. S.*, 1871.

be that blue and yellow do not make green, but a pinkish tint, when neither prevails in the combination. It was plain that no addition of *red* to this could produce a neutral tint. The fact that green cannot be formed by a mixture of blue and yellow was pointed out by E. C. Wünch[1] and by Young,[2] but the contrary was still believed by the highest optical authorities. The reason why mixtures of blue and yellow pigments are often green was soon after explained by Helmholtz; and Clerk-Maxwell, who was one of Forbes' pupils and witnessed his experiments, was led by them to make experiments on the mixtures of the colours of the solar spectrum, which showed that a yellow, equal to that of the spectrum, can be produced by the mixture of green and red light.

Of some other physical investigations by Forbes, such as those upon the *Trevelyan experiment*, and upon *Ampère's experiment* which exhibits the repulsion between elements of the same straight current, with various other questions connected with heat and electricity, we need not say much. There can be little doubt that Forbes did not so clearly apprehend the nature of the questions at issue as to be enabled to bring his unquestioned experimental skill properly to bear upon them. Consequently he was led to oversights, and, in one case at least, to serious mistakes. Some of this may be traced to a singularly erroneous idea, as to the nature of the relation between heat and electricity, which seems to have haunted him all his life, and occasionally to have led him to waste a great deal of time on experiments which physical science enables us to say could not possibly lead to anything in the direction in which he was seeking to go. He possessed, indeed, untiring energy coupled with rare sagacity and experimental skill, and everything he published may be fully relied on as a faithful description of what he observed. But it must be recollected that he

[1] 'Versuche und Beobachtungen über die Farben des Lichts' Leipzig, 1792.

[2] Lecture XXXVII.

was to a great extent prevented from fully appreciating such a theory, for instance, as that of Ampère, by the fact that he does not seem to have had any special talent for higher mathematics. It is greatly to his credit that, almost without assistance, he mastered sufficient mathematical knowledge to lay amply that solid foundation which is absolutely essential for the true scientific teaching of every branch of physics. But he did not wield the higher analysis with facility in original investigations. In fact, considering the natural bent of his mind to such studies as mineralogy, geology, and meteorology, it is wonderful to find him so very high in the list of original experimenters; and it would not have appeared at all strange had he been totally ignorant of all but the elements of mathematics.

Forbes' *Correspondence* (which his executors have placed in my hands) cannot fail to prove, some day, of very great importance to the future historian of the progress of physical science, alike in this country and abroad, during the last forty years. Every letter he received, whether from a great master or from a mere amateur observer or critic, he carefully docketed and preserved. Every letter he wrote was transferred by copying ink into his letter-book. Sufficient extracts from his own letters have been given in the preceding chapters. Among his correspondents, not to mention any still alive, we find such names as Brewster, Cauchy, Ellis, Faraday, Herschel, Melloni, Murchison, Powell, Rendu, Whewell, &c. &c.; and several of them give detailed remarks upon subjects at which he was working, or which happened to form the novelty of the day. Unfortunately, so far as this collection relates to great parts of Forbes' own work, it is yet too soon to try to avail ourselves of it, except in a very incomplete manner. The reasons will be obvious to those who think of the numerous controversies into which he was dragged. For, though many extracts, and even entire letters, may be selected which are free from controversy, yet in general these would give but an imperfect notion of the import of the whole. Others, again, cannot be published at present because the writers supply him with details of that mysterious wire-pulling which seems to be inseparable from every transaction involving honours (scientific, in common with all others, it is humiliating to have to confess). The value of this unique series is, however, so great, and its preservation so complete, that it is to be hoped that it may be safely deposited (under seal) in the care of some scientific society or institution, to be opened only when all the actors have passed from the scene. The hasty glance which I have bestowed upon these less scientific letters has shown me that Forbes attached great importance to mere honorary distinctions, as well as to the opinions of others regarding the value of his discoveries. It has opened up a view of a to me totally unexpected feature of his character, and enables me to understand many of his remarks which had previously appeared either meaningless or inconsistent. But, throughout the whole of the collection, the fearless honesty of the man shines out conspicuous. And his correspondence also shows that, though he was exact to a halfpenny in every pecuniary transaction, he could on fitting occasion be almost lavish in bestowing money upon struggling worth. It must be left to another generation, however, to decide whether the details of all this shall be made public.

In 1849 Forbes sent to the editors of the *Philosophical Magazine* the following short letter :—

'*On the alleged Evidence for a Physical Connection between Stars forming Binary or Multiple Groups, arising from their Proximity alone.*

'GENTLEMEN, 'MELROSE, *July* 13, 1849.

'In conformity with usage and with the spirit of your journal, I may perhaps be permitted to suggest a doubt as to the legitimacy of certain reasonings with respect to the evidence for the *physical connection* of binary or multiple stars arising from the mere fact of their juxtaposition, as stated and applied by some of the most eminent writers on sidereal astronomy. I should probably have hesitated to oppose my solitary opinion to that entertained by the eminent writers whom I am about to quote, had I not found it to be entirely supported by the eminent authority of two friends to whom I separately proposed it.

'Nearly a century ago Mitchell computed the chances to be 500,000 to 1 against the stars composing the group of the Pleiades being *fortuitously* concentrated within the small apparent space which they occupy; and he thence infers the probability of a physical connection between them. Struve has pushed this consideration much further. In his classification of double stars he has applied the same argument to estimate the improbability of the occurrence of even *single pairs* of stars in close proximity. He "calculates the odds at 9,570 to 1 against any two stars from the first to the seventh magnitude inclusive, falling (if fortuitously scattered) within 4″ of each other. Now the number of such binary combinations actually observed at the date of this calculation was already 91, and many more have been added to the list. Again, he calculates that the odds against any two stars fortuitously scattered, falling within 32″ of a third, so as to constitute a triple star, is not less than 173,524 to 1. Now, four such combinations occur in the heavens." Sir John Herschel, from whose *Outlines of Astronomy* I take this statement of Struve's results, adds, "the conclusion of a physical connection of some kind or other is therefore unavoidable."[1]

'Now, I confess my inability to attach any idea to what would be the distribution of stars or of anything else, if "fortuitously scattered;" much more must I regard, with doubt and hesitation, an attempt to assign a numerical value to the antecedent probability of any given arrangement or grouping whatever. An equable spacing of the stars over the sky would seem to me to be far more inconsistent with a total absence of law or

[1] '*Outlines of Astronomy*, p. 564. If I recollect aright, the passage does not occur in the edition in Lardner's *Cyclopædia*.'

principle than the existence of spaces of comparative condensation, including binary or more numerous groups, as well as of regions of great paucity of stars. Thus, to take a familiar instance: No bad representation of stars and their distribution may be made by sparking viscid white paint from a coarse brush upon a dark ground. It is impossible to conceive a nearer approach to a "random scattering." But I am assured by an ingenious friend, who has used this contrivance in aid of pictorial effect, that such an artificial galaxy will present every variety of grouping, with double and treble points innumerable (as I have indeed myself witnessed); nor can I well see how, upon any reasonable theory of chance, it should be otherwise.

'I wish to restrict this letter to the end proposed, that of nakedly setting forth a serious difficulty in an inferential interpretation of nature, sanctioned by high and also cumulative authority. I shall not therefore attempt now to enquire more minutely into the history of the error, if error it be, nor to insist on the great importance of arguing correctly in cases which admit of so very extensive application.

'I remain, Gentlemen,
'Yours faithfully,
'JAMES D. FORBES.'

In 1850 he communicated to the British Association a paper on the same subject, which was not immediately published. Meanwhile, in July 1850, appeared in the *Edinburgh Review* an article by Sir J. Herschel, who thus writes to Forbes on the subject:—

'MY DEAR SIR, 'COLLINGWOOD, *July* 27, 1850.

'I had hoped to have been able to have sent you a copy of an article in the last No. of the *Edinburgh Review*, in which an objection advanced by you against the argument from probability of a physical connection between the individuals of a double star, drawn from mere juxtaposition, is attempted to be answered. But I am told, on application, that the type is broken up, and that I cannot be supplied with any copies. All I can do therefore is to call your attention to the passage in the article in question (page 36, *Edinburgh Review*, No. 185), as, though you are not *named* as the author of the argument in question, it will no doubt be recollected by many readers as resting on your authority.

'Whether you may think the answer to your objection a valid one, or not, it seemed worth while at least to set in a clearer light than I believe had been heretofore done, the gist of the argument itself; and at all events I hope you will not find that the writer of the article in question has in any way distorted or misrepresented your meaning.

'Believe me, my dear Sir, yours very truly,

'J. F. W. HERSCHEL.'

The passage in the *Review* is as follows:—

'Astronomy affords us a very remarkable example of this nature, which we adduce, by reason of a singular misconception of the true incidence of the argument from probability which has prevailed in a quarter where we should least have expected to meet it. The scattering of the stars over the heavens, does it offer any indication of law? In particular, in the apparent proximity of the stars called "double" do we recognize the influence of any *tendency to proximity*, pointing to a cause exceptional to the abstract law of probability resulting from equality of chances *as respects the area occupied by each star?* To place this question in a clear light, let us suppose that, neglecting stars below the seventh magnitude, we have measured the distance of each from its nearest neighbour, and calculated the squares of the sines of half these distances, which therefore stand to each other in the relative proportion of the areas occupied exclusively by each star. Suppose we fix upon a circular space of 4″ in radius as the unit of superficial area, and that we arrange all the results so obtained in groups, progressively increasing from 0 by the constant difference of one such unit. Now, the fact to which M. Struve originally called attention, and on which we believe all astronomers are agreed, is, that the first of these groups *is out of all proportion richer than any of the others*, and that the numbers degrade in the groups adjacent with excessive rapidity; so that, for example, calculating on the numbers given by Struve, we find the first group to contain 180 cases; the next three 68, or on an average 22 each; the next twelve 70, or six each on an average; and the next forty-eight only 94 in all, averaging two to each, while a general average would assign only one star to 540,000 such units of area. The case, then, is parallel to that of a target of vast size, marked out into 6,700 millions of equidistant rings, riddled with shot marks in the bull's eye, and with a tolerable sprinkling in the first 50 or 60 rings, beyond which the whole area offers nothing for remark indicative of any particular local tendency, though *dotted all over with marks*, in the sparing manner above described. Anyone who could view such a target, bearing in mind what is said above, must feel convinced that a totally different system of aiming had been followed in planting the interior and exterior balls. Such we conceive to be the nature of the argument for a physical connection between the individuals of a double star prior to the direct observation of their orbital motion round each other. To us it appears conclusive; and if objected to on the ground that every attempt to assign a numerical value to the antecedent probability of any given arrangement or grouping of fortuitously scattered bodies must be doubtful, we reply that if this be admitted as an argument, there remains no possibility of applying the theory of probabilities to any registered facts whatever. We set out with a certain hypothesis as to the chances, granting which, we calculate the probability not of one certain definite arrangement, which is of no importance whatever, but of certain ratios being found to subsist between the cases in certain predicaments on an average of great numbers.

Interrogating Nature, we find these ratios contradicted by appeal to her facts, and we pronounce accordingly on the hypothesis. It may, perhaps, be urged that the scattering of the stars is *un fait accompli*, and that their actual distribution being just as possible as any other, can have no *à priori* improbability. In reply to this we point to our target, and ask whether the same reasoning does not apply equally to that case? When we reason on the result of a trial which in the nature of things cannot be repeated, we must agree to place ourselves in idea at an epoch antecedent to it. On the inspection of a given state of numbers we are called on to hold up our hands on the affirmative or negative side of the question, Bias or no bias? In this case who can hesitate?'

On receipt of this letter, and on reading the above extract, Forbes seems at once to have set about rewriting for the *Philosophical Magazine* his British Association paper; and for that purpose to have impressed into his service as many of his scientific friends as possible. Several of them furnished him with numerous letters on various aspects of the question—mathematical and metaphysical—some of them adding remarks not very complimentary to his critic. From these, which are all extremely interesting and characteristic, we make as many extracts as our space, and the nature of their contents, will permit. We choose, in preference, those which bear most directly on the nature and extent of Forbes' own knowledge of such subjects.

From PROF. KELLAND.

'*September 5th*, 1850.

'I will do my best to answer two of your questions.

'1st. Is not

$$1 - \frac{p(p-1)\ldots(p-n+1)}{p^n}$$

approximately represented by

$$1 - \frac{\left(p - \frac{n}{2}\right)^n}{p^n}$$

and why not?

'Take a simple case, $n = 4$.

$$1 - \frac{p(p-1)(p-2)(p-3)}{p^4} = \frac{6}{p} - \frac{11}{p^2} + \frac{6}{p^3}$$

But $$1 - \frac{(p-2)^4}{p^4} = \frac{8}{p} - \frac{24}{p^2} + \frac{32}{p^3} - \frac{16}{p^4}$$

a totally different thing; having (when p is very large) to the former the ratio of 4 to 3.

'And why is this? Simply because the large term goes out, and you take a false mean for the standing terms. If p is very large, and n very small,

$$1 - \frac{\left(p - \frac{n-1}{2}\right)^n}{p^n}$$

will be very nearly correct, but only when n is *very* small compared with p.

'2nd. You ask for the logarithms of e, and of 4254602 and 4254372. Now, I am not very skilled in computing logarithms to 12 places, and I am afraid of committing an error. But I will, instead, give you the process (which you can repeat, for I fear errors of multiplication and division) of finding the value of [your expression itself, which involves the ratio only of these large numbers—the calculation follows]. I hope you will find this correct. It is more laborious than either difficult or complicated. It is, in fact, a mere matter of multiplication and division.'

From SIR WILLIAM THOMSON.

'*August* 15*th*, 1850.

'. . . . In the first place, if we consider what we mean by a "physical connection" between two stars, we find that unless we are to speculate upon the cause of their being created near one another, or at least to speculate as the nebular theorists have done, it means merely the stars being near enough to sensibly influence one another's motions by their mutual attraction, and not having *vis viva* enough in their relative motions to cause at any subsequent time an enormously greater separation than they have at present. Hence, what we see of double stars, and of groups of stars, suggests two questions, one purely geometrical (to which I alluded in Section A), the other mechanical (which has only occurred to me definitely since the meeting of the Association).

'1. Are the stars of any individual group excessively near one another as compared with their distances from the earth?

'2. Are the *vis vivas* of different stars of a group estimated with reference to the centre of gravity of the group, so small that they will always be excessively near one another (unless disturbed by a body foreign to the group) as compared with their distances from the earth?

'The first question gives rise to some very curious geometrical considerations, which I think (so far as I know of what has been written) have been overlooked by those who have entered upon the probability calculations. Let us for a moment conceive that the individuals of a double star subtend an infinitely small angle when seen from the earth. Then, either the stars must be close to one another, or they must be in a line passing through the earth. Now, in general, any arrangement that may be made of points *in space of three dimensions* will be such that no two lines, each joining two of them, will intersect. In a particular case two of these lines may intersect, but there may be no other two which do so. It will be only in more and more special cases that there are more intersections than that of one pair of lines, or that there are three or more of the lines which have a common intersection. Question 1 then resolves itself into two:—Is there any position in the universe where n (the number of double stars usually reckoned) lines joining pairs of stars (of such and such brightness as seen from that position) intersect, or so nearly intersect that they all pass through a space of which the lineal dimensions are small compared with the distances and the difference of the distances of the two individuals of each pair from that position? and is the earth at present in that position? These two last questions would be excessively improbably answered in the affirmative if we merely knew that there are stars and an earth in space; and therefore the supposition that the individuals of many of the double stars are far from one another as compared with their distances from the earth is *à priori* excessively improbable. I scarcely think the calculations which have been made have any reference to the probability of an answer to either of these subsidiary questions or of Question 1, but I am not confident in this opinion. They all, so far as I know, relate to subjects of which the scattering of sand upon a surface, from a sieve, which you adduced, is an illustration, but I doubt whether the three dimensions of the geometrical question have been properly taken account of.

'Question 2 gives rise to some excessively curious considerations regarding the motion of a system of points acting on one another according to Newton's law. Will it not be only in special cases that any two will ever come to be extremely near one another as compared with their distances from the others, unless they are so initially? Again, it is certainly only in a special case that a pair, at any time far asunder, shall come to be very near, and that another pair shall come to be very near at the same time; and that numerous pairs may come to be very near, two and two, at the same time, is still more special. Hence it is excessively improbable that Question 2 could be answered otherwise than in the affirmative. Would it not be altogether futile to attempt to estimate numerically this improbability? . . .'

From Bishop Terrot.

'Dear Professor, 'Edinburgh, *October* 10*th*, 1850.

'. . . I was from home when Prof. Kelland left your paper dated August 30. I returned home and found it on October 5; this date I have written on the back. I will take care both of the letter and paper.

'Of the latter I have read a few pages respecting Mitchell. You hold as I do that he makes the atrocious blunder of saying that the antecedent improbability of an event is the consequent probability of its having a physical cause, *e.g.* that if the odds *against* throwing aces with two dice be 1 to 35, then (the aces having turned up) the odds *for* the dice being loaded or some equivalent cause is 35 to 1.

'But then this blunder is so gross that I can scarcely believe it could be swallowed by Herschel or any of the men of science who have endorsed Mitchell's argument. This leaves a sort of vague suspicion upon my mind that we do not do Mitchell justice.

'I am gratified by your quotation from Laplace as to *regular* distributions—for I had written down something to the same purpose, with the addition that it is *regularity* which induces us to believe in design, and that as there is no unit of perfect regularity there can be no definite expression of imperfect regularity by means of such fractions as Mitchell gives.

'I am, my dear Sir, yours faithfully,

'C. H. TERROT, Bp.'

From SIR GEORGE AIRY.

'ROYAL OBSERVATORY, GREENWICH, *October* 11*th*, 1850.

'. . . "Article 23, beginning." This is the only place in which you have adverted to the *long run of many trials* as entering into a chance-problem: and I think the want of more energetic reference to it is a defect in your paper. I think that that consideration is the foundation of all calculations of probabilities.

'Article 28: "An axiom which cannot be admitted as certainly true." I think in discussing Mitchell on chances you *must* admit it as certainly true. It is the mathematical expression of chance. The denying it amounts simply to saying that there cannot be any chance, in a shape for mathematical consideration, in the whole matter; and it might as well be said in so many words.

'Article 30: "Tending ultimately to uniformity of

effect over given areas." I think you are quite wrong as below.

'Article 31 : "It is in fact convertible with this. Every portion, &c. will include as many stars as any other." Negatur. I think you are here most absolutely wrong. All that is meant by being as likely to be in one situation as another is this, that if the experiment of creating a starry firmament were repeated a great many times, and if you took the mean of the number of stars found, at all the times, in each particular square degree, the greater the number of such creations, the more nearly would the mean contents of each square degree approach to uniformity.

'This affects several things which follow. If I might presume to indicate what I think a radical fault in your paper, it is this misunderstanding of the doctrine of chances. Pray look to it well.

'Note A.—I believe (but will not assert) that there is an error in Mitchell, as you have remarked : if so, it also affects the investigation which I sent you.

'And so end my comments. . . .'

'ROYAL OBSERVATORY, GREENWICH, *October* 26*th*, 1850.

'. . . I have had no further opportunity of writing, and do not see the least prospect of leisure sufficient for calm consideration of Struve and Mitchell. Therefore, do not reckon on that. Meantime I send you my remarks applying to the three things which you say a Mitchellite must believe, and you can judge whether I am Mitchellite or no.

'1. "That there is a definite probability against two stars being within a certain distance."

'Certainly there is : errors in calculation always excepted. But this is to be taken in the sense which I have before explained, that if it was possible to have a great number of starry firmaments created, the average number of the double stars would be determined nearly by the law asserted.

'But if, after finding that the chance is 10,000 to 1

against a double star, anybody infers that the existence of a double star leads to this: "It is a probability of 10,000 to 1 that the existence of that double star is due to a cause unrecognized in our chance-calculations,"—I say that such person is under a mistake. There is no logical connection between the two. Possibly it would be so on the average of the great number of firmaments as aforesaid, but on this I am not able at this moment to speak.

'I think that the existence of that double star gives a small probability of law, but I am not in a state to give an estimate of the probability.

'2. "That there is a mathematical certainty of cause or design if two stars and the spectator are in the same line."

'Certainly not. The chance of it is, as we say, infinitely small (a bad word), which means that if a million of money depended on finding the thing, the chance when you are turned adrift in a new cosmical system is not worth the millionth part of a farthing, but it is as possible as ever, without special cause or design.

'There are, in this neighbourhood, four conspicuous objects sensibly in a straight line (three church-spires and the centre of a conspicuous clump of trees or "spinney"), which I will walk you to see on the next clear day that you are here: everybody who sees it remarks, "How curious," but nobody insinuates either that they were so set on purpose, or that it was impossible that they could have so come by the mere accident of four independent local selections. But I would offer a very large bet, I mean very large odds, that you will not find such a thing near Edinburgh, or even anywhere else near London.

'3. "That a uniform spacing of the stars is more probably the result of chance than any other observed order of distribution."

'Certainly not—as a consequence of any theory of probabilities. Surely Mitchell cannot have said anything so absurd. There is nothing in the result of probabilities that bears on it except in the sense that I have said, that you must take a great number of starry heavens, and take the average of all.

'The number of persons on the different parts of London Bridge at any moment, when there is no spectacle of any kind, is a pure matter of chance (*i.e.* depends on laws which we are not considering); yet, as I know well, they are most oddly grouped, without any sort of uniformity. But, taking the average of 100 different states in the same day, I apprehend that the mean would be a very uniform treading of the pavement. . . .'

From R. LESLIE ELLIS.

'ANGLESEA HOUSE, GREAT MALVERN,
Sept. 3rd, 1850.

'MY DEAR SIR,

'. . . I am glad to see by what you have sent me, for which pray accept my thanks, that you have begun to say something touching one of the popular applications of the theory of probabilities.

'The simple view of the matter seems to me to be, that whenever we perceive a relation among phenomena we are disposed to infer the existence of a relation among their causes, and that the strength of such an inference varies with an infinity of circumstances, and in particular with what may be called the *à priori* conceivability of the existence of the causal relation. If in the midst of the stream of people whom you meet in the Strand you see a group gathered together at a particular spot, you infer that they are interested about the same object, whatever that may be: the strength of the inference varying with the size of the group, the similarity of their appearance, &c.; and in this case you know that a common cause may very well have produced what you observe. To set this last condition aside, and to attempt to estimate the force of the inference by the theory of probabilities, is to leave out something which is essential, and to introduce something which is irrelevant to the matter in hand. Everything which exists is *à priori* infinitely improbable, as that this sheet of paper should be the precise size that it is of all the infinite variety of possible sizes, &c.

'Consequently the improbability that the actual angular distance between Aldebaran and Arcturus should be within certain limits may, if the limits be taken narrow enough, be made to exceed as much as we please the improbability that there should be 91 cases of stars lying within 4″ of each other. If, therefore, in the case of double stars we are bound to believe in physical connection *because* of the *à priori* improbability of what is observed to exist, I do not see how the wit of man can escape from the conclusion that we are bound also to believe in a physical connection between Arcturus and Aldebaran. If it be said that the cases are not parallel except *quoad* the numerical calculation, I answer that this is undoubtedly true; but then the remark admits that the numerical calculation is in itself no ground for inference in either case. And so, as they used to say in the schools, *cadit quæstio.*

'*Avec des chiffres on peut tout démontrer*, ought to be the motto of most of the philosophical applications of the theory of probabilities—which in its own nature and according to the plain view of it, is only a development of the theory of combinations. To attempt to constitute it into the philosophy of science, is, in effect, to destroy the philosophy of science altogether. . . .'

'MY DEAR FORBES, 'BELGRAVE PLACE, *October 8th.*

'. . . I am greatly pleased with your argument in general; it expresses what I have long thought. The foundation of all the confusion is the notion that the numerical expression of a chance expresses the force of expectation, whereas it only expresses the proportion of frequency with which such and such an event occurs on the long run. From this notion that chances express something mental or subjective, is derived the assumption that the force of belief touching past events admits of numerical evaluation as well as the force of expectation touching future. If this were true, it would be a legitimate inquiry to try to assign numerical values to the force of belief in any given cases. All this folly, for one cannot give it any other name, grows out of such statements as "certainty is equal to unity," and the like. It belongs to the school of Condillac and the sensationalists—they were in the ascendant when the theory of probabilities received its present form, and there has not yet been philosophy enough to expel it.

'I hope you will notice the exquisite *non sequitur* in Mitchell's argument. All that it can tend to prove is, that there is some reason why so many stars *appear* very near one another. Let this be granted, and that you have a numerical expression of the cogency of the conclusion. What connection is there between this and the conclusion that they *are* near—or that there is a physical connection, to use Herschel's phrase, among them? How if the observer is drunk? There will then be reason enough why he should see double stars, and a reason which applies to all the stars he looks at. Mitchell's conclusion, therefore, would be justified in the case of such an observer. Only the common reason here applying to all the stars is merely subjective and phenomenal. And to set aside this illustration, how are we in any case numerically to estimate the chance that the common reason really belongs to the essential nature of the question?' . . .

'BELGRAVE PLACE, BRIGHTON,
October 10*th* [1850].

'MY DEAR FORBES,

'. . . I send you the second sheet, on which I have less to remark than on the first, little as that was. I should rather have liked you to insist more on the point that such phrases as *mere chance*, *at random*, and the like have really no meaning at all except with reference to the knowledge of the observer and with his system of classifying phenomena. For everything which exists there is a definite reason why it is what it is and as it is; the only question being what analogy exists among the causes of analogous phenomena; in other words, what general propositions can be affirmed about them, or, which is again the same thing, what law or laws they fulfil. However the stars had been fixed in the visible heavens, each of them must have been fixed there by an adequate cause, fixing it just where it is. "Random scattering" is, to use Bacon's words, "nomen rei quæ non est, aut confusæ et male terminatæ." The question is, Do the phenomena suggest to us the idea that the causes which placed the stars as they are, are connected with certain regions of the sky rather than with others? If *e.g.* all the stars lay between the tropics, there could be no doubt as to the answer. As it is, there is a certain presumption from the observed lists of double stars. But in the first place this presumption does not admit of evaluation—no presumption does: and secondly, if it did, the evaluation could only relate to the phenomena, viz., to the *apparent* places of the luminous points we call stars. If we divide space into a multitude of cones whose vertices are in the observer's eye, there would, according to the probabilitarians, be a numerical expression of the presumption that certain of these cones favour the existence of stars rather than others. But while the earth was supposed the centre of the universe, people would have been contented to say that stars tended to lie in rays converging towards it. No one now would offer this explanation; but no one, on the other hand, gives a numerical estimate of its improbability. And yet it is only *because* we reject it that the supposed calculable presumption in favour of a tendency in the stars to apparent nearness leads us to any inference as to their real nearness. Nevertheless the transition from apparent nearness to real nearness is the very essence of the matter in dispute.

'Do not think that any of this or of my marginal notes is in the nature of contradiction or even criticism. More like the latter is a remark I will add, that the phrase "calculus of probabilities" seems objectionable (it occurs in the last part which reached me since I began to write): the word "calculus" in mathematics seems to express a mode of calculating—a method rather than a theory. One does not say the differential theory, but the differential calculus, and *e converso* I should rather say theory of probabilities than calculus. I like your phrase "the laws of random." . . .

'Yours very truly,

'R. L. ELLIS.'

To what extent these and other very numerous suggestions and criticisms were taken advantage of by Forbes in finally preparing his paper for publication we can only roughly guess, inasmuch as we are not in possession of a copy of the original proof submitted to the critics. Forbes himself says in a postscript that several modifications and additions had been made, and that he had profited by the kind criticisms of several friends to whom he submitted the argument. The paper as it stands, in the *Philosophical Magazine* for December 1850, is well worthy of careful perusal, but in no wise from a mathematical point of view. Experiments, indeed, performed by dropping rice-grains through a sieve upon a chess-board, and counting the number on each of the squares, are given in illustration; but, so far as we can see, the paper is much too diffuse, and one or two pregnant sentences from the letters we have quoted might here and there have been substituted with advantage for whole sections together. The conclusions, however, must be given in Forbes' own words:—

'(1.) The fundamental principle of Mitchell is erroneous; the probability expressed by it is an altogether different probability from what he asserts. His calculations are also apparently inaccurate, in some instances at least.[1]

'(2.) All the *numerical* deductions of his successors are equally baseless.

'(3.) Were Mitchell's principle just, a perfectly uniform and symmetrical disposition of the stars over the

[1] 'It would probably be more correct to say that all Mitchell's calculations are wrongly deduced from his own premises.'

sky would (if possible) be that which could alone afford no evidence of causation, or any interference with the laws of "random"—a result palpably absurd.

'(4.) Special collocations, whether (α) distinguished by their symmetry, or (β) distinguished by an excessive crowding together of stars, or the reverse, inevitably force on the reasoning mind a more or less vague impression of causation,—an impression necessarily vague, having nothing absolute, but depending on the previous knowledge and habits of thought of the individual, therefore incapable of being made the subject of exact (*i.e.* mathematical) reasoning.

'(5.) The form of error into which those have been led who had stated numerical probabilities against given arrangements of stars being the result of accident, is twofold. First, a confusion between the *expectation* of a given event in the mind of a person speculating about its occurrence, and an *inherent improbability* of an event happening in one particular way when there are many ways equally possible. Secondly, a too limited and arbitrary conception of the utterly vague premiss of stars being "scattered by mere chance, as it might happen"—a statement void of any condition whatever.'

He next recapitulates in a succinct form some of the preceding arguments, and then proceeds (without naming Herschel as the author in question) to show that 'an ingenious writer in the *Edinburgh Review*' has represented him as making averments foreign to his thoughts: and has overlooked the precise objections contained in his letter (*ante*, p. 467), and the precise statements there quoted from Herschel's Astronomy. 'The reviewer,' he says, 'charges me with a singular misconception of the true incidence of the argument from probability. I can only say that my error, if such it be, is one sanctioned by Sir John Herschel, to whom I referred as my authority.' . . .

'The necessity of such definite citation has become evident from the misapprehensions to which I have been exposed. If Mitchell's deductions had remained buried in the heavy quartos of the "Philosophical Transactions," I should not have thought that the refutation of them was so important.'

The extracts and quotations we have given will suffice to show that Forbes, from a mere common-sense point of view, though with quite unnecessary elaboration, hit upon a real blot in Mitchell's argument, and rightly denounced its revival in Sir John Herschel's justly celebrated text-book. But they also show that in dealing with the subject, he fell, at first at least, into mistakes quite as grave as those he was endeavouring to expose.

The following account of Forbes' geological work is quoted, by permission, from an excellent obituary notice by Professor Geikie, in the 'Proceedings of the Geological Society of Edinburgh.'

'Principal Forbes was born just twelve years after the death of the great Hutton, only seven years after the publication of the "Illustrations of the Huttonian Theory," and he was already a boy of ten when Playfair died. Many of his friends had been personally acquainted with these philosophers, and the memory of the fierce Plutonian and Neptunian war was still fresh in their minds when he began to give himself to scientific pursuits. These early influences are traceable all through his life. He was profoundly impressed with the originality and truth of the views propounded by Hutton, and illustrated by Playfair. In one of his earliest papers he refers to "the splendid Huttonian theory which Playfair undertook so ably to illustrate."[1] In one of his latest writings he speaks with enthusiasm of the "precious lessons" which one of his friends had drawn from the lips of Playfair and of Hall.[2] I shall never cease to remember with gratitude that it was he who introduced me, when a boy, to the writings of these masters. He used to speak of Playfair's "Illustrations of the Huttonian Theory," as one of the best books ever written upon the first principles of geological science.

[1] *Edin. Jour. Science*, New Series, i. p. 274.

[2] *Proc. Roy. Soc. Edinburgh*, v. 63.

'Forbes' first effort in scientific literature, so far as I have been able to discover, was written in his eighteenth year. It took the shape of a letter to the editor of the "Edinburgh Journal of Science (1828)," and was dated from Rome, with the title "Remarks on Mount Vesuvius." It formed, however, merely the prelude to a series of eight papers, entitled "Physical Notices of the Bay of Naples," which were sent at intervals to the same journal.

'This series of papers is one of singular merit, considered as the work of so young a man. He describes with considerable detail his personal explorations in the volcanic districts, digesting at the same time the published information on the subject, and presenting a clear narrative of the physical features of that interesting region. Of special excellence is the fifth paper of the series, in which he enters learnedly into the history of the Temple of Jupiter Serapis at Pozzuoli. After enumerating the different hypotheses which had been proposed in explanation of the remarkable geological features of the ruin, he adopts and enforces that which has since been universally acquiesced in, and has become familiar from the writings of Sir Charles Lyell. He shows very clearly that the evidence points to oscillation of the land with respect to the level of the sea, and that other proofs of the same fact are furnished by adjacent parts of the Mediterranean shores. In his concluding paper he offers a *résumé* of the information he had been able to gather relative to the formations of the volcanic Neapolitan district.

'No one can read these, the earliest productions of the late Principal, without recognizing in them evidence of that scrupulous carefulness and caution which distinguished their writer from first to last. They show, too, the pains which he always took to make himself thoroughly acquainted with the literature of a subject before venturing to write upon it himself. Nor was his reading confined to his own language; it extended to the ancient and modern tongues in which the subjects he happened to be studying were discussed, or through which collateral information might be gathered.

'Another of his early contributions to geology is in the form of a short letter to Professor Jameson, on the occurrence of a large greenstone boulder in the Pentland Hills.[1] It is dated from Colinton House, 3d August, 1829, when its writer was a little over twenty years of age. It gives an account of the position of the boulder, its composition, dimensions, and specific gravity. But the chief interest it possesses lies in the broad generalization which the young observer drew from the facts he had so carefully noted. The boulder lay upon the side of a small, steep ravine, and its position there was such as to lead him to regard the induction as undeniable, "that the excavation of the valley must have taken place subsequently to the deposition of this boulder." He remarks further, that this inference as to the lateness of the erosion of valleys is forced upon us by many other instances, which intimate the gradual degradation of the soil. Those who have watched the progress of geological discussion in recent years will see at how early a period our departed friend had acquired clear views upon this subject, and had based them upon the results of actual observation. This early paper is further interesting, inasmuch as it serves to indicate the special field of geology into which Forbes' natural instincts turned him, and in which he was destined in later years to reap so abundant a harvest. He had often read, and treasured in his memory, the eloquent passages in which Playfair, following in the path of Hutton, had expounded the erosion of valleys, and the universal decay and waste of the continents. He saw that the happy suggestions and sagacious inferences of these philosophers ought to be regarded in the light rather of an outline of what remained to be discovered, than as the epitome of a completed philosophy. Whatever related to the forces which work upon the surface of the earth and effect geological changes had a special charm for him. It was this tendency which led

[1] *Edin. New Phil. Jour.*, vol. vii. p. 259 (1829).

him to wander, with more than a tourist's curiosity, among the glaciers of Switzerland; which first suggested to him the idea of working out, by accurate observation, the real cause of glacier-motion, still, in his opinion, undiscovered, and which brought him back, year after year, to these great mountains, where he toiled with a devotion that told at last upon his physical frame.

'The lessons which he had laboriously learnt among the living ice-rivers of the Alps bore fruit when he came again to wander among the more mountainous regions of his own country. In the year 1840 Agassiz had made the startling announcement that the British Islands had once been deeply buried under a vast mantle of snow and ice, and that the traces of its seaward motion were yet fresh and clear upon the sides of the countless valleys among the uplands. Following up the observations of the Swiss naturalist, Buckland and Lyell had pointed out the former existence of glaciers in the Highlands and other parts of the country. When, however, we look back upon the early discussion of this subject, we are forced to admit that conclusions were often based upon very hasty and imperfect observation. In particular, glacier-moraines were often recognized in places where no geologist would now be able to find them. Much as Forbes knew of the geological effects of ice, his natural caution kept him from taking part in this discussion for a time, until he was able to produce more accurately determined data than had, in many cases at least, been available. In the year 1845 he visited the Isle of Skye, and his eye, already trained to recognize the traces of vanished glaciers in Switzerland, was at once struck by the identity of the forms assumed by the rocks at Loch Scavaig with the *roches moutonnées* of the Alps. Further investigation led him to obtain complete demonstration of the former presence of a group of glaciers descending from the rugged scarps of the Cuchullin Hills. He walked over mountain and glen, filling in a rough sketch-map of the glacier valleys as he went along, and in December of the same year he read a narrative of his observations to the Royal Society of Edinburgh. This was the most detailed and satisfactory account which had yet been given of the proofs that the Highlands of Britain once nourished groups of glaciers.[1]

'In the year 1851 Professor Forbes undertook a journey to Norway, partly to make observations of the great solar eclipse, and partly drawn by his love of physical geography, and notably of glaciers. It was his design to compare the phenomena of glaciers in Northern Europe with those already so familiar in Switzerland. This he has done in a masterly way. His pages contain, in a clear and succinct form, the sum of all that was known at the time regarding the snow-line and the existing glaciers of Norway. I have myself gone over much of the ground he has described, and can bear witness to the accuracy of his sketches, alike of pencil and of pen. His two chapters on the physical geography of Norway have always appeared to me to be a masterpiece of careful yet rapid observation, broad generalization, and clear description.

'But though the tendency of his researches in geology was mainly towards the investigation of the phenomena connected with changes in the outline of the surface, he did not neglect the study of minerals and rocks, in which he had been trained under Jameson. Previous to 1836, with the view of learning more of the history of ancient geological upheavals, he had examined "the trap-rocks of our own island, the ophites of the Pyrenees, and the serpentines of Anglesea and the Lizard; the porphyries of Northern Italy, the granite veins of Mounts Bay and Glen Tilt; the ancient volcanoes of Auvergne, the Eifel, the Siebengebirge, and of Rome; and the modern volcano of Vesuvius."[2] In December 1835 he gave to the Royal Society of Edinburgh a narrative of his researches in Central France, dwelling more especially on the analogies between the volcanic rocks of that district and the trappean masses of his own country. Throughout his narratives of foreign travel, also, we everywhere meet

[1] *Edin. New Phil. Jour.*, vol. xl. p. 76. [2] Ibid., vol. xxi. p. 2.

with indications that, though busied with what had become his own more special branch of the science, he remained no indifferent observer of the rocks among which his journeys led him. He retained his fondness for mineralogy up to the end. When I last saw him at St. Andrews, he showed me a collection of veined agates which he had accumulated in the course of years, and with which he used often to beguile a little leisure in trying to speculate upon the manner in which the concentric siliceous coatings might have been formed.

'But it is not from the nature or the number of Principal Forbes' contributions to geology that his interest in our science is to be measured, or that we can learn how much he really did for its promotion in this country. As Secretary to the Royal Society of Edinburgh—an office which he held for many years—as Professor here, and finally as Principal at St. Andrews, he had numerous opportunities, which he was ever anxious to use, of encouraging, by a kind word or deed, those who were devoting themselves to geological pursuits. He had watched, with a sadness which he used often to express to his friends, how the halo which shone round Scottish geology in his youth had slowly faded. The last time that he addressed the Royal Society of Edinburgh, this feeling found vent in these expressive words:—"Of all the changes which have befallen Scottish science during the last half century, that which I most deeply deplore, and, at the same time, wonder at, is the progressive decay of our once illustrious Geological School."[1] I may add that among the last letters which he wrote before leaving this country for Cannes was one to myself, in which he referred anew to the desirability of reviving the active prosecution of geology in Scotland.'

My own, almost schoolboy, recollections of Forbes as a lecturer are still quite vivid. His clear, cold, unimpassioned style suited admirably the eternal verity of the laws he enunciated, explained, and illustrated by

[1] *Proc. Roy. Soc. Edin.*, vol. v. p. 18.

well-chosen and invariably successful experiments. I cannot recollect one instance in which he adopted, even for a few moments, the "sensational" style which is now so terribly common: and which is too often employed, after the manner of the cuttle-fish, to enable the lecturer to escape from a difficulty in which his own ignorance has involved him.

In mathematical applications, far from attempting to gloss over a difficulty, he was almost painfully elaborate, even in the simplest matters. For the bulk of his class his lectures were admirably adapted: and there can be no doubt that, had there been in his day a Physical Laboratory attached to his chair, he would have sent out many students trained, like his illustrious pupils Clerk-Maxwell and Balfour Stewart, to perfection in nicety of manipulation and accuracy of recording results.

From the students' point of view he was regarded as too strict a disciplinarian, visiting with what we looked upon as uncalled-for severity very slight infractions of order. This, however, was but the natural outward expression of his own intense earnestness of purpose and sense of duty. All of us, who came to know him well, found underlying it a grand substratum of geniality and kindly interest.

Like Brewster and Herschel, and even to some extent Faraday—in physical science the great luminaries of the generation which is just passing from amongst us—Forbes did not fully receive the grand doctrine of our own generation—the conservation of energy.

CHAPTER XV.

FORBES' CONTRIBUTIONS TO OUR KNOWLEDGE OF GLACIERS.

To the non-scientific public there can be no doubt that Forbes was known chiefly by his observations on glaciers: and, bearing this in mind, it is curious to find that, from the scientific point of view, this part of his work will bear comparison in importance with the remainder.

It is not at the first reading of Forbes' works that we can gather an adequate idea of the comparative value of his contributions to the Natural History of Glaciers: for in them, with scrupulous care, he gave at least full, sometimes indeed excessive, credit to everyone who to his knowledge had previously published on the subject.[1] We cannot too strongly insist upon this point, for he has actually been charged with depreciating or suppressing the claims of others! Such charges could not have been made by writers who had taken the trouble to read carefully what they criticised; for this characteristic of almost morbidly scrupulous justness must strike everyone who is acquainted with any of Forbes' writings on

[1] The curious and rare work of Bordier he seems not to have seen. I find among his correspondence a letter from M. Studer, which contains the following passage, and which probably turned his attention from the subject:—'A-propos du petit bouquin de Bordier, je ne manquerai pas de chercher à vous le procurer, si l'occasion se présente. Mais il est rare, même chez nous. Je l'avais trouvé dans la bibliothèque de mon père. C'est surtout Bourrit qui parait avoir offusqué ce pauvre pasteur, par l'attaque dans le discours prélim. de ses *Glaciers de Savoie*, 1773.'

subjects which had been previously treated (however superficially or even erroneously) by others. And nowhere do we find it more conspicuous than in his *Travels through the Alps*, where he collects the scattered letters which he had for some time written to Prof. Jameson, and adds to them the results of further protracted and careful observation.

He was the first who *deserved* to succeed in the study of this very curious problem, for he was the first to go to a glacier determined to measure and to examine what actually takes place, unbiassed by a belief in and a desire to promote any particular theory. Others preceded him, but they too often went prepared to gather only what could be made to fit in to the particular theory of which they were the partisans, or they went altogether unprepared with sound ideas of what to measure, and suitable instruments wherewith to measure; and, in consequence, they missed, distorted, or misinterpreted many of those points which Forbes saw almost at once, and which very soon arranged themselves in his mind in mutual interdependence. We may say at once that Forbes did not completely solve this extremely interesting problem. He nowhere asserts that he has done so: and, indeed, some of his theoretical conclusions are inconsistent with modern knowledge. But there can be no question that his insight into the problem, arrived at in at most a few weeks' measurement and study, was profound compared with that of any of his contemporaries, who seem to have brought to it little but a happy facility in taking for granted, and a recklessness of the laws of dynamics which may be fitly characterized as sublime. One of the few men who seem, in any point of consequence, to have had even one clear and accurate idea on the subject before Forbes is Mgr. Rendu, late Bishop of Annéçy: but this was so mixed up with error that it does not appear likely that in his hands it could ever have led to anything definite; for Rendu holds and enunciates, sometimes in the same sentence, facts and errors utterly incompatible with them. To this we shall recur, but it may be well to state (as a simple and complete answer to those who charge Forbes with plagiarism from Rendu) that no one more heartily congratulated Forbes on his success than this worthy prelate. Witness the following letter :—

'MONSIEUR, 'CHAMBÉRY, 17 *Août*, 1844.

'Je vous rémercie infiniment de la lettre si aimable que vous avez eu l'obligeance de m'adresser, et plus encore de la promesse que vous me faites de venir me voir. Il y a longtemps que je desirais faire votre connaissance, ou plutôt que je vous connais par vos travaux. Venez donc et descendez à l'evêché, où votre logement sera prêt. J'y suis de retour le 21, et le 29 les membres de la Société Géologique seront à Annéçy pour leurs travaux et pour l'inauguration de la statue de Berthollet : il faut que vous soyez de cette fête ; venez avant si vous le pouvez.

'Votre théorie de la marche des glaciers finira par être la seule admise, parcequ'elle est, selon moi, la seule vraie. La dernière note que vous avez publiée dans la "Bibliothèque Universelle" est claire, et fait démonstration pour les lignes de stratification. Combien je regrette que vous n'ayez pas été ici dans ce moment, la question aurait fait un pas de plus.

'Agréez les sentiments de la parfaite estime avec laquelle j'ai l'honneur d'être, Monsieur,

'Votre très humble et très obéissant serviteur,

'✠ LOUIS EVÊQUE D'ANNÉCY.'

We may add to this the testimony of one of Forbes' Swiss friends, one specially qualified to speak on this particular point ; but we will prefix to it a letter by Forbes himself, which shows his anxiety to do justice :—

To M. B. STUDER.

'NEUFCHÂTEL, *August 26th*, 1843.

'I hope that you will find my book as free from polemics as possible ; and also that you will consider that I have earnestly endeavoured to render justice to my predecessors, and especially to M. Rendu (pages 356, 367, 368, and 382). You will see that I have quoted the most prominent passages of his book bearing on my subject. You will, I hope, look to his book itself, and let me know whether I have done him justice. In your letter to M. de Léonhard, in February, you allude to him as the author of a theory of fluid motion; but no one knows better than you do to distinguish between a man who states four or five theories, as Rendu does, *without giving any preference to one of them* (see his work, p. 93), and the man who proves one of these theories to be correct.'

'To take your old subject of comparison—gravitation—Pythagoras did as much before Newton. But it has been my real and sincere wish to bring M. Rendu's little-known merits, and ingenious, and often correct, views into notice; and I think you will admit that I have done so. I even quoted him in my letters to Jameson (see Appendix to my volume, p. 408), on your authority, before I had read his book myself.'

'TRÈS CHER AMI, 'BERNE, *le* 3 *Oct.* 1860.

'. . . Je me souviens très bien qu'en 41 ou 42 nous causâmes ensemble dans ma chambre sur le livre de Mgr. Rendu, et que nous fûmes d'accord qu'il fallait accorder à la glace quelque chose comme de la plasticité, mais de là à vos travaux et à leurs resultats il y a un pas immense. Vous connaissez sans doute le livre de Humbolat sur l'histoire de la découverte de l'Amérique. Là aussi l'on voit que, dès les temps les plus anciens, l'on pensait et écrivait sur la possibilité de cette découverte; depuis Colombo, il ne manquait pas de détracteurs de sa gloire qui se fondaient sur des passages d'anciens auteurs. Il est même prouvé que Colomb n'eut pas la première idée de son voyage, qu'il la tint d'un astronome toscan. Cependant toutes ces attaques contre son mérite, quoique renouvellées jusqu'à nos jours, sont restées sans le moindre résultat, la gloire de celui qui le premier osa traverser l'Atlantique, reste à jamais inébranlable, et même l'imposition du nom d'un autre au continent qu'il a découvert, ne lui a pas porté atteinte. C'est que ce n'est pas d'avoir eu la première idée d'une explication, ou de l'avoir énoncée le premier, qui prouve pour la force d'esprit ou de caractère d'un homme, ces inspirations sont souvent dues au hasard et momentanées, mais c'est la persévérance à leur donner suite et l'habileté à trouver les moyens d'arriver au but, que l'on admire et glorifie. Les moyens dont vous vous êtes servi pour mésurer la vitesse des glaciers semblent très simples et comme devant se présenter de prime abord à chacun, et cependant personne avant vous n'a eu l'idée de s'en servir. C'est l'œuf de Colombo. L'on ne peut douter que plusieurs de vos devanciers, comme Rendu, et de meilleures physiciens que lui, n'aient senti la nécessité de posséder des données exactes sur ce mouvement, mais ils ne savaient pas comment s'y prendre pour les obtenir. Escher, qui m'avait consulté, planta des jalons sur la glacier d'Aletsch qui furent renversés au dégel, Agassiz fit exécuter tout un réseau trigonométrique qui compliqua la question, Hugi se tint aux estimations des guides. Je croix que la principale cause de la difficulté que l'on trouva à suivre une bonne méthode, fût la fausse idée que l'on s'était formée sur la nature de ce mouvement. On croyait que le glacier avançait par saccades, en suite des crevasses, et qu'entre deux sécousses, pendant des intervalles plus ou moins longs, des semaines peut-être, il restait en repos. Je crois que personne avant vous n'ait eu l'idée, pas même Rendu, qu'un glacier avançait d'un mouvement continu, et je n'oublierai jamais l'impression que me fit l'annonce de cette importante découverte. Peut-être que vous-même, vous ne croyiez pas à un mouvement aussi contraire à l'impression de nos sens, avant que le résultat de vos observations ne vous l'eussent révélé. . . . Votre tout dévoué,

'B. STUDER.'

When Forbes first attacked the problem, there were two rival theories which jointly held unchallenged sway over the subject, the *Dilatation Theory* and the (so-called) *Gravitation Theory*; both of which, as enunciated by their respective supporters, are inconsistent with simple dynamical principles. Before, however, we explain them, we shall first, as briefly as is consistent with clearness, describe the phenomena of glaciers as they appear to the superficial observer.

The masses of snow which fall on the higher Alps, and on other elevated mountain ridges, are only in small part melted or evaporated where they fall. They descend, sometimes as avalanches, into the higher valleys, which thus become choked with a mass of half snow, half ice, called *névé*, far too great to be removed by melting or evaporation in the course of a single season. Thus there would be a gradual accumulation from year to year, unless some third mode of escape presented itself. It is observed that at the lower ends of these valleys the *névé* escapes (often through deep and narrow gorges), but now in a state more resembling solid ice. This gradually works its way towards the plains, carrying with it on its surface sand, gravel, and even immense masses of rock, arranged in what are called *moraines*. Sometimes such masses are wholly imbedded in the glacier; sometimes they project from its lower or lateral surfaces, and in this case they grind out its channel as they proceed.

For it is obvious to the most casual observer that the glacier ice moves down its channel, though not at any great speed, and it was long ago recognized by Playfair[1] as the most efficient mode of transport of those so-called *erratic blocks* which had so much puzzled his predecessors. To the pedestrian on its surface the glacier presents, especially towards its sides, huge cracks called *crevasses*, whose direction is at least approximately transverse to its bed, and which are often of great breadth, and may sometimes extend downwards to the very bottom of the glacier.

[1] *Illustrations of the Huttonian Theory*, 1802, § 349. Also Jameson (1827). See paper by Forbes, *Edin. New Phil. Journal*, xl., p. 99.

We are not aware that anything of particular importance beyond this was known, in the sense of having been *observed,* not merely *seen,* till Forbes took up the subject, with the exception of Rendu's acute remark, which appears to have been made previously by Captain Basil Hall and others, that a glacier seems to flow in its channel like a sluggish stream.

'Rien ne me paraît plus clairement démontré que le mouvement progressif des glaciers vers le bas de la vallée, et rien en même temps ne me semble plus difficile à concevoir que la manière dont s'exécute ce mouvement si lent, si inégal, qui s'exécute sur des pentes différentes, sur un sol garni d'aspérités, et dans des canaux dont la largeur varie à chaque instant. C'est là, selon moi, le phénomène le moins explicable des glaciers. Marche-t-il ensemble comme un bloc de marbre sur un plan incliné? Avance-t-il par parties brisées comme les cailloux qui se suivent dans les couloirs des montagnes? S'affaisse-t-il sur lui-même pour couler le long des pentes, comme le ferait une lave à la fois ductile et liquide? Les parties qui se détachent vers les pentes rapides suffisent-elles à imprimer du mouvement à celles qui reposent sur une surface horizontale? Je l'ignore. Peut-être encore pourrait-on dire que dans les grands froids l'eau qui remplit les nombreuses crevasses transversales du glacier venant à se congeler, prend son accroissement de volume ordinaire, pousse les parois qui la contiennent, et produit ainsi un mouvement vers le bas du canal d'écoulement.'[1]

We shall presently see how much importance is to be attached to this one correct idea, and how altogether erroneous is the mass of other ideas in which it is here 'set.'

It was imagined, probably on the authority of guides and other untrained and totally untrustworthy observers, that the sides of a glacier move faster than the centre; and various other statements of a similar kind were

[1] Rendu, *Théorie des Glaciers*, p. 93. (Quoted in full by Forbes, *Travels in the Alps*, chap. xxi.).

adopted on equally valueless evidence; rendering it absolutely necessary to any real progress in the subject, that, as a preliminary, *facts* should be carefully ascertained. The following extract from Forbes' work (*Travels*, pp. 33–38) will give a clear notion of the state of affairs when he took up the subject:—

'The theory which appears at first sight most readily to account for the leading facts, is that maintained by De Saussure, that the valleys in which glaciers lie being always inclined, their weight is sufficient to urge them down the slope, pressed on by the accumulations of the winter snows above, and having their sliding progress assisted by the fusion of the ice in contact with the ground, resulting from the natural heat of the earth.[1]

'This cause of motion has been rejected as insufficient by M. de Charpentier, who has supported another, which (though like the last, suggested originally by an older author, Scheuchzer, as De Saussure's was by Gruner), having received a scientific form and detail in his hands, we will call "Charpentier's Theory of Dilatation," as the other may be called "Saussure's Gravitation Theory," or the sliding theory.

'De Charpentier's theory is this:—The snow is penetrated by water and gradually consolidated. It remains, however, even in the state of ice, always permeable to water, by means of innumerable fissures which traverse the mass; these are filled with fluid water during the heat

[1] 'I wish to quote De Saussure's own statement of his views, which is very distinct:—"Ces masses glacées entraînées par la pente du fond sur lequel elles reposent, dégagées par les eaux de la liaison qu'elles pourraient contracter avec ce même fond, soulevées même quelquefois par les eaux, doivent peu-à-peu glisser et descendre en suivant la pente des vallées ou des croupes qu'elles couvrent. C'est ce glissement lent, mais continu, des glaces sur leurs bases inclinées, qui les entraîne jusque dans les basses vallées, et qui entretient continuellement des amas de glaces dans les vallons assez chauds pour produire de grands arbres, et même de riches moissons."—*Voyages*, § 535. For De Saussure's very clear views respecting the action of the heat of the earth, see §§ 532—535, 739, &c.'

of the day, which the cold of the night freezes[1] in these fissures, producing by the expansion which freezing water undergoes in that process, an immense force, by which the glacier tends to move itself in the direction of least resistance—in other words, down the valley. This action is repeated every night during summer, in winter the glăcier being assumed to be perfectly stationary.[2]

'In the *Edinburgh Review* for April 1842 I have stated some leading objections to both of these theories, to which I refer the reader. I shall content myself with specifying one against each, which seems conclusive.

'1. If the glacier *slide* down its bed, why is not its motion continually accelerated? *i.e.*, why does it not result in an avalanche? And is it conceivable that a vast and irregular mass like a glacier, having a mean slope of only 8°, and often less than 5°, can *slide*, according to the common laws of gravity and friction, over a

[1] 'The following quotations make it quite plain that it is to the difference of the temperature of the day and night alone that the freezing of the water in the capillary fissures is attributed:—

'"Il résulte . . . que pendant les jours d'été les glaciers s'imbibent d'eau, et que celle-ci s'y congèle pendant les nuits."—CHARPENTIER, *Essai*, p. 11.

'"Dans la plupart des nuits durant l'été les glaciers augmentent de volume par la congélation de l'eau qu'ils ont absorbée pendant le jour." —*Ib.* p. 14.

'"Cette alternative de gelée et dégel, comme je viens de le dire, a lieu pendant la belle saison, surtout à l'époque des jours les plus chauds suivis de nuits fraîches."—P. 15. See also p. 23. Compare AGASSIZ, *Etudes*, pp. 165, 211.'

[2] '"Une troisième objection contre le mouvement des glaciers par leur propre poids, se tire de leur immobilité pendant l'hiver. Car c'est un fait reconnu et attesté par tous ceux qui demeurent dans leur voisinage, tels que les habitans de Chamounix, de Zermatt, de Saas, de Grindelwald, etc., que les glaciers restent *parfaitement stationnaires* dans cette saison, et ne commencent à se mouvoir qu'à la fonte des neiges."—CHARP., p. 36.

'"Le mouvement des glaciers suppose des alternances fréquentes de chaud et de froid. . . . Il en résulte que l'hiver est pour les glaciers l'époque de repos."—AGASSIZ, p. 175.

'"Pendant l'hiver toute sa masse (*c. à d.*, du glacier) est dans un état de rigidité permanente qui la maintient dans *une immobilité complète* jusqu'à l'époque du retour des variations de la température."—*Ib.* p. 212.'

bed of uneven rock, and through a channel so sinuous and irregular that a glacier is often embayed in a valley whence it can only escape by an aperture of half its actual width? On all mechanical principles we answer that it is impossible. We may add, that many small glaciers are seen to rest upon slopes of from 20° to 30° without taking an accelerated motion; and this is conformable to the known laws of friction. It is known, for instance, to architects that hewn stones, finely dressed with plane surfaces, will not slide over one another until the slope exceeds 30°.

'2. The dilatation theory is founded on a mistake as to physical fact. I am sorry to put it in this way, but it is unavoidable; and the respectable author of the only intelligible or precise account of the theory will, I hope, excuse me for pointing it out.

'"The maximum temperature which a glacier can have," observes M. de Charpentier, "is 0° Centigrade, or 32° Fahr., and the water in its fissures is kept liquid only by *the small quantity of heat* which reaches it by the surface water and by the surrounding air. Take away this sole cause of heat, *i.e.*, let the surface be frozen, and the water in the ice must congeal." Now, this is a pure fallacy; for the fact of the latent heat of water is entirely overlooked. The latent heat of water expresses the fact, that when that fluid is reduced to 32° it does not immediately solidify, but that the abstraction, not of "a small quantity of heat," but a very large quantity indeed, is necessary to convert the water at 32° into ice at 32°. Not a great deal less heat must be abstracted than the difference of the heat of boiling water and that at common temperatures. The fallacy, then, consists in this:—Admitting all the premises, the ice at 32° (it is allowed that in summer, during the period of infiltration, it cannot be lower) is traversed by fissures extending to a great depth (for otherwise the dilatation would be only superficial), filled with surface water at 32°. Night approaches, and the surface freezes, and water ceases to be conveyed to the interior. Then, says the theorist, the water already in the crevices and fissures of the ice, and in contact with ice, instantly freezes. Not at all; for where is it to deposit the heat of fluidity, without which it cannot, under any circumstances, assume the solid form? The ice surrounding it cannot take it; for, being already at 32°, it would melt it. It can only, therefore, be slowly conveyed away through the ice to the surface, on the supposition that the cold is sufficiently intense and prolonged to reduce the upper part of the ice considerably below 32°. The progress of cold and congelation in a glacier will therefore be, in general, similar to that in earth, which, it is well known, can be frozen to the depth of but a few inches in one night, however intense the cold. Such a degree and quantity of freezing as can be attributed to the cold of a summer's night must therefore be absolutely inefficient on the mass of the glacier.

'I will not stop to consider the attempt made by M. de Charpentier to show that the friction of any length of a glacier upon its bed may be overcome as easily as the shortest, from a consideration of the forces producing dilatation; but it is as indefensible on mechanical grounds as the preceding theory is on physical ones (*Essai*, p. 106). I quote from M. de Charpentier, not because his defence of the theory of dilatation is more assailable than that of others, but because his work is the only one in which an attempt is made to explain its physical principles with precision.

'I cannot admit, then, that either the sliding or dilatation theory can be true in the form which has hitherto been given to them. When I first began to study the subject minutely, under the auspices of M. Agassiz, in 1841, its difficulty and complication took me by surprise; and I soon saw that to arrive at any theory which, consistent with the rigour of physical science at the present time, would be worthy of the name, a very different method of investigation must be employed from that which was then in use by any person engaged in studying the glaciers.

'To a person accustomed to the rigour of reasonings about mechanical problems, the very first *data* for a solution were evidently wanting—namely, the amount of motion of a glacier in its different parts at different times. A few measures had indeed been made from time to time by MM. Hugi and Agassiz, of the advance of a great block on the glacier of the Aar from one year to another, but with such contradictory results as corresponded to the rudeness of the methods employed; for in some years the motion appeared to be *three times* as great as in others. I then pointed out to M. Agassiz, how, by the use of fixed telescopes, the minutest motions of the glacier might be determined,—a suggestion which he has, I believe, since put in practice. It seems very singular that ingenious men, with every facility for establishing facts for themselves, should have relied on conclusions vaguely gathered from uncertain data, or the hazarded assertions of the peasantry about matters in which they take not the slightest interest. The supposed immobility of the glaciers in winter,—the supposed greater velocity of the sides than the centre of the ice, were amongst the assumptions traditionally handed down, upon no sufficient authority; and I believe that I may safely affirm that not one observation of the rate of motion of a glacier, either on the average or at any particular season of the year, existed when I commenced my experiments in 1842. Far from being ready to admit, as my sanguine companions wished me to do in 1841, that the theory of glaciers was complete, and the cause of their motion certain, after patiently hearing all that they had to say, and reserving my opinion, I drew the conclusion that no theory which I had then heard of could account for the few facts admitted on all hands, and that the very structure and motions of glaciers remained still to be deduced from observation.'

Again (*Travels*, pp. 126—129):—

'If De Saussure's theory be true, the glacier moves onward without sensibly incorporating new matter into its substance—continually fed by the supplies from behind, which form a new and endless glacier. The mechanism may not inaptly be compared to that of the modern paper machine, which, from the gradually consolidated material of pulp (representing the *névé*), at length discharges, in a perpetual flow, the snowy web. The theory of De Charpentier, on the other hand, represents the fabrication of the glacier going on within the glacier itself, so that each part swells, and the dilatation of each is added to that which acted upon itself, in order to shove on the section of the ice immediately in advance. *In the former case, then, the distance between two determinate points of the glacier remains the same; in the latter, it will continually increase.* Again, *on the former hypothesis, the annual progress of any point of the glacier is independent of its position; on the latter, it increases with the distance from the origin (the transverse section of the ice being the same).* The solution of this important problem would be obtained by the correct measurement, at successive periods, of the spaces between points marked on insulated boulders on the glacier; or between the heads of pegs of considerable length, stuck into the matter of the ice, and by the determination of their annual progress.'[1]

'The more that I revolved the subject in my own mind, the more clearly was I persuaded that the motion of glaciers admitted of accurate determination, and must lead to definite conclusions.'

'We have seen that the *motion* of glaciers has been for much more than half a century universally admitted as a physical fact. It is, therefore, most unaccountable that the *quantity* of this motion has in hardly any case been approximately determined. I rather think that the whole of De Saussure's writings contain no one estimate of the annual progress of a glacier, and if we refer to other authors we obtain numbers which, from their variety and inaccuracy, throw little light on the question. Thus, Ebel gravely affirms[2] that the glaciers of Chamouni

[1] *Edinburgh Review*, April 1842, p. 77.
[2] *Guide du Voyageur*, art. 'Glacier.'

advance at the rate of fourteen feet a year, and those of Grindelwald twenty-five feet a year; whereas, as we shall see, such spaces are actually traversed by most glaciers in the course of a few days. This statement is quoted by Captain Hall[1] and other recent writers, and even by M. Rendu (now Bishop of Annéçy), the author of a most ingenious paper on glaciers, too little known.[2] Hugi perceived the errors arising from a confusion between the rate of *apparent* advance of an increasing glacier into a warm valley, whilst it is continually being shortened by melting, and the rate of motion of the ice itself.[3] He points out the correct method of observation; and although his work contains no accurate measures, he was perhaps the first who, by observing the position of a remarkable block upon the glacier of the Aar, indicated how such observations might be usefully made, instead of trusting (as appears to have been the former practice) to the vague reports of the peasantry. Hugi's observations on the Glacier of the Aar give a motion of 2,200 feet in nine years, or about 240 feet per annum.[4] Now, in contradiction to this, it would appear from M. Agassiz's observations, that from 1836 to 1839 it moved as far as the preceding nine years—that is, three times as fast.[5] There is reason, however, to think that M. Hugi's estimate is the more correct.'

'Bakewell[6] assigns 180 yards per annum as the motion of the Mer de Glace, and De la Beche[7] 200 yards, on Captain Sherwill's authority.[8] But both of these were

[1] *Patchwork*, i. 109.
[2] *Mém. de la Société Académique de Savoie*, x. 95.
[3] *Alpenreise*, p. 371. [4] Agassiz, *Etudes*, p. 150.
[5] Ibid. '[As conjectured in the text, Hugi's estimate has been almost exactly confirmed. Experience too has shown that the motion of glaciers is almost uniform from year to year. (See 9th Letter in the Appendix.) The enormous error of his successors on the Glacier of the Aar is therefore attributable to a want of the most ordinary attention to accuracy, and shows how little such considerations were deemed important by them.]'
[6] *Travels in the Tarentaise*, i. 365. [7] *Geological Manual*, p. 60.
[8] 100 yards, in *Philosophical Magazine*, Jan. 1831.

hearsay estimates by the guides. M. Rendu seems to have been more aware of the importance of the determination of the rate of motion of glaciers than any other author; but the best information which he could collect in 1841 did not much tend to clear up his doubts. He gives the following rates of motion of the Mer de Glace, or Glacier des Bois, without being able to decide upon which is the most trustworthy: 242 feet per annum; 442 feet per annum; a foot a day; 400 feet per annum; and 40 feet per annum, or *one-tenth* of the last!—a difference which he attributes to the different rates of motion of the centre and sides.[1] De Charpentier, so far as I recollect, offers no opinion in his work on glaciers as to what is to be considered as their rate of motion. I was not therefore wrong in supposing that the actual progress of a glacier was yet a new problem, when I commenced my observations on the Mer de Glace in 1842.'

'I had myself been witness to the position on the Glacier of the Aar, in 1841, of the stone whose place had been noted by Hugi fourteen years before, and it was manifest that it had moved several thousand feet. In conformity with the prevalent view of the motion of the ice being perceptible chiefly in summer, I made the hypothesis that the annual motion may be imagined to take place wholly during four months of the year, with its *maximum* intensity, and to stand still for the remainder. With this rude guide, and supposing the annual motion of some glaciers to approach 400 feet *per annum* (as a moderate estimate from the previous data), we might expect a motion of at least three feet *per diem* for a short time in the height of summer. There appeared no reason why a quantity ten times less should not be accurately measured, and I, therefore, felt confident that the laws of motion of the ice of any glacier in its various parts, and at different seasons, might be determined from a moderate number of *daily* observations.'

'I went to Switzerland, therefore, fully prepared, and not a little anxious to make an experiment which seemed

[1] *Mémoires, &c.* x. 95.

so fruitful in results, and though so obvious, still unattempted.'

These observations on the two competing theories are all, with the exception of a very few words, obviously and necessarily correct; and they show, better than any words of ours could do, how completely original was Forbes' work, the greater part of it having been performed under circumstances the most adverse to the starting of a new and correct hypothesis; as he was so constantly associated with a partisan of one or other of these incorrect theories.

We now come to his own statement (*Travels*, pp. 382—384) of his views on the subject:—

'The idea of comparing a glacier to a river is anything but new, and I would not be supposed to claim that comparison or analogy as an original one. Something very like the conception of fluid motion seems to have been in the minds of several writers, although I was not aware of it at the time that I made my theory. In particular, M. Rendu, whose mechanical views are in many respects more precise than those of his predecessors or contemporaries, speaks of "glaciers d'écoulement" as distinct from "glaciers reservoirs;" and in the quotation at the head of this chapter,[1] he evidently contemplates the *possibility* of the mutual pressures of the parts overcoming the rigidity.[2] He is the only writer of the glacier school who has insisted upon the plasticity of the ice, shown by moulding itself to the endlessly varying form and section of its bed, and he is also opposed to his leading contemporaries in his conjecture that the centre of the ice-stream would be found to move fastest. But M. Rendu has the candour not to treat his ingenious speculations as leading to any certain result, not being founded on experiments worthy of confidence. "The fact of the motion exists," he says, "the progression of glaciers is demonstrated, but the manner of it is *entirely unknown*.

[1] See *ante*, p. 498.

[2] 'See also p. 107 of his work for a comparison between a river and a glacier.'

Perhaps by long observations and well-made experiments on ice and snow, we may be able to apprehend it, *but these first elements are still wanting.*"[1]

'I feel bound also to quote the significant expressions of Captain Hall, pointing to the conception of a semi-fluid glacier. "When successive layers of snow," he says speaking of the Glacier de Miage, "often several hundreds of feet in thickness, come to be melted by the sun and by the innumerable torrents which are poured upon them from every side, to say nothing of the heavy rains of summer, they form a mass, not liquid indeed, but such as has a tendency to move down the highly inclined faces on which they lie, every part of which is not only well lubricated by running streams resulting from the melting snows on every side, but has been well polished by the friction of ages of antecedent glaciers. Every summer a certain but very slow advance is made by these huge, sluggish, slushy, half-snowy, half-icy accumulations."[2] It is plain, I think, that the author had an idea that liquid pressure might drag a mass over its rocky bed, which would not move upon it as a solid.'

'But such speculations could not pass into a theory until supported by the definite facts of which M. Rendu deplores the want. I too, like my predecessors, though independently of them, had compared the movement of glaciers to that of a ductile plastic mass, in 1841, when I spoke of the Glacier of the Rhone as "spreading itself out much as a pailful of thickish mortar would do in like circumstances,"[3] and again, when I likened the motion of glaciers to that of a great river, or of a lava stream.[4] But I

[1] '"Le fait du mouvement existe, la progression des glaciers est démontrée ; mais le mode est entièrement inconnu. Peut-être avec de longues observations, des expériences bien faites sur la glace et la neige viendra-t-on à bout de la saisir ; mais ces premiers éléments nous manquent encore." '—*Théorie des Glaciers*, p. 90.

[2] '*Patchwork*, vol. i. p. 104, *et seq*. The whole passage, which is too long to quote, gives an admirable picture of the glacier world.'

[3] '*Ed. Phil. Journal*, January 1842.'

[4] '*Edinburgh Review*, April 1842, p. 54. Both these articles were written in 1841.'

knew very well that such analogies had no claim to found a *theory*. I knew that the onus of the proof lay with the theorist,—(1), To show that (contrary to the then received opinion) the centre of a glacier moves fastest; and (2), to prove from direct experiment that the matter of a glacier is plastic on a great scale, a fact which seems so repugnant to first impressions as lately to have been urged in a most respectable quarter,[1] as rendering the doctrine of semi-fluid motion untenable. No one had a right to maintain the theory of fluid motion as more than a conjecture, until at least these preliminary obstacles were removed by direct observations.'

'These observations have been made, and the result is the viscous or plastic theory of glaciers, as depending essentially on the three following classes of facts, all of which were ascertained for the first time by observations in 1842, of which the proofs are contained in this work.'

'1. That the different portions of any transverse section of a glacier move with varying velocities, and fastest in the centre.'

'2. That those circumstances which increase the *fluidity* of a glacier—namely, heat and wet—invariably accelerate its motion.'

'3. That the structural surfaces occasioned by fissures which have traversed the interior of the ice, are also the surfaces of *maximum tension* in a semi-solid or plastic mass, lying in an inclined channel.'

Or again (*Travels*, p. 365), where he finally sums up his ideas:—

'My theory of Glacier Motion then is this: A GLACIER IS AN IMPERFECT FLUID, OR A VISCOUS BODY, WHICH IS URGED DOWN SLOPES OF A CERTAIN INCLINATION BY THE MUTUAL PRESSURE OF ITS PARTS.'

These, again, may be accepted even now as a nearly complete statement of the physical circumstances of the question, in so far at least as they describe accurately the observed phenomena, and account for many of them in a

[1] '*Bibliothèque Universelle*, January 1843.'

lucid and complete manner. It is only in recent years that the *vera causa* has been pointed out which makes Forbes' observational theory completely consistent with recognized physical principles. We shall add a few words on this subject after referring, more explicitly than Forbes' modesty permitted him to do, to the immense steps which he made towards the explanation of these very singular phenomena; steps whose immensity can only now be thoroughly appreciated, when we see round us, and even occasionally among good mathematicians, constant attempts to revive or to remodel one or other of the old exploded theories.

Forbes, as we have already said, did very completely the Natural History of these enigmatic creatures, the glaciers. His 'viscous theory' is rather the enunciation of a truth discovered by his own thoroughly scientific observations and geodetic and other measurements, than a physical theory.

'Some confusion of ideas might have been avoided on the part of writers who have professedly objected to Forbes' theory, while really objecting only (and we believe groundlessly) to his usage of the word viscosity, if they had paused to consider that no one physical explanation can hold for those several cases [mentioned in the text]; and that Forbes' theory is merely the proof by observation that glaciers have the property that mud (heterogeneous), mortar (heterogeneous), pitch (homogeneous), water (homogeneous), all have of changing shape indefinitely and continuously under the action of continued stress.'[1]

His detection of the 'veined structure' as something essential to the understanding of glacier motion, his tracing of its form and distribution in various parts of a glacier, and his remarks on the production of the dirt-bands, all show a very high order of inductive reasoning. His observations as to pressure and distortion *producing* the blue veins are very important. His explanation, though incomplete, of the convexity of crevasses towards

[1] Thomson and Tait's *Natural Philosophy*, § 741 (footnote).

the source of the glacier,[1] and their approximate perpendicularity to the veined structure, exhibits a sound though somewhat imperfect acquaintance with the nature of stresses in a viscous mass.

To say that Forbes thoroughly explained the behaviour of glaciers would be an exaggeration; but he must be allowed the great credit of being the Copernicus or Kepler of this science; the man who, though no doubt aided to some extent by a somewhat imperfect analogy, first swept away from the subject the mass of nonsense with which it had been overlaid, first clearly obtained from observation, not from dogma or *à priori* reasoning, its fundamental facts, and first put them together into a simple, complete, and connected whole: and who must therefore ever stand forward in the history of the question as one of its most effective and scientific promoters. The mode in which he arrived at these results is vividly described by himself in one of the preceding extracts; but no such mere detached passages can give the reader any proper idea of the value and interest of his *Travels through the Alps*, which is one of the most charming books it has been our good fortune to meet with. There is an utter absence of self, entire devotion to his subject, and a freshness and simplicity of manner thoroughly in keeping with the keen penetration of the observer. Other works of Forbes, connected with this subject, are *Norway and its Glaciers*, 1853, and *Occasional Papers*, 1859.

We who now see these things by the light which he was mainly instrumental in throwing upon them, and which has so immensely smoothed though not wholly removed their difficulties, might be in danger of underestimating the services Forbes rendered to this part of science, did we not reflect that 'absolute nonsense' is the fitting designation of the theories held by acute men

[1] This convexity had been taken by Agassiz as a proof that the sides moved faster than the central portions; while Rendu, erroneously assuming them to be concave to the source, wrongly deduced the truth that the central portions move faster than the sides.

when he commenced his work, as it is that of many views on the subject which are even now persistently maintained, and stoutly defended with the choicest weapons of the mathematician or the physicist.

It is difficult in thought to place ourselves in the position of Forbes at the time when his discoveries were made, but the more successfully we attempt to do so, the more are we led to admire the cautious boldness and the indistractable penetration of the young explorer.

This is not a scientific but a historical work, so that here is not the place to enter fully into the physical explanation of glacier motion, though it may be well to show briefly that in all probability J. Thomson's beautiful discovery of the dependence of the freezing-point on pressure, which has already explained much, will ultimately be found capable of explaining almost all that is at present known about the formation, structure, and motion of glaciers. This was not Forbes' opinion, so that we will first give a letter which he wrote a few years ago, with the full intention that it should some day be published, containing a brief but complete statement of his final views as to the physical explanation of the facts of observation whose *ensemble* was concisely enunciated as the VISCOUS THEORY. In one of his letter-books, in which are kept in copying ink duplicates of almost every note he ever wrote, it has the following heading:—

'*On some Debated Points connected with the Properties of Ice and the Motion of Glaciers.*

'MY DEAR PROF. TAIT, *January* 12*th*, 1865.

'. . . It occurs to me that I may as well note down the ideas I expressed to you the other day as to the properties of ice in relation to its plasticity. Though untoward circumstances have made me very shy of publishing anything which can possibly lead to controversy on this vexed subject, I am glad to put on record the opinions which I have for many years entertained.

'1. The "pure and simple" phenomenon of regelation discovered by Faraday, and as I hold adequately explained by me [*Occas. Papers*, p. 228], has never been *proved* to be an efficient agent in the functions of glaciers. During summer, when they move fastest, it may be doubted whether it comes into play at all, in consequence of the gorged state of the capillary fissures of the glacier and the constant trickling and drainage of water through them. At the same time it is not part of my case to deny that true regelation is a constant operation in a glacier. If so, it helps to form the blue bands.

'2. Dr. Tyndall's so-called proofs that it is through "fracture and regelation" that a glacier moulds itself to its bed, are to my mind no proofs at all. They will be found, I believe, to rest exclusively on experiments with ice under Bramah's press. Now (*a*) these experiments so much vaunted are in the first place not new. I find them recorded in my Journal of 1846 as having been successfully made by Dollfuss-Ausset. They were made later by Schlagintweit: but (*b*) in the second place, I attribute the plasticity of ice in this case to a cause perfectly distinct from true regelation. It is beyond all doubt due to the internal liquefaction and general softening of the bruised ice under intense pressure, on the principles discovered and enforced by the Messrs. Thomson.

'3. I am not as yet inclined to allow that there is any true analogy between this last case (*b*) and the moulding properties of a glacier; though here I fear that I differ (as he knows) from Prof. Thomson. In order to convert bruised ice into a homogeneous solid under Bramah's press, the pressure must be applied and relaxed with a certain suddenness.[1] I cannot admit that this takes place in a glacier. The changes of pressure within a glacier, in consequence of the slowness of its motion, must be almost infinitely gradual, and therefore plenty of time must elapse for the perfect restoration of

1 Principal Forbes forgets here that the 'sudden' relaxation, whether required or not, is always necessarily present in consequence of the diminution of volume whenever the ice melts.—P. G. T.

an equilibrium of heat, and therefore no appreciable internal fusion can occur.

'4. According to my view ice is only hard and crystalline at a temperature of 30° or even lower [*Occas. Papers*, p. 225]. At 32° it is a clammy, plastic solid, not unlike Cheshire cheese; and it behaves very like it under the press or the knife. Such a substance cannot be bruised and the bruised surfaces brought into contact with moderate pressure, without reattachment, as a simple result of "time and cohesion" [*Occas. Papers*, xxi., 201] without appealing to any other properties of ice or effects of pressure. Mr. Graham's opinion that ice is a "colloid," entertained on quite independent grounds, confirms this view.

'5. I consider ice at 32° to have a slight, but perfectly sensible, genuine molecular plasticity. When the urgency of the forces to which it is subjected causes this limit to be overpassed, the pasty solid is subject to a general bruise, producing innumerable internal fissures with finite displacements, and subsequent restoration of continuity under pressure. Hence the blue bands. When the urgency of gravity is still greater, of course the mass acts as a common solid, and crevasses open. [See *Occas. Papers*, p. 161, &c., and many other places.]

'It must have been remarked that I have never attempted to reply to Mr. Hopkins' mathematical demonstrations about glacier motions and the internal forces by which they are controlled: and for this very good reason, that I never felt myself mathematically strong enough to replace his alleged demonstrations by others which I could regard as any true representations of the problem presented by Nature.

'From general physical considerations, and from an experimental study of plastic bodies, I was well aware that Mr. Hopkins' mode of treatment was applicable only to a body possessing properties, or placed under conditions, which I was perfectly certain that a glacier had not or was not placed in. I regarded his solutions therefore (I refer to his early papers in the *Cambridge*

Trans. and *Phil. Magazine*) as irrelevant mathematical exercitations. His later writings I have not studied, as I perceived that he had made no enlargement of his physical basis of reasoning. A proper theory of glaciers must include a constant, or constants, expressive of the state of the internal friction, according to the magnitude of which, in relation to the forces acting to set the body in motion, it will break up as a solid or *flow* like a liquid. I never felt myself equal to attempting this very general problem, but I recollect well that in 1851, when Mr. Stokes' paper on the Friction of Fluids appeared, I urged him to consider the motion of glaciers from the same point of view. I am glad to think from what you tell me, that there is now a prospect of our having a real solution of this interesting problem.

'Tresca's paper on the quasi-fluid motion of solids under intense pressure (*Comptes Rendus*, Nov. 7, 1864) seems to have excited some notice through a paragraph in some English journal, to which two of my correspondents refer. . . . .

'Yours very sincerely,

'JAMES D. FORBES.'

We find the Thomsons' explanation of at least a great part of the puzzling phenomena of glaciers given in a charming popular form by Helmholtz in his *Populäre Wissenschaftliche Vorträge*, from which we have paraphrased the following passages, with which this chapter may fitly terminate.

'One of the fixed points of our thermometric scale, which we call zero or the freezing point, is determined by placing the thermometer in a mixture of ice and pure water. Water, so long as it is in contact with ice, cannot be cooled below the freezing point without becoming ice; ice cannot be heated above this point without melting. A mixture of ice and water must therefore remain stationary at the temperature zero. If one tries to warm such a mixture, the ice melts, but the temperature of the mixture does not rise above zero so long as there is any ice unmelted. By the applied heat ice at zero is thus converted into water at zero, while no observable rise of the thermometer takes place. Physicists say, therefore, that the applied heat has become latent, and that water at zero contains a certain amount more of latent heat than ice at the same temperature. Again, when we abstract heat from the mixture of ice and water; the water begins to freeze, but its temperature remains at zero as long as there is any of it unfrozen. Here water at zero has given off its latent heat, and has become ice at zero.

'Now a glacier is a mass of ice through which little rills of water are running, and which, therefore, has everywhere in its interior the temperature of the freezing point. Even the lower layers of the *névé* appear, at such heights as we find in the chain of the Alps, to have this same temperature throughout. For, even though fresh fallen snow at such elevations is generally colder than zero, the first hours of warm sunshine melt its surface and form water, which trickles into the lower cold layers, and continues to freeze in them until they are by degrees brought to the freezing point. This temperature thenceforth remains unchanged. For by warm sunshine the surface of the ice can be melted, but not raised above zero, and the winter's cold does not penetrate deep into the badly conducting snow and ice masses. Thus the interior of the *névé*, like that of the glaciers, retains unchanged the temperature of the freezing point.

'But the temperature of the freezing point of water can be altered by great pressure. This was first pointed out by James Thomson of Belfast, and almost at the same time by Clausius of Zürich,[1] as a consequence of the dynamical theory of heat: and by the same reason-

[1] Prof. Clausius has habitually claimed so much that he sometimes, as in the present instance, gets credit for things he has himself expressly disclaimed. In his paper on this question (*Pogg. Ann.*, September 1850), he declares his object merely to be to show that Thomson's result (*Trans. R. S. E.* 1849) is in accordance with his own mode of viewing the subject.—P. G. T.

ing the magnitude of the change could be rightly predicted. In fact the temperature of freezing must be lowered by $\frac{1}{144}$th of a degree Réaumur for each additional atmosphere or pressure of 15 lbs. on the square inch. The brother of the first named, William Thomson, verified by experiments this result to its numerical details by compressing in a very strong vessel a mixture of ice and water, which was found to become colder and colder as the pressure rose, and exactly as much as the dynamical theory of heat requires.

'Now, if by the application of pressure a mixture of water and ice becomes colder without heat having been withdrawn from it, this can only happen by some free heat becoming latent, in other words, by some of the ice in the mixture melting, and in this lies the reason that mechanical pressure produces a change in the freezing point. Ice occupies more space than the water from which it is produced. When water freezes in a closed vessel, it bursts not merely glass flasks, but even iron bombshells. Thus, therefore, because in the compressed mass of ice and water some ice melts, the volume of the whole becomes less, and thus the mass yields more to the pressure than it could have done without this change of freezing point. Here mechanical pressure, as happens in the majority of cases of interaction of different natural forces, favours the production of the change, melting, which is favourable to the development of its own action.

'In W. Thomson's experiment, water and ice were shut up together in a closed vessel from which nothing could escape. The situation is somewhat different, when, as is the case in a glacier, the water produced between the pieces of ice pressed together can get out through cracks. The ice indeed then is pressed, but not the water, which escapes. The pressed ice becomes then colder, in consequence of the lowering of its freezing point by pressure, but the freezing point of the water, which is not compressed, is not lowered. In these circumstances we have ice colder than zero in contact with water at the temperature zero. The consequence of this is, that continually there is freezing of water and formation of new ice round the ice which is pressed, while a portion of the latter goes on melting.

'This happens, for instance, when two pieces of ice are pressed together, for thus they are firmly united to one another by the water which freezes at their surface of contact, and united into one piece of ice. Under powerful pressure, which cools the ice more, this takes place rapidly; but even with very slight pressure it happens, if one gives it time enough. Faraday, who discovered this phenomenon, called it the regelation of ice. There has been much dispute about the explanation of this; I have given you that which appears to me the most satisfactory.

'This freezing together of two pieces of ice is easily exhibited with pieces of any form, but they should not be colder than zero; they act best when they are beginning to melt. One needs only to press them powerfully against one another for a few instants and they unite. The flatter are the two touching surfaces, the more firmly do they unite, but very slight pressure suffices, if one only leaves the two pieces of ice very long in contact.

'This property of melting ice is made use of by boys when they make snowballs and snow-men. It is known that this succeeds only when the snow is beginning to melt, or at least, when it is so little colder than zero that it can easily be raised to that temperature by the warmth of the hand. Very cold snow is a dry loose powder, and does not stick together.

'What children making snowballs do on a small scale, goes on on the most tremendous scale in glaciers. The deeper portions of the originally loose and powdery snow of the *névé* are pressed together by the overlying heaps of snow, often many hundred feet deep, and under this pressure agglomerate into denser and more solid layers. The fresh fallen snow originally consists of tender microscopic needles of ice. . . . Thus, whenever the sun plays upon the upper layers of the snow field, water trickles down and freezes again in the deeper layers where it meets with colder snow; the *névé* becomes first granular and brought up to the temperature of the freezing point, and as the weight of the overlying snow mass increases more and more, it changes itself at last by firmer union of its separate granules into a dense and hard mass of ice. . . . . Thus, as the snow in the glaciers is pressed together into dense ice, so the already formed irregular masses of ice are in many places united into dense clear ice. This is most conspicuously visible at the foot of a glacier cascade. There are glacier-falls, where an upper part of the glacier ends at a steep rocky wall, and its blocks of ice fall down like avalanches over the edge of this wall. The heap of smashed blocks of ice which in consequence piles itself up below, unites itself again at the foot of the rocky wall into a continuous mass of dense ice, which pursues its downward course as a glacier. . . . Thus you see how, to the eye of the physicist, the glacier, with its blocks of ice confusedly towering over one another, its waste, stony, and discoloured surfaces, its destruction-threatening crevasses, has become a majestic stream, which flows down more peacefully and regularly than any other; which, according to fixed laws, contracts, spreads itself out or, crashing and roaring, plunges down into the abyss. If we follow it in conclusion, to its termination, we find the water produced by its melting united into a strong brook, breaking out and flowing down from the icy gate of the glacier. No doubt such a brook, at first, when it comes into view from below the glacier, looks very dirty and muddy, for it brings down with it all the dust which the glacier has rubbed off from the rocks. One feels disenchanted to see the wonderfully beautiful and transparent ice changed to such muddy water; but in fact, the water of the glacier brook is itself just as beautiful and clear as the ice, although for the time its beauty is concealed. You must seek these brooks again, when they have passed through a lake and there deposited their rock-dust. The lakes of Geneva, of Thun &c., Lago Maggiore, the lakes of Como and Garda, are principally fed on glacier water; the clearness, and the beautiful blue or greenish blue colour, of their water are the delight of all travellers. Let us leave the beauty and ask after the utility, we shall find still more cause for wonder.'

[*Added December* 16*th*, 1872.]

A popular work, *The Forms of Water, &c.*, just published by Dr. Tyndall, reiterates an estimate of the actual and relative originality, extent, and value of Forbes' contributions to the theory of glaciers, so utterly different from that contained in the preceding chapter as to render it imperatively necessary for me to make a few additional remarks. I believe the question to be one entirely of facts, not of opinions; and I feel assured that the work referred to would have elicited a prompt reply from Forbes had he lived to see it.

In proof of the general correctness of the view I have taken above, and as the only approximation now possible to the answer Forbes would probably have given, I have reprinted (in the form of Appendices A and B below), two most masterly papers of his, which unfortunately had not a very wide circulation: but which minutely detail his relations to Rendu on the one hand, and to Guyot and Agassiz on the other. Though the first of these has been published for twelve years, and the second for thirty, no attempt has ever been made to controvert or to answer either.

Till it is shown that these papers, along with others from which extracts have been made above, are not the work of Forbes, in whose character unswerving rectitude was the grandest feature, I shall continue to receive their contents as facts. From these facts no conclusions are possible but those drawn in the preceding chapter.

P. G. TAIT.

APPENDIX A.

Reply to Professor Tyndall's Remarks, in his work 'On the Glaciers of the Alps,' relating to Rendu's 'Théorie des Glaciers.' By JAMES DAVID FORBES, D.C.L., LL.D., F.R.S., &c., Corresponding Member of the Institute of France, Principal of the United College in the University of St. Andrews, and late Professor of Natural Philosophy in the University of Edinburgh. (A. and C. Black, 1860.)

§ 1. *Introduction.*

PROFESSOR TYNDALL informs us, in the introductory chapter of his recent work *On the Glaciers of the Alps* of the circumstances which, in 1856, first directed his attention to the structure of the ice of glaciers, and subsequently to all their other phenomena. Mr. Sorby's ingenious experiments on the production of cleavage in soft substances by pressure, led him to vary those experiments; and a reference by Professor Huxley to the laminated structure of glaciers, as described by me, induced him first to look into my *Travels in the Alps of Savoy* and afterwards (still in 1856) to visit some of the Swiss glaciers. A series of papers suggested by these and subsequent experiments and excursions, have been successively printed by Professor Tyndall in the Proceedings of the Royal Institution, and in the Proceedings and Transactions of the Royal Society of London.

From the first I naturally bestowed a close attention on publications which promised to revive the then flagging interest in the Theory of Glaciers, which had been with me, between the years 1841 and 1851, matter of laborious study. I could not fail to observe that Professor Tyndall, in the warmth of pursuit incident to a new inquiry, had not, in the first instance, apprehended the full import of what had been done or asserted by me, and that he not unnaturally attached great, and I might think undue, importance to the observations which two or three brief though actively employed visits to the glaciers had enabled him to make. Perhaps I also thought that Professor Tyndall unnecessarily enhanced the distinctions which he conceived to exist between his view sand my own, even where he had adequately studied the latter.

Although, therefore, in the different memoirs above referred to, criticisms appeared on what I had discovered, or believed that I had discovered—and although I considered these to be more or less unfounded, and, moreover, that much which was substantially mine was not attributed to me,—I steadily avoided entering into any controversy on the subject.

A strong belief in the eventual justice awarded in such cases, though it may be temporarily or locally withheld, sustained me in this course. I contented myself with reprinting in a separate volume those scattered papers in which from time to time I developed my Theory subsequently to the publication in 1843 of my *Travels in the Alps of Savoy*. I only omitted those which were directly controversial; but I did not suppress a single word or sentence which could possibly be held as open to criticism, either on the ground of inconsistency of opinion or of misinterpretation of fact. A copious index was added to facilitate reference. I introduced this compilation[1] by a prefatory note, showing the bearing of these expositions of my views upon certain speculations of Professor Tyndall. This preface was anxiously worded so as (I trust) to give no just offence, and it could scarcely be deemed polemical.

Perhaps I should not have thought myself warranted in now departing from my previous course in this matter, but should have left any outstanding differences between Professor Tyndall and myself to be calmly decided by an impartial public, had it not been for one section of his new work, which is headed RENDU'S THEORY.[2]

[1] The title was, *Occasional Papers on the Theory of Glaciers, now first Collected.* A. and C. Black : Edin. 1859.

[2] Pages 299 to 308 of the *Glaciers of the Alps.*

§ 2. *Professor Tyndall's General Account of Rendu's 'Théorie des Glaciers.'*

In the course of these few pages, Professor Tyndall conveys to his reader the apprehension that the memoir of Monseigneur Rendu on the Glaciers of Savoy has been almost entirely overlooked by English writers and readers on this subject: that the descriptions of it have been inaccurate and even deceptive, the extracts partial and not characteristic, so that when, after several years of study of the science of glaciers, his own attention was distinctly called by a Swiss friend to Bishop Rendu's work,[1] he was surprised to find evidence of extensive knowledge, close and accurate reasoning, and an extraordinary faculty of observation; together with a constant effort after quantitative accuracy, and 'a presentiment concerning things as yet untouched by experiment which belongs only to the higher class of minds.'[2] Nor was he less struck to find that the memoir contained passages of 'cardinal import' which previous writers had 'overlooked,' and that it should devolve on himself to call attention to them 'nearly twenty years after their publication.'

Whether these merits of Bishop Rendu had really been overlooked, as is here supposed—how far the statements of 'cardinal import' referred to had been knowingly or otherwise allowed to pass into forgetfulness, and required to be revived after twenty years' oblivion—I will presently endeavour to show. But I must first state why I feel my credit involved in these allegations, so as to induce me to withdraw from the neutral attitude which I have generally adopted towards Professor Tyndall's criticisms.

The reasons of my present remonstrance are these:—

First, That though the allegation that Bishop Rendu's *Theory* has been undervalued, or represented only by insufficient and partial extracts from his writings, is for the most part made in general terms,[3] yet it is to be inferred from the entire section that *I* am the writer most to blame. Moreover, at page 304 of his work, Professor Tyndall, after quoting a passage from Rendu, states that it enables him to correct a 'grave misappre-

[1] See *Phil. Trans.*, 1859, p. 271. [2] *Glaciers of the Alps*, p. 299.

[3] As, for example, in the following note:—'In all that has been written upon glaciers in this country, the above passages from the writings of Rendu are unquoted; and many who mingled very warmly in the discussions of the subject were, until quite recently, ignorant of their existence. I was long in this condition myself, for I never supposed that passages which bear so directly upon a point so much discussed, and of such cardinal import, could have been overlooked; or that the task of calling attention to them should devolve upon myself nearly twenty years after their publication. Now that they are discovered, I conceive no difference of opinion can exist as to the propriety of placing them in their true position.'—TYNDALL's *Glaciers of the Alps*, p. 308, note.

hension' at page 128 of my *Travels in the Alps;* and on the preceding page (303), while he allows that certain passages of Rendu's work 'are well known, from the frequent and flattering references of Professor Forbes,' he adds, 'but there are others of *much greater importance* which have hitherto remained unknown in this country.' Through whose fault they remained unknown he does not explicitly say; *who* extracted with commendation the *less* important passages, leaving the *more* important ones in forgetfulness, he leaves it to the reader to judge. I regret to say that a similar tone prevails in many other passages.

Now, an allegation that an author or observer has suppressed his knowledge of the antecedent labours of others in the same field, that he has even omitted some facts, while he claims credit for candour by citing others of less importance, or which bear less directly upon his own claims to their discovery,—such an allegation, I say, is an odious one, whether made explicitly or by inevitable implication. It requires to be openly met by the person whose character is really in question much more than his originality. This is the first reason why I write the present reply.

Secondly, Had not the scarcity of my own work (the *Travels in the Alps,* which has been long out of print), and in England that of Rendu's Essay also, made the indispensable verification of Professor Tyndall's citations and inferences (which are or may be in the hands of everyone) a matter of difficulty, I might possibly have spared myself the disagreeable task of writing these pages. My principal object must be to furnish the reader with the means of deciding—(1.) Whether the character given by me of Rendu and his theory was so indefinite as to leave it to Professor Tyndall to proclaim his merit for the first time. (2.) Whether my citations from Rendu were not only different from those made by Professor Tyndall, but whether they omitted matter of *much greater importance* than they included, and, by suppression, were calculated to mislead. (3.) Whether the context of the passages cited by Professor Tyndall from my work, and from that of Rendu, does not affect or modify their significance.

Before proceeding to this, may I be allowed to make one other observation, generally applicable to such historical criticism as that to which Professor Tyndall devotes so many pages of his writings on the Glacier question?

It is a matter notorious in scientific discovery, that every theory of the least importance has been preluded by the anticipations of men of sagacity and penetration, who yet wanted the skill, or the perseverance, or the opportunity necessary to demonstrate their speculations to be true. *Isolated quotations* from authors who formed just conceptions of a possible or antecedently probable explanation of a complex phenomenon, convey to the reader (trusting to these alone) an inaccurate conception of the exact importance of these anticipations. Seen by the light of subsequent observations and discoveries, they are incontrovertible truths; but when viewed in the aspect in which they appeared to contemporary writers, or even to the author himself—when tried by the context of the work in which they are contained, they appear what they really are—happy conjectures, supported by general analogies and by a few obvious or reputed facts. The history of science, if attempted to be based on such expressions alone, would become a maze of mingled truth and fiction. Hooke and Borelli would assume the position of being authors of the theory of gravitation; Grimaldi and Hooke (again), of the undulatory theory; De Dominis and Descartes, of the discovery of the unequal refrangibility of the rays of the spectrum; Hero and Porta, of the steam-engine; Bacon, of the aberration of light; Boerhaave and Fahrenheit, of specific heat; Wright and Lambert, of the laws of sidereal astronomy; Brugmanns, of diamagnetism; and Higgins, of atomic chemistry.

Professor Tyndall has treated Rendu's happy suggestions in this spirit, and consequently claims for him the parentage of every idea which thrown out by him, has since been proved to be true,[1] while he casts a veil over his less fortunate conjectures.

He has treated my writings in precisely an opposite spirit. He forgets that when my labours commenced I found *no single precise or quantitative fact* respecting glaciers established in a reliable manner, and that the whole weight of scientific authority was then ranged on the side of one or other of two theories of glacier motion, neither of which is now considered worth refuting.[2] As an example of the minuteness of Professor Tyndall's

[1] As when he claims for him (at p. 300) the theoretical inferences of Messrs. Grove and Helmholz on the mutual convertibility of the physical forces, and my explanation of Regelation by assuming ice to have a proper temperature lower than 0° (Cent.). In the latter instance, the context evidently shows that the meaning is, that zero is the *limiting* temperature of the ice of glaciers. It may be colder, but never warmer. 'Les quantités de calorique,' he says, 'fournies à la glace sont employées à la fusion sans jamais pouvoir élever la Glace *au-dessus de zéro*.' (RENDU, p. 70.)

[2] I refer, of course, to the Gravitation or sliding Theory of De Saussure, and to the Dilatation Theory of De Charpentier. It is not, I presume, alleged that the circulation of Rendu's Essay had the slightest influence in converting the partisans of those theories. The latter—the theory that the motion of glaciers arises from the dilatation of water frozen in the crevices of the glacier—was maintained by De Charpentier after receiving Rendu's work, which he seems even to have regarded as countenancing his opinions; and was also maintained by M. Agassiz, after he had made for several years observations on the Glacier

criticism when my *Travels in the Alps* are concerned, he has thought it worth while to signalize *three times*, in as many different publications, the venial error of the True North being set off on the compass-card of my Map of the Mer de Glace on the wrong side of the *Magnetic* North, according to which (as it is stated in the text) the map had been laid down.[1] Yet he allows that the error was corrected by myself nearly ten years before he had any chance of detecting it.

§ 3. *My first References to Rendu's 'Théorie.'*

The *Théorie des Glaciers de la Savoie* of Bishop Rendu appeared in the 10th volume of the Memoirs of the Academy of Chambéry in 1841.[2] It comprises 120 rather widely printed octavo pages. I heard it spoken of in August 1841 by M. Agassiz or by his friends, on the Glacier of the Aar, certainly not in very respectful terms. It appeared to be regarded by them as the work of a visionary. After leaving the glaciers in the same year, I saw the *Théorie* in the hands of a Swiss friend, but so cursorily, that the circumstance would probably have escaped me, but for a reference to it in a note to an article in the *Edinburgh Review* for April 1842, which was written by me, from which it also appears that the Essay itself seemed to me to be by no means worthy of the ridicule which I had heard applied to it. There can be no reasonable doubt that my main if not sole interest in inspecting M. Rendu's book on this occasion, was to endeavour to find a reference to the 'veined structure' of glaciers, which was then the engrossing subject of discussion. I did not find any account of it there.[3]

That a glacier moves like a sluggish river, and under the same laws, was an idea which first clearly entered my mind as a

of the Aar, which he considered as demonstrative of its correctness. The former theory—that the glacier merely slides by gravity over its inclined bed—was still later defended by the aid of assumptions of a very arbitrary kind. At this stage of the subject only a stinted measure of approbation was by many persons awarded to the observations contained in my work, the *general fact* that glacier ice behaves like a plastic mass being then resisted, though necessarily involved in the special facts which could not be denied. These occurrences, well known to those who interested themselves in the subject fifteen years ago, might easily escape the notice of the reader of the present day.

[1] The magnetic variation is also correctly stated in the text (p. 123, 1st edit.) at 19° W.

[2] Since writing the above passage, I find in the preface to Dr. Charpentier's *Essai sur les Glaciers* (Lausanne, 1841), that he received a copy of Rendu's *Théorie* on the 24th October, 1840, and that it bore the date of 1840. This must have been on the separate copies.

[3] Compare *Travels in the Alps*, p. 29, and also extract (*a*), page 11 of the present *Reply* [p. 528].

definite probability on the evening of the 24th July, 1842, when from the heights of the Charmoz I saw the dirt-bands stretching across the breadth of the Mer de Glace at my feet, like floating scum on a partially stagnant stream, such as I have described at page 162 of my *Travels in the Alps;* and from that hour the viscous or plastic theory was to me a conviction and a reality.[1] I will not at this date hazard an assertion as to the exact period at which I coupled with this thought the hitherto slighted speculation of the Canon of Chambéry. I will therefore, in my statement, rely only on documentary evidence. On the 22nd August I made the following remark in a letter to Professor Jameson, written from Zermatt, in which I expounded the leading features of my own theory:—'It is not difficult to foresee, that if my view should prove correct, a theory of glaciers may be formed which, without coinciding either with that of Saussure or Charpentier, shall yet have something in common with both. Whether that of M. Rendu may not avail something, I am unable to say, not yet having been able to procure his work.'[2]

Thus far, at least, there is no tendency manifest on my part to suppress the fact that M. Rendu had a theory, and that it might probably bear upon mine.

Knowing just so much as I then did of M. Rendu's book, had I wished to maintain a colourable ignorance of its contents, I might have proceeded to develop tranquilly my own theory. But I find (again appealing to documentary evidence for what would otherwise have unquestionably escaped my memory), that both in Switzerland and at Turin (whither I went in the summer of 1842) I sought for Rendu's work in vain, and requested Abbé Baruffi at the latter place to do me the favour of writing to his brother ecclesiastic to obtain a copy for me.

Notwithstanding all this, I returned to Edinburgh in the autumn of 1842, and commenced writing my 'Travels,' with the uncomfortable feeling that I had now procured and read every book about glaciers which I knew or had heard of, except the one whose contents it might prove to be most important for me to know.

[1] See my Third Letter on Glaciers, dated 22nd August 1842, where the following passages occur:—'One afternoon I happened to ascend higher than usual above the level of the Mer de Glace, and was struck by the appearance of discoloured bands traversing its surface nearly in the form indicated in fig. 4. . . . . These dirt-bands perfectly resemble those of froth and scum which every one has seen upon the surface of slowly moving foul water; and their figure at once gives the idea of *fluid motion,* freest in the middle and obstructed by friction towards the sides and bottom.'—*Occasional Papers on the Theory of Glaciers* (1859), pp. 21, 23.

[2] *Edin. Phil. Journal,* Oct. 1842; *Travels in the Alps,* 1st ed. Appendix, p. 407, and *Occasional Papers* (1859), p. 24.

As a last resource, on the 12th December, 1842, I wrote a letter to Monseigneur Rendu (then Bishop-elect of Annéçy), of which I have fortunately preserved a copy, which is now before me. In it I stated the particulars which (on the evidence of this letter, and not from memory) I have mentioned above;—the cursory inspection for a few minutes of a copy in the hands of a Swiss friend,—my inquiries for the work itself, both in Switzerland and at Turin,—the application to M. Baruffi, and my desire to give an account of M. Rendu's views in my volume then in preparation. Of course I ended by requesting the Bishop to excuse the intrusion of a stranger, and to forward a copy of his book to Edinburgh with the least possible delay,[1] and by offering him mine in return.

I received from the Bishop a polite and friendly answer, dated the 20th December, 1842, which is also before me, promising the despatch of the volume of Academical Memoirs containing his Essay through a Paris bookseller. I am unable to state the exact date at which it reached me—probably in February 1843; but this is not of importance. I shall now indicate with precision the notice which I took in my work (already in great part written before the arrival of Rendu's Memoir[2]) of the Essay which I had taken so much pains to procure.

§ 4. *References to Rendu in my 'Travels in the Alps'* (1843), *and in my later publications.*

The following extracts from my *Travels in the Alps* will speak for themselves. The pages referred to are those of the first edition (published in July 1843). No alterations whatever were made in these extracts in the second edition (1845), only the pages where a few of the later ones occur are slightly different.

(*a*) *Travels*, pp. 28, 29.—' . . . In the writings of Agassiz, Godefroy, De Charpentier, and Rendu, devoted exclusively to glaciers, and published in 1840 and 1841, there is an equal silence as to the real nature of glacier structure, &c.'[3]

---

[1] I then believed the book or pamphlet to be a separate publication. This accounts for the difficulty which I everywhere experienced in obtaining it through bookselling channels.

[2] See extract (*e*) below. I may here note that the chapters were not composed in the order in which they are printed.

[3] [Professor Tyndall (p. 301) ascribes to Rendu an allusion to the veined structure. A reference to p. 60 of the *Théorie*, however, must satisfy any impartial person that he is speaking of the true stratification of the *névé*, as would have been apparent had Professor Tyndall not stopped short in his quotation. After the word 'snow' in his citation this passage follows, identifying the description with those of De Saussure and other older writers, applied to the higher glaciers: 'Il y a dans l'ensemble de grandes et de

(*b*) *Travels*, p. 127.—'Ebel gravely affirms[1] that the glaciers of Chamouni advance at the rate of 14 feet a year, and those of Grindelwald 25 feet a-year; whereas, as we shall see, such spaces are actually traversed by most glaciers in the course of a few days. This statement is quoted by Captain Hall[2] and other recent writers, and even by M. Rendu (now Bishop of Annécy), the author of a most ingenious paper on glaciers, too little known.'[3]

(*c*) *Travels*, p. 128.—'M. Rendu seems to have been more aware of the importance of the determination of the rate of motion of glaciers than any other author; but the best information which he could collect in 1841 did not much tend to clear up his doubts. He gives the following rates of motion of the Mer de Glace, or Glacier des Bois, without being able to decide which is the most trustworthy: 242 feet per annum; 442 feet per annum; a foot a day; 400 feet per annum; and 40 feet per annum, or *one-tenth* of the last!—a difference which he attributes to the different rates of motion of the centre and the sides.'[4]

We shall see below (page 16) that Professor Tyndall, in citing the preceding passage from my *Travels* has made two important omissions.

(*d*) *Travels*, p. 356.—Motto or heading to Chapter XXI. [which is entitled 'AN ATTEMPT TO EXPLAIN THE LEADING PHENOMENA OF GLACIERS.'] —'Rien ne me paraît plus clairement démontré que le mouvement progressif des glaciers vers le bas de la vallée, et rien en même temps ne me semble plus difficile à concevoir que la manière dont s'exécute ce mouvement silent, si inégal, qui s'exécute sur des pentes différentes, sur un sol garni d'aspérités et dans des canaux dont la largeur varie à chaque instant. C'est là, selon moi, le phénomène le moins explicable des glaciers. Marche-t-il ensemble comme un bloc de marbre sur un plan incliné? Avance-t-il par parties brisées comme les cailloux qui se suivent dans les couloirs des montagnes? S'affaisse-t-il sur lui-même pour couler le long des pentes, comme le ferait une lave à la fois ductile et liquide? Les parties qui se détachent vers les pentes rapides suffisent-elles à imprimer du mouvement à celles qui reposent sur une surface horizontale? Je l'ignore. Peut-être encore pourrait-on dire que dans les grands froids l'eau qui remplit les nombreuses crevasses transversales du glacier venant à se congéler, prend son accroissement de volume ordinaire, pousse les parois qui la contiennent, et produit ainsi un mouvement vers le bas du canal d'écoulement.'—RENDU, *Théorie des Glaciers*, p. 93.

(*e*) *Travels*, p. 367.—'When a glacier passes from a narrow gorge into a wide valley, it spreads itself, in accommodation to its new circumstances, as a viscous substance would do; and when embayed between rocks, it finds its outlet through a narrower channel than that by which it entered. This remarkable feature of glacier motion, already several times adverted to, had not been brought prominently forward until stated by M. Rendu, now Bishop of Annéçy, who has described it very clearly in these words: "Il y a une foule de faits qui sembleraient faire croire que la substance des glaciers jouit d'une espèce de ductilité qui lui permet de se modéler sur la localité qu'elle occupe, de s'amincir, de se rétrécir, et de s'étendre, comme le ferait une pâte molle.

petites assises. Les premières représentent sans doute les années, et les secondes représentent peut-être une saison, un jour de pluie, une chute de neige, un violent coup de soleil.'—RENDU, p. 61. This is evidently no account of the veined structure. (1860.)]

[1] *Guide du Voyageur*, art. 'Glacier.'
[2] *Patchwork*, i. 109.
[3] *Mém. de la Société Académique de Savoie*, x. 95.
[4] *Mémoires*, &c. x. 95.

Cependant quand on agit sur un morceau de glace, qu'on le frappe, on lui trouve une rigidité qui est en opposition directe avec les apparences dont nous venons de parler. Peut-être que les expériences faites sur de plus grandes masses donneraient d'autres résultats." [1] Now it is by observations on the glacier itself that we can best make experiments on *great masses* of ice, as here suggested.'

(*f*) *Travels*, page 382.—'The idea of comparing a glacier to a river is anything but new, and I would not be supposed to claim that comparison or analogy as an original one. Something very like the conception of fluid motion seems to have been in the minds of several writers, although I was not aware of it at the time that I made my theory. In particular, M. Rendu, whose mechanical views are in many respects more precise than those of his predecessors or contemporaries, speaks of "glaciers d'écoulement" as distinct from "glaciers réservoirs;" and in the quotation at the head of this chapter, he evidently contemplates the *possibility* of the mutual pressures of the parts overcoming the rigidity.[2] He is the only writer of the Glacier school who has insisted on the plasticity of the ice, shown by moulding itself to the endlessly varying form and section of its bed; and he is also opposed to his leading contemporaries in his conjecture that the centre of the ice-stream would be found to move fastest. But M. Rendu has the candour not to treat his ingenious speculations as leading to any certain result, not being founded on experiments worthy of confidence. "The fact of the motion exists," he says; "the progression of glaciers is demonstrated, but the manner of it is *entirely unknown*. Perhaps by long observations and well-made experiments on ice and snow, we may be able to apprehend it, but these first elements are still wanting." [3] . . . . [Here follows a quotation from Captain Basil Hall.] But such speculations could not pass into a theory until supported by the definite facts of which M. Rendu deplores the want.'

(*g*) *Travels*, p. 385.—'I have no doubt, however, that the convex surface of the glacier (which resembles that of mercury in a barometer tube) is due to this hydrostatic pressure acting upwards with most energy near the centre. It is the "renflement" of Rendu, the "surface bombée" of Agassiz.'

Such are the quotations from and references to M. Rendu in the work which embodied the results of my own experiments and speculations. In my later writings I did not suffer the bishop's name to drop out of sight.

[1] '*Théorie des Glaciers de la Savoie*, p. 84. Whilst I am anxious to show how far the sagacious views of M. Rendu coincide with, as they also preceded my own, it is fair to mention, that all my experiments were made, and indeed by far the greater part of the present volume was written, before I succeeded in obtaining access to M. Rendu's work, in the tenth volume of the Memoirs of the Academy of Chambéry, which I owe at length to the kindness of the right reverend author.' *

[2] 'See also page 107 of his work for a comparison between a glacier and a river.'

[3] '"Le fait du mouvement existe, la progression des glaciers est démontrée; mais le mode est entièrement inconnu. Peut-être avec de longues observations, des expériences bien faites sur la glace et la neige, viendra-t-on à bout de le saisir: mais ces premiers éléments nous manquent encore."'

* [I have explained already (page 9) that though I had cursorily seen M. Rendu's book in the possession of a friend, I never had access to it for consultation, either while deducing my theory from observations on the glaciers themselves, or afterwards, when elaborating it at home, until the time above referred to, which was probably in February 1843. (1860.)]

(*h*) *Eighth Letter on Glaciers*, 1844 (reprinted 1859, in *Occasional Papers*, &c., p. 62), I say: 'Facts like this seem to show with evidence, what intelligent men, such as Bishop Rendu, had only supposed previously to the first exact measures in 1842, that the ice of glaciers, rigid as it appears, has, in fact, a certain "ductility" or "viscosity" which permits it to model itself to the ground over which it is forced by gravity,' &c.

(*i*) Paper in *Philosophical Transactions* for 1846 (read 1845; reprinted 1859, in *Occasional Papers*, &c., p. 84). 'M. Rendu, Bishop of Annéçy, in his excellent Essay on Glaciers, refers in one passage (and I believe in one only) to the possible analogy with a lava stream "[le glacier] s'affaisse-t-il sur lui-même pour couler le long des pentes comme le ferait une lave à la fois ductile et liquide?"'[1]

(*k*) Paper in *Phil. Trans.* 1846 (*Occas. Papers*, p. 160). 'Still less could I have anticipated that when the plastic changes of form had been measured and compared, and calculated, and mapped, and confirmed by independent observers, that we should still have had men of science appealing to the fragility of an icicle as an unanswerable argument! More philosophical, surely, was the appeal of the Bishop of Annéçy, from what we already know to what we may one day learn if willing to be taught. "Quand on agit sur un morceau de glace, qu'on le frappe, on lui trouve une rigidité qui est en opposition directe avec les apparences dont nous venons de parler. *Peut-être que les expériences faites sur de plus grandes masses donneraient d'autres résultats.*"'[2]

(*l*) *Thirteenth Letter on Glaciers*, 1846 (reprinted 1859, *Occasional Papers*, p. 201). 'Bishop Rendu, whom I had the pleasure of visiting at Annéçy' [in 1846], 'remarked a familiar circumstance which illustrates the same thing. We often see, in the coldest weather, that opaque snow is converted into translucent ice by the sliding of boys on its surface; friction and pressure alone, without the slightest thaw, effect the change which must take place still more readily in a glacier,' &c.[3]

I leave these extracts to the consideration of the impartial reader, that he may decide, by a comparison of them with the extracts in Professor Tyndall's work,—or, still better, with Rendu's Essay itself (where, of course, these sagacious views are mixed with a certain alloy of ingenious but inaccurate conjecture)—whether they do not constitute an adequate and impartial recognition of the Bishop's *Théorie*, and at the same time invite and facilitate reference to it.

[1] '*Théorie des Glaciers. Mém. de l'Académie de Savoie*, tome x. p. 93, published in 1841.'

[2] '*Théorie des Glaciers de la Savoie*, p. 84; quoted in my *Travels*, p. 367, second edition.'

[3] [Had I been aware of the following passage in Rendu's *Théorie*, p. 75, quoted by Professor Tyndall (p. 301), I would certainly have cited it:—'D'où [*i.e.*, from an observation by De Saussure on the Seracs] l'on peut conclure que le temps favorisant l'action de l'affinité et en même temps la pression des couches les unes sur les autres, rapprochent les petits cristaux qui forment la neige, et en les amenant au contact, les font passer à l'état de glace.' Nor did the Bishop refer me to his work when he made his remark about 'les petits polissons,' as he called them. (1860.)]

§ 5. *Professor Tyndall's Extracts from Rendu examined.—The First Extract.*

I have already adverted to some of Professor Tyndall's slighter and more general claims on behalf of Rendu. But it is at page 303 of Professor Tyndall's work that the more serious allegation is made,—that while certain passages from Rendu are well known, from the frequent and flattering references in my work, 'others of much greater importance, which have hitherto remained unknown in this country,' which, having 'discovered,' he proceeds to divulge.

The passages to which attention is thus so significantly directed are only two in number, both relating to the comparison of a glacier with a river. I will consider them separately.

Quotation A (*Glaciers of the Alps*, p. 303).—It will be desirable, I believe, to copy the original passage from Rendu, restored to the original French, which has been quoted and translated by Professor Tyndall:—

'J'ai cherché à apprécier la quantité de son mouvement; mais je n'ai pu recueillir que des données un peu vagues. J'ai interrogé mes guides sur la position d'un énorme rocher qui est au bord du glacier, mais encore sur la glace et par conséquent soumis à son mouvement. Les guides m'ont montré l'endroit où il était l'année précédente, et celui où il était il y a trois, quatre et cinq ans ; bien plus, ils m'ont montré l'endroit où il se trouvera dans un an, deux ans, etc., tant ils croient être certains de la régularité de ce mouvement. Cependant leurs rapports n'étaient pas toujours précisément d'accord, et leurs indications de temps et de distances manquent toujours de cette précision de mesure et de quantité sans laquelle on est obligé de marcher à tâtons dans les sciences physiques. En réduisant ces différentes indications à une moyenne, je trouvai que l'avancement total devait être d'environ 40 pieds par année. Dans mon dernier voyage j'ai obtenu des renseignements plus certains que j'ai consignés dans le chapitre précédent, et l'énorme différence qui se trouve entre les deux résultats provient de se que les dernières observations ont été faites au milieu du glacier d'écoulement, qui marche avec plus de rapidité, tandis que les premières ont été faites sur le bord, où la glace est retenue par le frottement des parrois [*sic*] rocheuses.'—RENDU, pp. 94, 95.

From the emphasis laid on the last sentence by the italics and capitals of Professor Tyndall in citing this passage, it would appear that what he most insists on as novel, and hitherto overlooked, is the more rapid motion of the central ice, and the retention of the lateral ice by friction; and in evidence of the 'grave misapprehension' to which he alleges that Rendu's statements have been exposed, he cites[1] from my *Travels*, p. 128, the reference which I have made to Rendu's estimates of the motion of the Mer de Glace, which has been already

[1] *Glaciers of the Alps*, p. 304.

quoted in Extract (*c*), page 12 of this Reply. But I regret to state that he has given it in an incomplete shape, to the manifest perversion of the meaning as respects the credit given to Rendu. In the first place, the introductory sentence is omitted. It is this:—'*M. Rendu seems to have been more aware of the importance of the determination of the rate of motion of glaciers than any other author; but the best information which he could collect in* 1841 *did not much tend to clear up his doubts.*' But the following omission is more serious; for it suppresses half a sentence, including the very point under discussion. Professor Tyndall closes the quotation (see p. 12) with the words, '40 feet per annum, or *one-tenth* of the last!' and he places four dots, thus, ..., in lieu of the concluding words of the sentence, which in my book are as follows: '—A DIFFERENCE WHICH HE ATTRIBUTES TO THE DIFFERENT RATES OF MOTION OF THE CENTRE AND SIDES' (accompanied by a foot-note reference to the page of Rendu's *Théorie*). I cannot but look upon this suppression of the end of a quoted sentence, including the very point under discussion, as an evidence of strong prejudice in the writer, against which I feel myself called upon to protest.

The idea that I had *suppressed* Rendu's correct opinion,—that the centre of a glacier moves faster than the sides, and its consequent analogy to the movement of a river,—could have no plausibility to anyone who had before him the account of Rendu's theory contained in my 'Travels,' and reprinted in this Reply; as, for instance, in Extract (*f*), page 13. I there recall (while in the act of stating my own opinions) that M. Rendu 'is the only writer of the Glacier school who has insisted upon the plasticity of the ice, as shown by moulding itself to the endlessly varying forms and sections of its bed; *and he is also opposed to his leading contemporaries in his conjecture that the centre of the ice-stream moved fastest.* But M. Rendu has the candour not to treat his ingenious speculations as leading to any certain result, not being founded on experiments worthy of confidence.'

§ 6. *Value of the Estimates of Glacier Motion previous to* 1842 *discussed.*

Professor Tyndall next proceeds (at p. 304) to discuss the question whether I was correct in stating (*Travels*, p. 128) that 'the actual progress of a glacier was yet a new problem when I commenced my observations on the Mer de Glace in 1842.' He cites a table containing the earlier estimates from Ebel to Rendu, and he seems to challenge a proof of their inaccuracy; but he avoids stating whether *any one of these* rests on evidence which would have been satisfactory to himself.[1] Nearly all of them depend on the reported estimates of the guides, who are too well known (with rare exceptions) to affect an accuracy of knowledge in such matters of which they are wholly destitute, and even (I am sorry to say) to frame their replies in the sense which they believe will be most in conformity with the ideas of their employers for the time.[2]

The estimates of MM. Hugi and Agassiz, which might seem to form exceptions, left the question in almost a worse position than ever, for the velocities which they measured in the *same* region of the *same* glacier differed in no less a proportion than *one* to *three*.[3] The chief error proved to be in the larger number,

[1] If the reader will turn to pp. 37, 127, 128 of my *Travels,* he will find reasons given for my scepticism in several instances. Thus, at p. 37—'It seems very singular that ingenious men, with every facility for establishing facts for themselves, should have relied on conclusions vaguely gathered from uncertain data, or the hazarded assertions of the peasantry about matters in which they take not the slightest interest.' These remarks, though not referring primarily to the statements of Rendu, are, as we shall see, applicable to them also.

[2] Before the time of Hugi (see *Alpenreise,* p. 371, cited in my *Travels,* p. 127) it was a common error to confound the actual movement of the ice with the amount of the advance in cold seasons of the lower termination of a glacier into the inhabited valleys,—a circumstance evidently of infinitely more moment than the other to the native peasantry.

[3] Cited also on the same page of my *Travels.* The table quoted by Professor Tyndall at p. 305 omits M. Agassiz' most irreconcilable estimate, that of 2,200 feet in three years, or 733 feet per annum—Hugi's estimate having been only 244 feet per annum. M. Agassiz accepted this enormous irregularity of motion (since admitted to be a mere mistake) in these words :—'Il résulte de ces faits que pendant les trois dernières années le glacier a fait autant de chemin que pendant les dix [neuf?] premières. Ce qui semble indiquer une marche de plus en plus rapide à mesure qu'il avance vers la vallée.' (*Etudes sur les Glaciers,* p. 151.) Yet this last conjecture is annihilated by the statement which follows in the next sentence but one, that the motion for one year (1840) *posterior* to all the others was only 200 feet! In such confusion of fact and inference was this subject when I entered upon it.

I must here add, that Professor Tyndall has referred in several places to M. Agassiz having made measurements similar to mine (in 1842), contemporaneously or nearly so. That Professor Tyndall should have remained unaware (as he indicates at p. 274) of the measurements made by M. Agassiz in 1842, until Professor Wheatstone referred him (apparently quite recently) to the *Comptes Rendus,* is to me a matter of the utmost surprise. Not only are the observations detailed, and that work cited, in M. Agassiz' *Nouvelles Etudes,* which Professor Tyndall might have been expected to consult, but they are specially referred to in an article in the *Westminster Review,* for April 1857, in which Professor Tyndall can hardly have failed to recognize the hand of a zealous supporter of his own. In that article, while justice is at the same time done to Rendu, every possible advantage is allowed to M. Agassiz; but the writer records an impartial decision in favour of my priority (*Westminster Review,* New Series, Vol. XI. p. 427). The measurements of M. Agassiz, moreover (as I have indicated at p. 37 of my *Travels in the Alps*), were made

and was, I believe, due to M. Agassiz having trusted these important measurements to his guides, without having given them any previous training. So little can even the *measures* of untutored peasants be relied on.

Rendu's Estimates.—From M. Rendu's work, it would appear that *one solitary measurement of any description made by himself upon a glacier is recorded.* These are his words: 'J'avais fixé en 1838 la position de deux blocs de rochers qui étaient à la surface du Glacier des Bois; une année après je suis allé mesurer le chemin qu'ils avaient parcouru; l'un avait avancé d'environ quatre cent pieds, et l'autre avait disparu,' &c.[1] Will it be maintained that this rough estimate of 'about 400 feet' is to be the geometrical basis of the whole theory of glaciers?

Rendu's other estimates repose, like those of most of the earlier writers, on the reports of the guides, and he does not hesitate to let us know in many places the uncertainty which he attached to them. He describes them as 'des données un peu vagues,' 'pas toujours précisément d'accord,' 'manquent de précision de mesure et de quantité.'[2] Indeed, with reference to the reported observation of an annual velocity of 40 feet per annum near the side of the Glacier des Bois, near Montanvert, there was enough on the face of the statement to inspire doubt. In the first place, it was *six times less* than the velocity of a block, described as 'sur le bord,' which he had quoted (at p. 85), on the same authority, at 242 feet. In the next place, it was the result of the *recollected* (not recorded) position of the block, one, two, three, four, and five years previously. No one who knows the ideas of the peasants of Chamouni on such subjects twenty years ago can give credit to such an alleged precision of memory, and I am confident that in this Professor Tyndall will agree with me. Lastly, it implied, if true, an amount of plasticity in the ice which *ought* to have excited an amount of scientific scepticism which could only be overcome by direct and authentic measurements. No such were produced.

Now, on this last point Professor Tyndall quotes Rendu's facility of belief in admitting that the central velocity might be at the Montanvert *ten times* that of the lateral velocity of the ice, in favourable contrast to my obvious doubt of the fact expressed by the mark of admiration (!) which I attached to the citation

at my suggestion, and were chiefly pursued by methods specifically indicated by me to him in 1841; methods similar to those which I myself put in practice. It is also well known that no such observations had been undertaken by M. Agassiz until after the period of my visit to the glacier of the Aar. I had also publicly urged the necessity for such observations being undertaken, in the *Edinburgh Review* for April 1842.

[1] *Théorie*, p. 86. [2] Ibid. pp. 94, 95.

of the fact by Rendu,[1] and which I find to have been faithfully transferred from the margin of my own copy of the *Théorie*, where I had thus recorded my incredulity by a pencil mark still extant, on a first perusal of the passage in 1843.

In truth, I disbelieve the alleged observation of 40 feet per annum for five years as much *now* as I did them. My own observations on that or any other part of the Mer de Glace lend no countenance to it. The smallest annual rate which I have estimated for that region is 486 feet, or in any other region of the glacier, 260 feet. *Névés* and snow-beds have of course nothing to do with the question.[2] But, what is much better evidence in the present instance, Professor Tyndall himself, and Mr. Hirst, his surveyor, who made it their business to measure the velocity of this very region of the glacier[3] in no less than four transverse lines, and at numerous stations in each line, have nowhere recorded a velocity of less than 6½ inches a day in summer, which, taking a mean with Professor Tyndall's own estimate of the ratio of winter to summer motion,[4] must give us, at the very lowest reasonable estimate, from 140 to 150 feet of advance per annum.

This station 'stood close to the moraine,' and the result is interesting as giving us a velocity almost entirely due to the *drag* of the ice over the bank of the glacier. Now, the motion reported by the guides to M. Rendu was *three and a half times* less than this velocity of ice in contact with the soil, which was, in fact, the minimum possible for that section of the glacier.

[1] See extract, marked (*c*), p. 12.

[2] In a passage at p. 306 of the *Glaciers of the Alps*, the author states that the measurements of glacier motion, made with my own hands, vary from less than 42 feet a year to 848 feet a year. The reader might very naturally infer from the context that these measurements indicated the effect of plasticity, and that they both referred to the Mer de Glace. As he might have difficulty in correcting this view, or in discovering where I have recorded a velocity of only 42 feet a year, and as references to my writings are here withheld, I may mention in this place, that the only observation by me which at all warrants such a statement, appears to be one which I made on a high and dry little glacier, in the state of *névé*, 8,000 feet above the sea, niched in a cavity near the Simplon Pass (*Occasional Papers*, p. 122). The velocity which I found in summer was 1·4 inches in 24 hours, which, multiplied by 365, gives 42 feet and a fraction. It is needless to add, that there was *no plastic connection* between this little glacier and the Mer de Glace, to which apparently the movement of 848 feet belongs. (See Map in *Johnston's Physical Atlas.*)

[3] See *Philosophical Transactions* for 1859, p. 263, and *Glaciers of the Alps*, p. 277.

[4] I will allow with Professor Tyndall that the winter motion is $\frac{16}{30}$ of the summer motion, and I will take the mean of the two, or $\frac{3}{4}$ nearly of the summer motion, to be the average of the year. Now, at 6½ inches per diem, the annual motion would be 198 feet at the summer rate, or nearly 150 feet at the average rate.

Even had the guides of M. Rendu, favoured by some exceptional circumstances, commenced observations on a block so extravagantly retarded by the glacier-wall, it is physically certain that before two seasons, much more before *five*, it would have been stranded on the fixed moraine, conformably to the rule in such cases.[1]

But I have not quite done with the quotation from Rendu (*A*, page 15). Professor Tyndall says, at page 306 of his work, that I came tardily in 1845 to admit a differential motion of a glacier even greater than Rendu admitted in 1841. The distinction in the two instances lies in this, that what Rendu admitted as applicable to the Mer de Glace at Montanvert, on very indifferent evidence, I admitted in a different case or cases, and on satisfactory geometrical proof. It is certainly the first time that it has been urged against me that I was not a thoroughgoing believer in my own theory. Most persons probably thought that I was too facile in accepting evidence in favour of it. The fact undoubtedly is, as Professor Tyndall says, that in 1844 and 1845 I was perfectly satisfied, that in immense glacier streams of small slope and low velocity, like those of Aletsch and the lower Aar, the central velocity might diminish towards the side in almost any given ratio. The evidence on which I accepted this astonishing proof of plasticity was in the first case (Aletsch) my own observations with the theodolite, in the second (Aar) similar measures by M. Agassiz' surveyors. But I did not do so without calling attention (which Professor Tyndall omits to state) to the fundamental difference betwen these glaciers and the Mer de Glace, which prevents us from expecting so striking a confirmation in the latter case. 'A glacier,' I then stated, 'like the Mer de Glace of Chamouni has so considerable a velocity (on an average at least three times that of the glacier of the Aar), that the ice is impetuously borne along and torn from the sides at the expense of innumerable lacerations and crevasses. So that, in the whole extent of the middle and lower regions of that glacier, in no place do the ice and ground meet without the former being more or less fissured by rents. But the contrary is the case on the great glaciers, which move on small slopes, and with smaller velocities,' &c.[2]

I am sorry to have been obliged to dilate so much upon the

[1] This happened after twelve months to the Block D 7, which I commenced observing in September 1842, under these circumstances, and near the same locality. See my *Travels*, second edition, chap. vii., and *Occas. Papers*, p. 36.

[2] *Ninth Letter on Glaciers* (1845), *Occas. Papers*, p. 73.

first quotation which Professor Tyndall has made from Rendu, and the inferences which he has drawn from it. It embraces, however, the real gist of the matter, and the only other quotation to which I must refer need not occupy us long.

§ 7. *Professor Tyndall's Second Extract from Rendu examined.*

Quotation B (*Glaciers of the Alps*, pp. 305-6).—It appears unnecessary to repeat the first portion of Professor Tyndall's second quotation from Rendu, as it merely contains in general terms an assertion of the analogy of the movement of a glacier to a river, the modifications of its velocity and depth depending on the width and slope of the valley. No one who reads the extracts which I have already given from my 'Travels' will doubt that I have given M. Rendu entire credit for this generally accurate anticipation. I also ascribed due merit to our countryman Captain Basil Hall (whose prior claim is not alluded to by Professor Tyndall) for his sagacious anticipation, 'pointing to the conception of a semi-fluid glacier.'[1] I further explicitly stated, that 'the idea of comparing a glacier to a river is anything but new, and I would not be supposed to claim that comparison or analogy as an original one;' that 'such analogies had no claim to found a *theory*;' that 'the *onus* of the proof lay with the theorist.'[2] The latter sentences of Professor Tyndall's quotation from Rendu, restored to the original French, are as follow:—

'Ce n'est pas tout, il y a entre le Glacier des Bois et un fleuve une ressemblance tellement complète qu'il est impossible de trouver dans celui-ci une circonstance qui ne soit pas dans l'autre. Dans les courants d'eau la vitesse n'est pas uniforme dans toute la largeur ni dans toute la profondeur; le frottement du fond, celui des bords, l'action des obstacles, font varier cette vitesse, qui n'est entière que vers le milieu de la surface.'—RENDU, p. 96.

The quoted extract terminates abruptly (*again* with four dots, ...), and there is a manifest incompleteness in the sense. For the introduction, 'Ce n'est pas tout,' shows that the writer is going to explain an additional analogy of the glacier to a river. But in the preceding extract it is evident that only the conditions of *river* motion are noted, and that the analogues of glacier motion, which ought to follow, do not appear.

On turning up Rendu's *Théorie* for the context, I found the missing member of the analogy. It is as follows:—

'Or, la seule inspection du glacier suffit pour prouver que la vitesse du milieu est plus grande que celles des rives. La surface entière est coupée par

[1] *Travels*, first edition, pp. 382, 383. [2] Ibid.

des crevasses, qui sont en général transversales à sa direction. Si le mouvement était le même dans toute la masse, ces crevasses, qui coupent la surface en ondées parallèles, formeraient une ligne droit qui serait toujours à peu près perpendiculaire aux deux rives ; mais il n'en est point ainsi ; la ligne générale est une courbe dont la convexité s'avance vers le bas de la vallée, ce qui ne peut être attribué qu'à l'excès de vitesse que les glaces ont sur ce point.' —RENDU, *Théorie*, pp. 96, 97.

From this extract we find—very unexpectedly to the reader of Professor Tyndall's account of Rendu's theory—that the Bishop relies for his conviction of the river-like motion of the ice on an observation (if observation it was) altogether fallacious, and from which a larger experience, and even a more impartial study of the Mer de Glace alone, must have led him to draw a consequence diametrically opposed to the plastic or river hypothesis. It is now admitted by *all* parties, including Professor Tyndall,[1] that, as a rule, the crevasses of a glacier stretch across it in curves *convex towards the origin*. M. Agassiz and others, misled in their deductions from this fact, which *they* recorded correctly, espoused a false hypothesis. M. Rendu, having a just hypothesis, observed the facts inaccurately, or must have relied on some altogether local or apparent exception to a general law.

The explanation of the whole matter is probably this: The Bishop perceived the general river-like analogy of glaciers, which had already struck Captain Hall and many other persons. He hastened forward to the conclusion, and viewed the facts with a favourable bias. He seized on the inaccurate estimate of the guides as to lateral retardation, without considering its improbability, or relying on measures made either by himself or by any educated person ; and he *fancied* that he saw in the direction of the crevasses in some portion of the glacier the desired confirmation of the same view. This appears indeed to have been his main reliance ; for in the comparison to a river, just quoted, there is no mention of the guides' estimate ; and in a parallel passage further on (which I have referred to in my Extract, marked *f*), where he recalls in similar terms the points of resemblance to a river, he again recurs, for proof of the more rapid central motion, *to the form of the crevasses, and to that alone.*[2]

I hope that I have now satisfactorily shown that the two quotations from Rendu, to which Professor Tyndall attributes so much significance, do not constitute any new claim in favour of

[1] *Glaciers of the Alps*, pp. 322-324.

[2] 'Les ondées de glace marquées par les crevasses transversales semblent se suivre ; mais elles paraissent avoir plus de rapidité vers le milieu, ce qui donne à chaque ondée la forme d'un demi-cercle dont la convexité est tournée vers le bas de la vallée.' Nor is there a word of the guides' estimate of the lateral retardation. (*Théorie*, p. 107.)

their author's theory. Their drift or bearing is simply that the movement of a glacier is analogous to that of a river generally; and, in particular, that Rendu was of opinion that, as in a river, the centre moves faster than the sides, and for the same reason. I have shown, by extracts from my own writings, that in both these respects I have given full credit to Rendu for his just anticipations. But I have also had to show that the grounds of his belief were inadequate, as they were also unsatisfactory to his contemporaries; that they relied, on the one hand, on reported estimates of velocity, at best vague, probably erroneous; and, on the other hand, on a fallacious observation, which, if made aright, should have led the author to an *opposite* result. That thus the claim of Rendu, viewed by the light of Professor Tyndall's extracts, amounts to no more than I had previously cordially admitted, and had also been (I may say) the first to proclaim,—that of having made a sagacious anticipation of a true theory from limited observations of no great precision.

§. 8. *Reception of Rendu's 'Théorie' at Home and Abroad—Conclusion.*

In two passages which I have already quoted, Professor Tyndall has made the specific complaint, that '*in this country*' certain passages of Rendu's writings have been unquoted, certain original views which are patent upon the face of them ignored. Can Professor Tyndall cite *any foreign work* in which greater justice has been done to them, or in which the passages relied on by him have been cited? He has not done so. Yet separately printed copies of the *Théorie des Glaciers* were circulated by the author among men of science. Indeed, I have the Bishop's own information that he had none remaining in December 1842. How is this? The truth is, that the effectual promulgation of Rendu's opinions and merits came *from this country*, and through my own writings.

That Rendu's theory did not rest on bases sufficient to convince the foreign glacialists and physical philosophers of his time, it would not be difficult to prove from the writings of his contemporaries, not one of whom gave any heed to the doctrine of plasticity.[1] We even find his essay quoted with approbation

[1] Thus, in a review of Rendu's *Théorie* in the *Bibliothèque Universelle* for February 1841, signed by an eminent professor at Geneva, I find no reference to his speculations on the quasi-fluid motion of the glacier, or to its plasticity, though he admits the more rapid motion of the centre on the basis of the inaccurately observed curvature of the crevasses. On the contrary, he describes Rendu as leaving the cause of the descent of glaciers indeterminate. 'Quant au mouvement progressif des glaciers d'écoulement, M. Rendu en

by De Charpentier, who was zealously attached to the Dilatation Theory, which was at the time in vogue. Indeed, there are passages in it which leave it quite open to the author to be considered as a supporter of this view.[1] But such a review of contemporary writers would not only lead me too far, but would at best be but secondary proof of what I have undertaken to explain on its own merits.

There is, however, one reviewal of the question, as between Rendu's work and mine, which I cannot in justice to myself omit, because it anticipated the very claim now raised by Professor Tyndall, at least *twelve years* before he had entered on the subject at all, and therefore cannot possibly be regarded as having any personal application. In quoting it, I emphatically disclaim any imputation of unworthy motives to Professor Tyndall. At the same time, it is impossible to read so striking a prediction of the course which he has taken without noticing that it gives a not-unneeded warning, how mistaken zeal on behalf of even a deserving client may take too strongly a forensic tone, and may even wear the appearance of detraction and hostility to another. The *Edinburgh Review* was one of the first in which my *Travels* were noticed, and the writer thus stated the case of M. Rendu and my own in the number for July 1844, p. 149:—

'Amid this penury of thought, it did occur to M. Rendu, Bishop of Annéçy, that "glaciers might roll down declivities like a ductile and liquid lava."[2] He speaks of *glaciers d'écoulement*, in contradistinction to *glaciers reservoirs*. He was the first, likewise, to conjecture that the central portions of glaciers move faster than the lateral ones, as in fluid motion; and he distinctly states, "that there was a number of facts which would make us believe that the substance of glaciers possesses a kind of ductility, which permits them to mould themselves on the locality which they occupy, to become thinner, and to contract and expand themselves in the same manner as a soft paste would do."[3] This idea, however, just though it be, was too bold to meet with reception even from philosophers; and when the author of it himself states, "that the rigidity of a mass of ice, when struck, is in direct opposition to this theory (though experiments made on greater masses may give other results)," we cannot but view it as one of those glimmerings of truth which would have soon died away, had it not been immediately surrounded by that pure atmosphere of oxygen, in which our author has made it flash into a brilliant light. Although he has, with great candour, published and given currency to the sagacious conjectures of Rendu, yet we fear that those spiteful critics who delight in depreciating inventions and discoveries, will seek to confer the honour of the new theory on the Sardinian prelate. No real philosopher, however, will countenance such a claim. Professor Forbes's theory was established by numerous and direct ex-

---

donne des preuves irrécusables ; c'était d'ailleurs un fait déjà acquis à la science ; mais il ne sait quelle est la cause qui peut leur donner ce mouvement.'—*Bibl. Univ.* xxxi. 375.

[1] See for example the last sentence of Extract (*d*), page 12.

[2] *Théorie des Glaciers de Savoie.* Chambéry, 1840, p. 93. [3] Ibid. p. 84.

periments before he saw the Bishop's work ;[1] and even if he had seen it, the possession of a conjecture which its own author distrusted, cannot, in the slightest degree, affect either the originality or the merit of his discovery.'

This extract alone appears to afford ample evidence, that Rendu's belief in the more rapid central motion of the glacier-stream was known 'in this country' by other persons as well as by myself; and also that he was allowed to have applied it to illustrate the analogy which he had conceived to the motion of a fluid body.

Moreover, in the *North British Review* for August 1859, from which Professor Tyndall quotes at page 306 of his *Glaciers of the Alps,* the following passage *immediately precedes* that which he has selected:—

'. . . While we give all honour to the Bishop for an idea which possesses so much ingenuity and truth, we must admit that it was a mere conjecture, hardly admitted as of any value by the author himself, and that it might never have been heard of had not Professor Forbes, with great candour, given it currency at the same time that he published his own theory, of which it was but the germ.'

Those who can recollect the silent disregard with which the pamphlet of Rendu was at first received—the ridicule, even, with which my attempts to base on rigorous demonstration the quasi-fluid behaviour of the glacier-stream were assailed, and the deaf ear turned for so many years to all the arguments by which I could enforce the analogy which to Professor Tyndall himself appears now so plain,—may afford to smile at the notion, that the evidence brought forward in the *Théorie* of the Bishop of Annéçy could have silenced 'the sneers of a presumptuous criticism,' to which, in the language of the same reviewer, my arguments were exposed, or that they would have excited attention now, save by the reflected light which a long and persevering discussion of the subject has thrown upon it.

To conclude: In the whole of this matter I am conscious of having acted an honourable and a loyal part towards Monseigneur Rendu. I unhesitatingly quoted those passages of his Essay in which his statements approached nearest to my own doctrine— the ductility of a glacier as a whole,—with which I felt my credit and originality most closely bound up.[2] So fully

[1] [I had indeed *seen* it, but little more. See page 9 of this Reply (1860).]

[2] It will have been noticed that my Extracts from Rendu's writings bore exclusively upon such of his views as have been upon the whole *confirmed*, and which may be regarded as his happiest prognostics. Nowhere have I allowed myself to detract from his merit by quoting, as a set-off, the few indifferent observations or inaccurate deductions or hazarded hypotheses which his work contains. and of which it cannot be supposed that I was ignorant. If I have been compelled to allude to one or two of these in the present Reply, it is because they have bèen extorted by the partial citations and unwarrantable claims which have been unwisely set up on his behalf.

am I satisfied of this, that on reading Professor Tyndall's section on Rendu's Theory, my first impulse would have been to appeal to the Bishop himself, had this been possible. I know that he only desired credit for having put forth sagacious anticipations of what was afterwards shown to be true in the face of many palpable objections. Far from thinking that I had deprived him of any share of his just glory, I am sure he felt that his claims had been brought into notice (even in Switzerland), and had acquired respect, chiefly through the influence of my writings. The Bishop's death made the desired appeal impracticable, but looking into the correspondence with which he occasionally honoured me, I found sufficient evidence of what I now state. I will therefore conclude this Reply with a short extract from a very friendly letter, dated from Chambéry (where a scientific meeting was then held), 17th August, 1844, which I am sure that he would have permitted me to use, and which certainly shows no trace of a thought that I had appropriated any of his laurels:—

'Votre théorie de la marche des glaciers finira par être la seule admise, parcequ'elle est, selon moi, la seule vraie. La dernière note que vous avez publiée dans la *Bibliothèque Universelle* est claire, et fait démonstration pour les lignes de stratification. Combien je regrette que vous n'ayez pas été ici dans ce moment; la question aurait fait un pas de plus.'

APPENDIX B.

Historical Remarks on the first Discovery of the real Structure of Glacier Ice. By PROFESSOR FORBES, Corresponding Member of the Royal Institute of France. (*Edin. New Phil. Journal*, 1843.)

I FEEL myself most reluctantly called upon to state some circumstances respecting the discovery of a fact in the theory of Glaciers which M. Agassiz has declared, in a paper printed in the last number of the *Edinburgh Philosophical Journal,* to be erroneously claimed by me.

The first account of '*a remarkable structure of the ice of glaciers,*' by myself, was printed in this Journal for January 1842. A history of this discovery, entirely opposed to mine, appears at pages 265 and 266 of the last number. By the kind permission of the Editor, I have now the opportunity allowed me of stating how the facts really stand, and at the same time of explaining the circumstances under which the publication of the original paper, claiming the discovery, took place,—circumstances which delicacy prevented me from mentioning at the time, but which it now appears essential to make known.

Private report proverbially exaggerates and misrepresents the history of transactions little interesting to any but those immediately concerned. I believe that my own conduct and its motives have been misunderstood, with reference to the matter in question. A few extracts from the ample correspondence of which I am possessed, in illustration of every step of the transaction, will, I hope, suffice to place the matter clearly before such readers as shall feel sufficient interest to follow them. I pledge myself to their accuracy, and to their being fairly extracted in conformity with the tenour of the letters to which they belong. If any doubt shall be raised on this point, I shall have only the disagreeable alternative of publishing the entire correspondence, the length of which would render it unsuitable for the pages of a scientific journal. But I repeat my belief that the extracts I shall make, and the narrative with which I shall connect them, will put the matter in a light sufficiently clear; and for the facts which I shall have to state, I am conscious of their admitting of no colouring or denial.

In the *first* place, I shall briefly state the circumstances under which the observation of THE VEINED STRUCTURE IN THE ICE OF GLACIERS[1] was made.

In the *second* place, I shall explain the circumstances under which I made it public.

In the *third* place, I shall discuss shortly the claims to priority of observation which have subsequently been made.

I.

In 1840, M. Agassiz invited me to make a tour with him the next summer amongst the glaciers of the Oberland, Vallais, and Savoy. I understood the invitation to extend simply to our mutual companionship on a journey of mutual interest. Of third parties there was no mention; and it was with diffidence that I requested permission for my friend and fellow-traveller, Mr. Heath, Fellow and tutor of Trinity College, Cambridge, to increase the number. It was only after all preliminaries were arranged, and after I had agreed, in order to accommodate M. Agassiz, to change the direction in which I proposed to commence our intended tour, that I learned that he had several friends in company with him; and it was not until my arrival at the Grimsel, on the 8th of August, that I learned that the plan of a tour, into which I had originally gone, had been abandoned by my fellow-traveller, for reasons which he did not assign, and that I was expected to unite with the party he had formed at Neufchâtel, to spend some time on the glacier of the Aar, instead of prosecuting the journey originally proposed. I cheerfully acquiesced, however, in the arrangement, which promised to give me a good insight into the structure of glaciers, which I proposed farther to study by prosecuting alone, or with Mr. Heath, my originally projected tour to Monte Rosa and Mont Blanc.

It is to be remembered that the Glacier of the Aar was the one which M. Agassiz had already repeatedly visited in former

[1] See *Edinburgh Philosophical Review*, January 1842, p. 89.

years, and on which he had constructed a sort of hut, in which he had lived for some time.

His other friends not having all arrived, M. Agassiz, Mr. Heath, and myself, accompanied by (I believe) a single guide, ascended the glacier on the 9th August, 1841.

Fact 1. We had not walked for half an hour on the ice, when I directed the attention of my companions to what I called a *vertical stratification* pervading the ice. It appeared to me so plain, that it scarcely occurred to me that it could be new to M. Agassiz, who had so often traversed the same ground.

Fact 2. M. Agassiz, having his attention called to the fact, stated that he thought I was deceived in considering that it penetrated the ice; that, indeed, the surface of the glacier

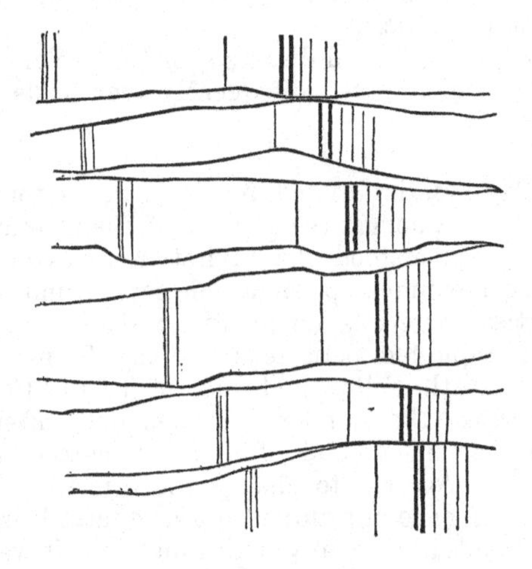

seemed to him much changed since last year, but that he had observed *superficial linear markings* of the same kind on (I think) the Glacier des Bois.

Fact 3. At each new *crevasse* we came to, I took pains to show him that the apparent strata penetrated into the mass of the glacier; but he seemed incredulous until I noticed a deep hollow in the ice close to the left margin of the medial moraine between Hugi's and Agassiz' cabins, at least twenty feet deep, to which I called M. Agassiz' attention, in proof of the position I had maintained.

Fact 4. To this he assented, but expressed his belief that it would only be found in the neighbourhood of the moraine, and not throughout the breadth of the glacier.

Fact 5. In the course of the same afternoon, we ascertained, by conjoint inspection, that the structure in question was traceable *all across* the glacier of the Finster Aar.

Fact 6. M. Agassiz, unwilling to admit that he could formerly have overlooked so palpable a structure, expressed a frequent doubt whether this structure had not been superinduced since his last visit.

Fact 7. I took the following means of proving that this could not be the case. I showed him some *crevasses,* and asked him how old he supposed them to be? he answered, several years; they certainly had not opened since last summer (1840). I showed that the veined structure *crossed these crevasses,* and was dislocated by them, as in the margin, and, therefore, must have been anterior to their formation.

Let us hear the evidence of Mr. Heath and M. Agassiz, the only witnesses present besides the guide.

Mr. Heath wrote to me thus, on sending him the above statement of facts:—

EXTRACT FIRST.—*Rev. J. M. Heath to Professor Forbes (printed by Mr. Heath's permission).*

'TRINITY COLLEGE, *8th March*, 1842.

'. . . But those who were there this summer have very different evidence that this was a new fact. I remember when it was first remarked, Agassiz said he had seen it before, but not to such an extent. That it had a peculiar relation to the medial moraines, and would not be found in the centre of the glacier; that it was *only superficial*, and owing, as he believed, to the sand, which placed itself in parallel straight lines, and produced these incisions by melting the ice. The afternoon was taken up in what I then thought a very superfluous endeavour to make out whether it was superficial or not, and I believe he maintained the contrary opinion until the discovery of the great hole of which you have given a drawing.'

It will be observed, then, that the whole question lies in this, Whether the lined appearance of the ice was due to an inequality of melting, occasioned by a linear arrangement of sand on the surface, washed from the moraines, and intercepting here and there the sun's rays?—or, Whether it was occasioned by the unequal action of the weather on alternating vertical bands of friable and of compact ice, of which the glacier is composed. M. Agassiz appears, upon Mr. Heath's testimony and my own, to have taken the former view, whilst I took the latter. According to him, the ice was striated *on its surface*, because the sand lay in lines; according to me, the sand lay in lines, because the ice has a veined structure *throughout its mass.*

M. Agassiz, the other witness, admitted as much himself, when I requested him to say whether the above-cited facts were accurately stated or not. In a letter to me, dated 29th March, 1842, he says:—

EXTRACT SECOND.—*Professor Agassiz to Professor Forbes, 29th March,* 1842.

'Comme vous en convenez vous-même lorsque nous discutâmes pour la première fois les bandes de glace de teintes diverses que l'on observe dans le glacier, *je vous dis que j'en avais remarqué* DES TRACES SUPERFICIELLES au Glacier des Bois en 1838, ce qui est mentionné dans mon livre p. 121, à l'occasion des moraines médianes.'

It appears, then, that Mr Heath's memory and my own agree thus far precisely with M. Agassiz. Let us see whether the reference to the 'Etudes sur les Glaciers,' published in 1840, gives any farther evidence.

EXTRACT THIRD.—*Agassiz, 'Etudes sur les Glaciers,' pp.* 121-2.

'Les trainées régulières et parallèles de grains de sable que l'on poursuit quelquefois sur de très grandes étendues, le long des moraines médianes, me paraissent être un effêt de la dilatation de la surface chargée de débris, combiné avec le mouvement progressif de toute la masse. Les petits grains de sable épars, n'agissant pas comme les gros blocs,[1] tendent à former des séries (Qu. *stries?*) longitudinales et parallèles qui se transforment quelquefois en rainures, et qui servent même souvent de lit aux petits filets d'eau qui coulent le long des moraines. Nulle part je n'ai observé ce phénomène d'une manière aussi frappante que sur la Mer de Glace de Chamonix en 1838 ; je l'ai également remarqué sur le Glacier de l'Aar, et ce qui m'a confirmé dans l'explication que j'en donne, c'est qu'ici on remarque sur le côté gauche de la grande moraine une petite moraine qui lui est parallèle, et qui me paraît détachée de la même manière que les traînées de sable dont je viens de parler se détachent des moraines en général.'

It appears then, that, after three years of observation of the glaciers, M. Agassiz still entertained, in 1841, the same view of the cause of a fact which he had observed in 1838, and published in 1840. The *fact* was the superficial arrangement of lines of sand near the moraines of glaciers, which, according to him, arose from some molecular dilatation of the ice, which he does not very clearly explain ; and its *effect* was *sometimes* to produce grooves (*rainures*), by the heat of the sun acting on the sand thus arranged.

The fact which I pointed out to him on the 9th of August had no reference to the arrangement of sand on the ice, but consisted in a *texture* which the ice itself presented throughout its mass, of harder and softer layers, whose wasting, when it occurred in the neighbourhood of the moraines where the glacier was covered with sand, occasioned hollow grooves, into which, for obvious reasons, the sand was speedily washed, and there it lay. M. Agassiz was very naturally and properly slow to admit, in explanation of a fact which had for three years been before his eyes, the existence of a prevalent structure to

[1] This refers to the well-known action of large blocks of stone in defending the surface of the ice from evaporation ; here, on the other hand, the sand sunk in the ice.

which he had not adverted. Accordingly, his convictions were proportionably gradual; and, as Mr. Heath observes, 'the afternoon was taken up in what I then thought a very superfluous endeavour to make out whether it was superficial or not.'

Two days after the discovery of the structure, namely, on the 11th of August, we were joined by Professor Studer, the distinguished geologist of Berne, and by other friends of M. Agassiz. The structure in question having been discussed, it is important to know the impression which it left as to novelty or originality upon the mind of so competent a judge. M. Studer writes to me:—

EXTRACT FOURTH.—*Professor Studer to Professor Forbes, 19th March,* 1842. *Extracted by M. Studer's permission.*

'M. Desor[1] m'a écrit il y a quelques semaines de cette contestation de priorité : je lui ai répondu que je ne me mêlerais pas de cette affaire, mais que bien certainement vous m'aviez fait remarquer pour la première fois la structure en question, et que j'avais cru en effet que son importance avait échappée à Agassiz, comme à tous ses devanciers.'

I will only cite one other testimony as to the origin of the discovery on the Glacier of the Aar, also by an eye-witness, Mr. Robertson of Newton House, near Elgin, a friend of M. Agassiz, whom I did not know before, and whom I have not seen since, but who, having learnt the nature of the contest as to priority which has occurred, generously and voluntarily sent me the following statement of facts, which I have likewise his permission to publish.

EXTRACT FIFTH.—*Mr. Robertson, of Newton, to Professor Forbes.*

'NEWTON, *May 4th,* 1842.

'Before joining you on the 13th August last year, I was pretty familiar, from reading, with all the ordinary phenomena of glaciers, and, on my walk to the "Cabane," examined each as it presented itself. Among others I observed the *superficial* indications of the ribboned structure ; and, during the first half-hour after my arrival, I recollect perfectly, in walking from the "Crevasse" at the end of the Finster Aar glacier (where you had been preparing the experiment on the absorption of ice with red wine) to the left flank of the Lauter Aar (where we exposed, with a hatchet, the contact of the ice and rock, in order to see the sand, &c., between them), having asked Agassiz how it was produced? He told me that the surface of the glacier had completely changed since last year, when he had *scarcely* observed it,—that it was an effect of the moraines, and probably caused by the greater variations of temperature to which they were subject as compared to the rest of the glacier, and that it had nothing to do with stratification. I remember also asking whether the horizontal lines at the end of the glacier were those of stratification? and was told "undoubtedly."

'On our return to the "Cabane," I pointed out the structure very well marked, at some distance from the moraines, and, on cross-questioning Agassiz, saw that he was far from satisfied with his theory.

[1] A friend of M. Agassiz.

'I have thus abundant evidence, independent of your ample testimony, to show that, at the date I have mentioned, my friend Agassiz was unaware of the general occurrence of the ribboned structure, through the mass of glaciers; and, in writing to him some days ago, mentioned my conviction that the discovery, certainly the most important of the recent ones, was due to you. I shall be glad to find that, as I believe is the case, M. Desor alone, and not M. Agassiz, could call it in question.'

The 'stratification' alluded to at the close of the first paragraph of the preceding letter, refers to the twisted planes of structure which I have described in my paper, and which are, in fact, continuous with the veins which, throughout the greater mass of the glacier, run parallel to its sides, when these sides are steep and continuous. The complex form of the surfaces of the shells into which a glacier is divided by these bands of compact and friable ice, I was first able to discover, during a visit to the glacier of the Rhone, on the 23rd August, 1842. I was accompanied by Mr. Heath and Mr. Calverley Trevelyan, but not by M. Agassiz or any of his party. In the course of a very careful examination of the glacier, I succeeded in satisfying myself completely of the conoidal form of the veined surface, and in explaining the apparent frontal stratification, which I have since confirmed in every point.[1] On our return to the Grimsel, I explained my views to M. Agassiz, who copied the sketch I had made, which corresponds exactly to that in the *Edinburgh Philosophical Journal*, January 1842, p. 89. A month later, I explained this system of curves of structure of the glacier of the Rhone to M. Studer at Berne. His penetration immediately perceived its importance, and he expressed great satisfaction at the insulated fact which I had pointed out to him on the glacier of the Aar being thus generalized.[2] We both agreed that its explanation must involve, in a good measure, the true theory of glaciers. In a letter to Professor Bronn of Heidelberg, dated 1st October, 1841, a week after I had quitted Berne, M. Studer gives an accurate account of my observations, being the first publication on the subject.[3]

[1] See Letters to Professor Jameson in this Journal for October 1842, p. 346.

[2] M. Studer, after quitting the glacier of the Aar, had recognized the structure on several others in the canton of Vallais. I should add that I pointed out the veined structure to M. Agassiz on the glacier of Gauli, in the Urbachthal, on the 20th August, and it was afterwards noticed by both of us on the Oberaar Glacier, and that of Aletsch. So that no reasonable doubt remained, at least, on my mind, that, having been observed on no less than five contiguous glaciers, it was a general and not a particular phenomenon. This meets M. Agassiz' statement, that I not only 'erroneously claimed the discovery,' but 'assigned to it a generality which the facts observed by myself did not at all justify.'—*Ed. Phil. Jour.*, p. 265.

[3] Leonhard's *Jahrbuch*, 1841.

II.

I now come to state shortly the circumstances which led to the publication of my paper describing this new structure of glacier ice; and about which there seems to have prevailed a misapprehension which I am anxious to remove.

It has been supposed that I resisted every offer to take a share in a joint publication of the proceedings of the summer, in order to bring forth a separate notice of the structure which I had observed; that even whilst in Switzerland, I contemplated such a separate publication; and having reached England, hastened to anticipate M. Agassiz.

The facts are precisely the reverse. The idea of publishing either this or any original observation of my own, on a subject so new and so unexpectedly difficult as I found the glacier theory to be, had certainly not entered my imagination during any part of my stay abroad. A *précis* of the labours of *others*, in the form of a Review of the writings of Venetz, De Charpentier, and Agassiz, such as subsequently appeared in the 'Edinburgh Review,' I certainly contemplated, thinking, that if I pursued the subject another year, such a preliminary study would be the fittest introduction to any original investigations. But I can safely say, that the way and manner in which my observations on the glacier structure should be brought out, was not a matter of the slightest concern to me, until an unexpected circumstance brought it to my mind.

I must mention, however, what passed between M. Agassiz and myself relatively to a joint publication, when I was at Neufchâtel in the middle of September 1841. I will state this in the words which I employed in writing to a friend a few months after the transaction took place.

EXTRACT SIXTH.—*From a Letter from Professor Forbes to a Friend, dated* 1st *April,* 1842.

'M. Agassiz never asked me, so far as I recollect, to publish with him on the subject of the Glaciers. He once proposed to me to communicate the observations I had made on Solar Radiation on the Glacier of the Aar, to form part of the description of the journey, of which the narrative part was to be written by Desor.

'This I declined, on the ground that these observations formed part of a series of experiments, long since commenced, and which must be treated of in connection.[1]

'I was very well aware, however, that a declaration of my opinion on the Glacier Theory was what was desired; and M. Desor took upon him to intimate this to me at Neufchâtel, in these words:—"M. Forbes ne

[1] They accordingly form part of a very extensive inquiry since communicated to the Royal Society of London.

veut pas se compromettre, mais nous le compromettrons"—which you will think rather a singular way of securing support to a scientific dogma. The following reasons determined me against taking any part in a joint publication:—

'1*st*, That however willing I might be to have my name associated with that of Agassiz, in any common work, experience led me entirely to decline such an association with M. Desor.

'2*nd*, That the utmost extent to which I could then conscientiously have gone in support of the Glacier Theory, would, I know, not have satisfied M. Agassiz.

'3*rd*, That from the perusal of Charpentier's work, and from communications with those best acquainted with the history of the Theory in Switzerland, I had begun to perceive, that were I to take any part in the discussion going on between Agassiz and Charpentier, it must be in favour of the latter, and not of the former, as an original observer and just reasoner.'

I reached home in the month of October 1841, and soon commenced the Historical Review of the Glacier Question which I had projected. Whilst I was thus engaged, the *Comptes Rendus* of the Academy of Sciences at Paris for the 18th of October reached me. In it I found a letter from M. Agassiz to Baron Humboldt, containing the following passage, with reference to the observations made on the Glacier of the Aar.

EXTRACT SEVENTH.—*From Professor Agassiz to Baron Humboldt.*

'Le fait le plus nouveau que j'ai remarqué, c'est la présence dans la masse de la glace de rubans verticaux de glace bleue, alternant avec des bandes de glace blanche d'un quart de ligne à plusieurs pouces de large, s'étendant sur toute la longueur du glacier, c'est à dire, à plusieurs lieues de longueur, et pénétrant à une profondeur d'au moins 120 pieds puisque j'ai observé encore ce phénomène au fond du trou de sonde.'

On reading this letter, from which all mention even of my presence on the Glacier of the Aar is excluded, my first impression was of surprise and pain. That I could not suffer so direct a plagiarism to remain unchallenged never appeared to me to admit of doubt; *le fait le plus nouveau que* J'AI *remarqué*, was an assertion as articulate as it was unfounded. How to take notice of it was a point of more difficulty. I felt fully the delicacy of my position. Towards M. Agassiz I felt the warmest friendship; sympathy with his zeal, and gratitude for his kindness and hospitality. This he well knew: during several weeks of the closest intimacy, we had been perpetually engaged in discussions connected with his theoretical views, and also respecting facts. I believe it may safely be stated, that neither of us ever for a moment lost temper in these amicable disputes, which often lasted for hours together, and which were uninterrupted either by our walks or our meals. His enthusiasm and good temper in these discussions delighted me, even where he failed to convince me of the constancy of his alleged facts, or the cogency of his reasons. We parted at Neufchâtel with even more cordiality (at least on my part), than we had met at the Grimsel; and my letters written afterwards testify that I freely acknowledged my obligations. Accordingly, in vindicating the originality of my observation, I resolved to take the plan which seemed to me most likely to secure a continuance of a friendship I so much valued. Were I to write to complain directly of want of justice on his part, though I did not doubt his willingness to correct his error, I felt that it would place him in a somewhat painful position, after so direct an assertion of his own rights. I preferred a different, and, I think, a natural course. Knowing well the facts stated in the commencement of this paper, and feeling that M. Agassiz must be equally aware of their truth, I resolved to make no reclamation, and especially to testify in my letters no irritation at the part which he had taken; but simply in a short and matter-of-fact communication to the Royal Society of Edinburgh, which I lost no time in transmitting to him, to state my own version of the affair, and claim my discovery without the slightest allusion to its having been erroneously claimed elsewhere. This was the origin of the paper which, at Professor Jameson's request, was communicated afterwards to his Journal; and anyone who looks at it in this view, will, I think, admit that it was well calculated to answer the end proposed. It has not a trace of a controversial character, but I well knew that when it should meet M. Agassiz' eye, it would be understood as an intimation that when he should next publish, I expected my claims to original observation to be more carefully regarded, though, in consideration of our friendship, and of the informal character of the communication to M. de Humboldt, I was both willing and happy to dispense with any apology.

At the same time that I communicated my paper to M. Agassiz, I sent it to Mr. Heath, the only other party to the observation of the 9th August—referring to the letter to Humboldt as the cause of the publication, and requesting his friendly opinion as to whether I had acted prudently in thus asserting my claim, and whether he considered all that I had stated to be justly my due. To this letter I received the following reply, which is here printed with Mr. Heath's kind permission.

EXTRACT EIGHTH.—*From the Rev. J. M. Heath to Professor Forbes.*

'TRINITY COLLEGE, *Feb. 25th*, 1842.

'I am very much obliged to you for the extract from the *Philosophical Journal*. I saw the paper in the Journal before you sent it me, and I most cordially approved of its appearing. I did not know, what you seem to say in your letter, that Agassiz has claimed the main observation as his own . . . . I will witness that 1*st*, He knew nothing about it: 2*nd*, When he did see it, he said it was superficial, and caused by superficial *sand;* 3*rd*, That he was the last to believe that it went to any depth. I think your account very true, and not claiming one iot more than fully belongs to you.'

I certainly anticipated that my forbearance with respect to M. Agassiz would have been rightly interpreted, and that silent acquiescence would have acknowledged the justice of my claim.

The event proved otherwise. The particular steps which were taken by M. Agassiz to vindicate what he professed to consider his due, arbitrarily and unexpectedly claimed in this paper of mine, were singularly in contrast to my conduct in the matter of Humboldt's letter, and to the usage in such cases. But *that* I am willing to pass over for the present, and I will now refer to the new claims of priority which he ultimately substituted for his own.

III.

We now pass on to the other claims to the priority of the observation.

About the same time that M. Agassiz claimed the observation of the Lamellar Structure of Glaciers, in the letter to Humboldt, he communicated verbally to the societies of Geneva and Neufchâtel the same fact; and though my information is not specific on this point, I presume that my name was not mentioned in connection with it. This I learn from my friend Professor Guyot of Neufchâtel, who, immediately on hearing the account of the observations on the Glacier of the Aar, recollected having observed and described something similar, three years before, on the Glacier of the Gries. The note containing this observation, and others connected with glaciers, had been read in 1838 in the presence of M. Agassiz, to the meeting of the Geological Society of France at Porrentruy, but was published neither at large nor in abstract. It appears to have dropped not only out of the records of the meeting, but from the minds of those who were present, since M. Agassiz, whom it was specially calculated to interest, takes no notice of it in his book, published two years later, containing his own observations, already quoted, on the superficial striæ; which he could not in common fairness have published without mentioning M. Guyot's contemporaneous and far more important observation of the *structure,* of which these striæ are only the outward indication, had he been acquainted with its true bearing, or, in truth, had he recollected it at all. Be this as it may, it seems that M. Guyot himself never repeated the observation, and so far as it appears, never even *spoke* of it, between the meeting at Porrentruy in 1838, and his hearing, first at Geneva in October 1841, then at Neufchâtel in November, M. Agassiz' account of his 'new fact.' M. Guyot has most honourably testified to me[1] that *not one word had ever passed between himself and me,* which could have informed me of what he already knew on the subject; and, also, that he twice traversed the Glacier of the Aar, on the 18th and 19th of August, 1841, without recognizing the structure which he had himself described. I mention this, because M. Agassiz has thought it necessary to assume that the Glacier of the Aar was more distinctly veined in 1841 than in any of the previous years that he visited it, in order to account for his not having noticed it until he returned to the glacier in my company. In the *Edinburgh Philosophical Journal* for October last, page 266, he says:—'During the months of August and September, 1841, this phenomenon was so well developed in the Glacier of the Aar, that *it could not fail to strike every observer.*'

M. Guyot's next step was a perfectly natural and just one. Finding that his original observation had been totally forgotten, he reproduced his paper from his bureau, where it still remained in MS., and read it afresh before the Société des Sciences Naturelles at Neufchâtel on the 1st December, 1841, just five days before I was similarly engaged, not merely in claiming for myself, before the Royal Society of Edinburgh, the priority of observation to M. Agassiz, but likewise proving that he had his attention directed by me to the structure in question. The transaction with M. Guyot did not come to my knowledge until long after.

Meanwhile, M. Agassiz sent no direct answer or complaint upon the receipt of my Paper on the Structure of Glaciers. I will not now advert to the means taken, through third parties, to discredit my statements, on the one hand, and on the other, to obtain from me a renunciation of my claim under a threat of exposure. Having no exposure to fear, I contented myself with sending to M. Agassiz a statement of the various *facts*, cited in the commencement of this paper, connected with the discovery on the 9th of August, requesting to know whether any of them, or which, were denied. A tardy and involved reply (29th March, 1842) contained a denial of none of them, but (as we have seen, *Extract Second*) an exact confirmation of what both Mr. Heath and I recollected him then to have stated respecting his own observations. But the real cause of the marked embarrassment of his reply I was not at the time aware of. He

[1] In a Letter dated 3rd June, 1842, in answer to mine in *Extract Tenth.*

had now no apology for ignorance of M. Guyot's claim to prior observation, yet feeling that his own dissatisfaction with my publication was solely grounded upon my having claimed for myself something which rightfully belonged to *him* (M. Agassiz) —'*le fait le plus nouveau*' of 1841—'les observations *les plus précieuses de la campagne*;' he naturally felt an embarrassment at being obliged to admit that similar facts and observations had been described in his hearing at Porrentruy three years before. Unable to maintain any longer his own originality, in his letter of the 29th March, 1842 (afterwards privately printed), he endeavours to impeach mine; and, describing what passed on the 9th August, in the words already quoted in Extract Second, he adds :—

EXTRACT NINTH.—*From Professor Agassiz to Professor Forbes.*

'Je suis certain d'avoir ajouté que M. Guyot les avait vues la même année (1838), à une profondeur notable sur le Glacier du Gries.'

To prove the negative fact that M. Agassiz did not cite M. Guyot upon the occasion, I can only state (1) that neither Mr. Heath nor myself recollect his name to have been mentioned, although we perfectly collected M. Agassiz' meaning as to his having observed the linear arrangement of the sand on the surface. (2) That had the occurrence of this structure to any depth been a recognized fact subsisting previously in the mind of M. Agassiz, whether from his own observations or those of another, Mr. Heath and I would not have spent the whole afternoon in what then seemed to Mr. Heath 'the very superfluous endeavour to make out whether it was superficial or not.' (3) What seems decisive in the matter, M. Agassiz claimed the observation *as his own* in the letter to Humboldt, written in October; nor does he appear to have made any allusion to M. Guyot in his communication on the same subject to the Société de Physique at Geneva, which occasioned M. Guyot to mention his prior observation.

Between M. Guyot and myself there remains nothing to explain. That gentleman has never contested the originality of my observation, and I have never pretended to doubt the reality of his, which, far from being made known to the world by the publication of the proceedings at Porrentruy, seems to have slipt entirely from the memory of the persons present (including, I am informed, MM. Studer and Agassiz), whilst every written proof of it remained in manuscript. Accordingly, so soon as I had satisfactory evidence of the nature of M. Guyot's communication, I hastened to write to him, and assure him that I admitted his observation to be identical with mine. This I did in the following terms :—

EXTRACT TENTH.—*From Professor Forbes to Professor Guyot of Neufchâtel.*

'EDINBURGH, *April 28th,* 1842.

'MY DEAR SIR,—In a printed letter which M. Agassiz has forwarded to me, I find a memorandum (printed for the first time) from your manuscript, containing an account of the structure of the Glacier of the Gries, observed in 1838, and stated to have been read at a meeting of naturalists at Porrentruy, in that year.

'I have no hesitation in saying that that note describes clearly a structure similar to that which I observed and pointed out to M. Agassiz and Mr. Heath on the Glacier of the Aar, on the 9th of August last.

'Whilst, then, I am most ready to do you full justice in respect to the originality and clearness of your observation, you will, I doubt not, as freely admit, that not having the pleasure of your acquaintance at the time of my observing and ascertaining the existence and modifications of this structure on the Aar Glacier, and never having heard, to the best of my recollection, during the course of my stay in Switzerland, of your having made such an observation, I could not, in any respect, have borrowed it from you. As no printed record of your communication then existed, I could not, of course, have learned of it from books. You will also, I doubt not, candidly admit, that your having failed to publish your observation in any even the most abridged abstract, your having omitted to press it as a fact important in the theory of glaciers upon any of your Swiss friends, and especially on M. Agassiz, who was writing a book on the subject, shows that the observation had not excited either on your part or that of your auditors at Porrentruy, any very lively interest. The fact itself would probably have been soon lost to science, if it had not been revived last summer by re-discovery, and by a strong indication of its generality and importance in the theories now agitated.

* * * * *

'Everyone in the slightest degree conversant with questions of this kind will see, on reading M. Agassiz' letter, that your observations, communicated three years before at a provincial meeting, not published even in the vaguest form in the minutes of the proceedings, nor alluded to in their writings by any one of the contemporary authors who are stated to have been present, leave my claim to have made the observation independently, and first insisted on its importance and generality, quite unimpeached.

* * * * *

'My firm belief is, that M. Agassiz had totally forgotten this passage in the verbal proceedings at Porrentruy. I believe him to be incapable of the sustained duplicity of affected ignorance and surprise when I first pointed out the fact to his notice on the 9th of August. I believe his present newly displayed zeal for your originality in this matter to be occasioned solely by finding it impracticable to maintain the charge against me of plagiarism and ingratitude towards *himself,* which he at first alone urged.

'The dilemma in which M. Agassiz has placed himself appears to be this :—

'Either he was acquainted with this structure of ice on the 9th of August, or he was not.

'If he was not acquainted with it, he learned it from me; for he has never attempted to maintain that he showed it to me.

'If he was acquainted with it he learned it from you. And if he learned it from either of us, how does he claim it as his own in the letter to Humboldt, and in one other private letter at least, not yet published?

'I am, my dear Sir, yours very truly,

'PROFESSOR GUYOT.' 'JAMES D. FORBES.'

There are few sciences which have not offered parallel cases of insulated observations which lie dormant for many years, before, by being generalized and made units of a class of facts, they form the basis of theoretical induction. This is what I claim to have done :—to have re-discovered M. Guyot's unpublished and all but forgotten fact ;—to have generalized it so as to show that it was common to most if not all glaciers ;—to have explained the law of its occurrence in one glacier (the lamellar surfaces of the glacier of the Rhone) ;—and to have applied it to account for two appearances formerly ascribed to other and imaginary causes, the distribution of sand on the surface of the ice, and the supposed stratification of the terminal face of some glaciers.

I might here close my observations on the question of priority, but I will add a sentence or two, in order to avoid all cavil.

Can it be necessary to state that M. Agassiz has found a friend—M. Dubois—obliging enough to state, in April 1842, that M. Agassiz had described to a meeting at Bâle, in 1838, a structure similar to that noticed by M. Guyot, at Porrentruy, in the same year ? Is it possible that discoveries in science can be made without the consciousness of those who make them ? or that a discovery made in 1838 shall be wholly misrepresented by the discoverer himself in 1840 (see *Extract Third*), claimed anew for himself in 1841, and when re-claimed in the same year by two other persons, the discoverer recollects to have heard at Porrentruy the very fact which his friends assure him they heard him claim for himself at Bâle the same year ? Yet such is the *newest* claim of M. Agassiz to an observation, which a discussion respecting priority of six months' duration failed to recall to his mind, but which he is now persuaded that he made, upon the friendly testimony of M. Dubois of Montpéreux, in the following words :—

EXTRACT ELEVENTH.—*M. Dubois' Certificate.*

'Je soussigné conjointement avec M. Arnold Escher de la Linth, Secrétaire de la Section de Géologie de la Société Helvétique des Sciences Naturelles lors de la Réunion à Bâle, certifie que dans les notes recueillies pendant la Séance du 14 Septembre, 1838, il se trouve mentionné page 12 que M. Agassiz a signalé le fait de la structure lamellaire des glaciers, et qu'il en trouvait la cause dans l'accumulation des matières congélables qui se déposent à la surface du glacier. La note est accompagnée d'un dessin représentant cette structure. 'FREDERICK DUBOIS.

'*Péreux*, 27 *Avril*, 1842.'

Now, this was on the 14th September, 1838. M. Guyot made his communication on the 6th September in the same year, consequently M. Agassiz could not have revisited the glaciers in the interval. The communication at Bâle was therefore, no doubt, a repetition of the communication at Porrentruy made eight days before, and the drawing of Agassiz was probably done from memory after the drawing of Guyot. At least, I am at a loss to explain these seemingly independent communications in any other way, nor will I even put the question, whether the structure described was a *vertical* structure at all. I do not suspect M. Agassiz of the reserve of having made no mention at Porrentruy that the fact of Guyot had been ascertained by himself, and then of having gone immediately to claim it as original at Bâle. I apprehend rather that the secretaries at Bâle (to whose MS. notes we are indebted alone for any knowledge of this transaction, forgotten even by the principal actor in it) had supposed, from M. Agassiz' *verbal* communication (*de vive voix*), that whilst relating what his friend M. Guyot had seen, he was really giving an account of his own observations.

I mention this as the explanation most natural and most favourable to M. Agassiz. But I would ask, if facts and theories are to be introduced *thus* into the history of science, where is the palm of discovery ever to be bestowed? Surely a man must have very little skill as an observer, and have exercised still less thought to render his observations worth recording, if he cannot recognize his own discovery when pointed out to him, but is obliged to take the authority of his friends, at the end of three years, that he ever knew it! Such evidence is barely tolerated in the case of posthumous claims. I suppose that this is the first instance of its being gravely urged during life. That I may not be imagined to have brought forward this claim more strongly than its author has done, I quote from his letter to myself.

EXTRACT TWELFTH.—*Professor Agassiz to Professor Forbes.*

'MONSIEUR,—Je reçois la lettre suivante de M. Dubois de Montpéroux . . . . dont je crois devoir vous donner copie afin de vous prouver que de mon côté j'avais aussi remarqué dès 1838, la structure lamellaire d'une partie des glaciers, alors même que faute de plus amples détails, je n'en ai mentionné dans mon livre que les apparences superficielles. Vous verrez par là que si vous avez pû croire avoir fait une découverte à ce sujet, ce n'a pû être qu'en mécomprenant ce que j'ai pû vous dire [1] sur la profondeur à laquelle ces lames déscendent, et qui n'avaient été remarquées qu'à une faible profondeur avant 1841, et par moi seulement dans le voisinage des moraines.

[Here follows the Letter, and *Extract Eleventh.*]

---

[1] See *Extract Second* for M. Agassiz' own account of what he *did* tell me of his previous observations.

'Jugez maintenant si j'ai dû être surpris de vos réclamations[1] et si j'étais en droit d'y répondre comme je l'ai fait. N'ayant pas l'habitude de tenir un journal régulier des moindres particularités des observations que je fais, et adressant dans nos sociétés scientifiques toutes mes communications de vive voix, ces faits ne me sont revenus à moi-même avec les circonstances accessoires que lorsque mes amis me les ont rappelés.'

* * * * *

'*Neufchâtel,* 28 *Avril,* 1842.'

After receiving the preceding letter, I gave up all thoughts of attempting to convince M. Agassiz respecting the history of this or indeed of any, scientific question. In the course of a few months, he had entertained four different opinions respecting the authorship of the discovery in question, and still, I suppose, has some doubt as to whether he discovered it himself in 1838, or only in 1841; or whether he learned it from M. Guyot at Porrentruy, or from me at the Glacier of the Aar.

The structure in question, which is common to every glacier in which I have looked for it, is in some so exceedingly striking, that it would seem impossible to escape notice. Such, for instance, is the case with a glacier of great beauty and extent, and which is remarkable from being almost touched by a frequented mule-road, whence the structure is admirably seen, —I mean the glacier of La Brenva, near Courmayeur. That it has not been described by any of the modern writers on glaciers, De Saussure, De Charpentier, Hugi, Agassiz, or Godefroy, is certainly a most convincing proof of how long the most evident and important facts may remain practically unnoticed. It can hardly be doubted that it must have been casually seen by these intelligent persons, who have traversed such a vast extent of glacier surface; but certainly every principle of interpretation leads us to the conclusion that it was not *observed* in such a way as facts must be to enter within the pale of science, since no trace of it is to be found in any of their writings on this very subject. I have it on the authority of three eminent persons in England, France, and Switzerland,—all men of science, much travelled, and much observing,—that, upon reading my account, they recognized what they could distinctly recall having seen on the glaciers which they had visited, though they never attempted to generalize the observation, or to attach theoretical importance to it.

[1] Of course I maintain that he had no right whatever to be surprised, since it appears from the following sentence that he was equally ignorant with myself of what he had himself done in 1838, until receiving M. Dubois' letter, dated the day before this was written :—'*Ces faits ne me sont revenus à moi-même avec les circonstances accessoires que lorsque mes amis me les ont rappelés.*'

In like manner the older observers, whose more vague language and antiquated terms make their meaning capable of several interpretations, may very possibly have described this appearance, without its having been handed down to their successors. I have not yet seen any evidence that they have done so, but I stated last winter to the Royal Society of Edinburgh, that I should not feel the least surprise if such an anticipation were discovered. How easy it is to find meanings in undefined phrases, *after* a well-marked truth has been announced, may be judged of from the interpretation given even by a very able and candid judge, of a passage in Godefroy's *Notice sur les Glaciers,* p. 12, as referring to the present question, but which a closer examination shows has no relation to it whatever.

I cannot conclude with any observation so just, or so much to the point, as that which Professor Studer has added to the testimony, of which I have already quoted a part (*Extract Fourth*), in a letter to myself. 'C'est toujours l'histoire de l'œuf de Colombo; je ne doute pas que de Saussure, de Charpentier, Agassiz et tant d'autres, parmi lesquels je me placerai moi-même, comme vous vous y êtes placé aussi, n'aient vu cette division verticale de la glace bien avant notre dernier voyage au Grimsel: —comme Newton aura souvent vû tomber des pommes sans songer à la lune. Dans toutes les découvertes il ne suffit pas de voir les choses, ou bien la science ne ferait pas des progrès aussi lents.'

APPENDIX C.

AUGUSTE BALMAT.

From ALFRED WILLS, ESQ.

LE NID D'AIGLE, SIXT, *October 1st*, 1872.

No account of the travels of our late friend Principal Forbes would be complete without a notice of Auguste Balmat, and I therefore gladly comply with your request that I would send you some reminiscences of him. During the (then) Professor's prolonged observations on the Mer de Glace, Balmat was constantly in his service, and by his intelligence and faithfulness undoubtedly contributed to their success. Professor Forbes, on the other hand, helped largely to make him the priceless treasure he was when I knew him—a perfect guide.

I made the acquaintance of Balmat in 1852, by which time he had already gained some celebrity as a guide. I was greatly struck by the absence of false modesty or bashful timidity from his bearing; and although he might sometimes be guilty of a conventional solecism, in all graver respects he was sure to speak and act like a gentleman. At a subsequent period, he stayed for many weeks at my house, always as a member of our family circle, where he won everyone's good will by the ease and courtesy of his demeanour.

His appearance was remarkable. He was of the middle height, but the great breadth of his shoulders and the muscular development of his back and chest, gave him the appearance of not being so tall as he really was. His head was large, erect, and well set, while his broad and lofty forehead and squarely cut features gave to his face an appearance of firmness and decision, ill according with the only serious blemish in his character—a too great readiness to yield to the importunities of others. He was of singular strength and activity, could carry the heaviest burdens without apparent effort, and not only in distance but in pace, could easily outstrip most persons with whom he was brought into contact, whether of his own calling or not.

Professor Forbes, to whom he had been recommended in 1842 by M. Lanvers, the Curé of Chamounix, thought highly of his intelligence and capacity, and recommended him to various scientific men, who in their turn had introduced him to others. Hence it happened that, as he accompanied now a geologist, now a botanist, now a student of glacial phenomena, Balmat picked up from each something of his own particular knowledge; and being a man of retentive and accurate memory, he rarely forgot a fact which had been brought to his notice, or an observation made in his presence. He acquired also the habit of constant and patient attention to the phenomena of nature, and the power of applying the large store of facts he accumulated to the elucidation of any question under discussion. He evinced on many occasions a capacity for close and accurate reasoning, very remarkable in a man without any early education. The flattering attentions, however, which in his later years he was constantly receiving, never led him to over-estimate the extent or value of his acquirements, but left him, to the last hour of his life, the most modest and unassuming of men.

He had listened with especial interest to all that he could learn about the movements, structure, and peculiarities of Glaciers, and upon this department of nature he had exerted all his powers of observation and reasoning. He became, in consequence, everywhere at home upon the ice, and when on a perfectly new glacier appeared to have the same unerring judgment as to the best way of attacking its séracs, or winding through its labyrinths of crevasses, that he exhibited on the Mer de Glace, or the Glacier des Bossons. The Mer de Glace was at that epoch a famous school of ice-craft, and Balmat had learned in no mechanical manner. The contempt which he always evinced for the little pyramids of stones which were, and are still, used to mark the track through a piece of crevassed glacier—'la ressource des mauvais guides,' as he always said—was perfectly unaffected. I was a sort of pet pupil of his in ice-work, and he always begged me emphatically to pay no heed to such things, but to trust to the infallible indications afforded by the inclination and volume of the glacier, the form of its bed, and the direction and mass of its affluents.

It must not be supposed that he was deficient in the sort of mechanical instinct, which often seems to direct aright, whether on ice or rocks, the man whom nature has intended for a guide, through darkness, storm, and bewildering mist; but he supplemented this instinctive process, to which so many guides trust entirely, by the closest and most accurate observation. I have been with him on the 'calotte' of Mont Blanc, in cloud and snow-storm, when we could not see many paces before us, and when, to my eyes, every part of the mountain seemed just like every other part. I well remember the unhesitating confidence with which he pronounced that the trackless waste was here rather too steep, there rather too flat, there again too rounded, for the passage we were seeking, and led us through worse than darkness directly to the narrow strip, by which alone access to the Mur de la Côte could be won. His calm, unperturbed manner, his absolute freedom in moments of danger from either panic or hurry, were of inestimable service in the chief guide of an expedition, and were admirably calculated to maintain the sense of security in others. I never saw a sign of jealousy towards him on the part of the other guides, who seemed tacitly to acknowledge his pre-eminence. He had a great and genuine love for nature. I speak from very many opportunities of observation, and it is my deliberate belief, that he was as keenly alive to the beauties of scenery, and drew from them as refined an enjoyment, as any of the numerous highly educated persons with whom he travelled.

It was characteristic of him, that he never thought any of the duties of his calling beneath him. To the last day of his active career as a guide, he would lavish every care and attention which his kindly nature could prompt, to render the tour of Mont Blanc, or a trip to the Montanvert, easy and pleasant to a lady. He would carry any weight, or go to any distance, to do a service to any traveller he had undertaken to guide, and never betrayed the smallest sense that such drudgery was beneath the notice of a man who corresponded with half the men of science in Europe, and who, had he been mercenarily inclined, might have made his fortune by putting his own price on his services. But this he never did. It was a part of the simplicity of his nature that, especially in the later years of his life, he fell into a habit—very rare for a Savoyard—of being unduly reluctant to receive payment for the many services he rendered, in all sorts of capacities, to all sorts of people.

In such a nature, gratitude and patience were as inherent as self-sacrifice. I have seen him under the severest physical trials, —once in great danger of losing his sight from inflammation of the eyes, once in great danger of losing his hands from frostbite,[1] and again, in his last long and painful illness,—and I never knew which to admire the most, the patience with which very severe physical suffering was borne, or the touching gratitude with which any attention, however slight, was received. To say that he would lay down his life for a friend, would give but a feeble notion of his generous and devoted nature. He would have laid it down—he did venture it freely on more than one occasion—in answer to the simplest claims of humanity or charity. If he could help or please another, if he could ascertain a fact which could aid scientific investigation, he would never count the cost, or dream of either sparing or benefiting himself. To render a service, or to give a pleasure, was his highest delight.

It is not to be wondered at that such a person should have carried good-nature to the point of weakness, and he was accordingly practised upon by many designing and inconsiderate persons. 'If I could but teach you one monosyllable—"*Non!*"' I often said to him, 'I should render you the greatest of earthly services.' A nature like his was made to be imposed upon, and unprincipled people, particularly in his later years, when his reputation and character had given a value to his name, prevailed upon him too easily to sign papers, and otherwise to become responsible for their engagements. In the zenith of his apparent prosperity, without a single personal indulgence, without one expensive habit or questionable practice of his own, he fell into hopeless difficulty and embarrassment, which so preyed upon a sensitive and remorseful nature, that he literally died of

[1] On this occasion he accompanied Dr. Tyndall and myself to the summit of Mont Blanc, for the purpose of burying a self-registering thermometer of Dr. Tyndall's as deeply as possible in the ice of the 'calotte.' It was then late in September, and the operation was performed in weather of a fearful character, the thermometer marking 22° of Fahrenheit below the freezing-point, and the wind blowing a hurricane. In his eagerness to assist in the difficult process of cutting a hole four feet deep in the solid glacier, Balmat had used his hands for shovelling out the ice and snow, and both hands were soon found to be badly frost-bitten, and quite black. I shall never forget the intense agony he suffered, when, after half-an-hour's rubbing and beating, circulation began to be restored; but throughout it all, and during our perilous descent, he was not one whit less thoughtful for the safety and comfort of everyone else, than he was when in the height of health and personal enjoyment. He was for some time in danger of losing his hands, but as he said, he could have borne even that calamity the better, as it would have been met with in the cause of science.

He resolutely declined all remuneration for his services, and the Lecture Hall at Leeds rung with well-deserved applause, when, shortly afterwards, at the meeting of the British Association, Dr. Tyndall recounted to the first *savants* of Europe, to most of whom Balmat was personally known, the danger he had undergone, and the courage and disinterestedness he had displayed.

a broken heart at the age of fifty-four, stricken to death before one trace of old age could be detected, or one physical or mental power had begun to fail.

He died here, in my house. In the difficult and wearisome negotiations through which I overcame the prejudices which my proposal to settle here had excited, as well as in the details of the planning and construction of my Alpine home, he was of unspeakable help to me. He refused the most tempting and almost extravagant offers of remuneration and distinction, by which one *soi-disant* foreign *savant* tried to induce him to desert me, at a time when it would have gone hard with me had he done so. I had earnestly hoped that I should have been able to offer him a congenial employment and an honourable retirement in his declining years. '*Dîs aliter visum*'—and my hands tended him in mortal illness, and closed his eyes. In him died the most skilful, brave, and intelligent of mountaineers; a man whose natural powers of observation and induction were such, that had he enjoyed the blessings of early training and high culture, he could scarcely, I think, have failed to achieve eminence among men of science. As it is, a little cross of iron in the churchyard of Sixt marks his humble resting-place, and his memory lives in the affectionate recollection of myself and of those few friends who knew him well, and can never forget his gentle and self-sacrificing nature.

APPENDIX D.

LIST OF PRINCIPAL FORBES' SCIENTIFIC WRITINGS.

Extended, by Professor Geikie, from the Royal Society's Catalogue.

1. Remarks on Mount Vesuvius. Edinb. Journ. Science, VII., 1827, p. 11. [*Anonymous.*]

2. Physical Notices of the Bay of Naples :—No. 1. On Mount Vesuvius. Edinb. Journ. Sci. IX., 1828, pp. 189-213. [*Publ. anonymously.*]

3. Physical Notices of the Bay of Naples :—No. 2. On the Burnt Cities of Herculaneum, Pompeii, and Stabeæ; with Note on Mount Vesuvius. Edinb. Journ. Sci. X., 1829, pp. 108-136.

4. Physical Notices of the Bay of Naples :—No. 3. On the District of Pausilippo and the Lago d'Agnano. Edinb. Journ. Sci. X., 1829, pp. 245-267; Leonhard, Zeitschrift, 1829, pp. 717-729.

5. On the Defects of the Sympiesometer, as applied to the Measurement of Heights. Edinb. Journ. Sci. X., 1829, pp. 334-346.

6. Physical Notices of the Bay of Naples :—No. 4. On the Solfatara of Pozzuoli. Edinb. Journ. Sci. I., 1829, pp. 124-141 Froriep, Notizen, XXV., 1829, col. 113-121, 132-138.

7. Physical Notices of the Bay of Naples :—No. 5. On the Temple of Jupiter Serapis at Pozzuoli and the Phenomena which it exhibits. Edinb. Journ. Sci. I., 1829, pp. 260-286; Boné, Journ. de Géol. I., 1830, pp. 354-373.

8. Notice of a large Greenstone Boulder in the Pentland Hills. Edinb. New Phil. Journ. VII., 1829, pp. 259-261.

9. Description of a new Anemometer. [1829.] Edinb. Journ. Sci. II., 1830, pp. 31-43.

10. Physical Notices of the Bay of Naples :—No. 6. District of the Bay of Baja. Edinb. Journ. Sci. II., 1830, pp. 75-102.

11. Physical Notices of the Bay of Naples:—No. 7. On the Islands of Procida and Ischia. Edinb. Journ. Sci. II., 1830, pp. 326-350; Froriep, Notizen, XXVII., 1830, col. 243-250, 257-264.

12. Physical Notices of the Bay of Naples:—No. 8. Concluding View of the Volcanic Formations of the District. Edinb. Journ. Sci. III., 1831, pp. 246-278.

13. Memoir on Barometric Instruments acting by compression, considered particularly in their application to the Measurement of Heights: including some Trigonometrical Determinations. [1830.] Edinb. Journ. Sci. IV., 1831, pp. 91-122, 329-351.

14. Observations respecting Professor Leslie's Formula for the Decrease of Heat in the Atmosphere; and his Opinions respecting the Polar Temperature. Edinb. Journ. Sci. V., 1831, pp. 17-23.

15. Report upon the recent Progress and present State of Meteorology. Brit. Assoc. Rep. 1831-32, pp. 196-258.

16. Notice sur la Détermination des Positions géographiques du Prieuré de Chamouny, et de l'Hospice du Grand Saint Bernard. Bibl. Univ. LI., 1832, pp. 113-117.

17. Notice respecting a Vitrified Fort at Carradale in Argyleshire. Edinb. Journ. Sci. VI., 1832, pp. 94-100.

18. On the Horary Oscillations of the Barometer near Edinburgh, deduced from 4,410 Observations; with an Inquiry into the Law of Geographical Distribution of the Phenomenon. [1831.] Edinb. Journ. Sci. VI., 1832, pp. 261-286; Edinb. Roy. Soc. Trans. XII., 1834, pp. 153-190.

19. Account of some Experiments in which an Electric Spark was elicited from a natural Magnet. Phil. Mag. I., 1832, pp. 49-53; Edinb. Roy. Soc. Trans. XII., 1834, pp. 197-205.

20. Researches on the Conducting Power of the Metals for Heat and Electricity. Proc. Roy. Soc., Edinb., I., 1833, p. 5.

21. Account of some Optical Phenomena observed upon the Righi. Proc. Roy. Soc., Edinb., I., 1833, p. 10.

22. Experimental Researches regarding certain Vibrations which take place between Metallic Masses having different Temperatures. [1833.] Edinb. Roy. Soc. Trans. XII., 1834, pp. 429-461; Poggen. Annal. XXXIII., 1834, pp. 553-556.

23. Notice of Experiments on the Diminution of Intensity sustained by the Sun's Rays in passing through the Atmosphere. Proc. Roy. Soc., Edinb., I., 1834, p. 55.

24. On the Vibration of Heated Metals. Phil. Mag. IV., 1834, pp. 15-28, 182-194.

25. An Account of some Experiments on the Electricity of Tourmaline and other Minerals when exposed to Heat. [1832.]

Phil. Mag. V., 1834, pp. 133-143; Edinb. Roy. Soc. Trans. XIII., 1836, pp. 27-38.

26. Note respecting the Application of the Compressibility of Water to practical purposes. Edinb. New Phil. Journ. XIX., 1835, pp. 36-39; Edinb. Trans. Scot. Soc. Arts, I., 1841, pp. 53-56.

27. On the Refraction and Polarization of Heat. Phil. Mag. VI., 1835, pp. 134-142, 205-214, 284-291, 366-371; Edinb. Roy. Soc. Trans. XIII., 1836, pp. 131-168.

28. Note relative to the Polarization of Heat. Phil. Mag. VII., 1835, pp. 349-352.

29. Sur la Polarisation de la Chaleur. Paris, Comptes Rendus, II., 1836, p. 156; VI., 1838, pp. 705-707.

30. Experiments on the Weight, Height, and Strength of Men at different Ages. Brit. Assoc. Rep. 1836 (*pt.* 2), pp. 38, 39; Franklin Inst. Journ. XXII., 1838, pp. 115-118; Phil. Mag. X., 1837, pp. 197-200; Quetelet, Corresp. Math. IX., 1837, pp. 205-209.

31. On the Geology of Auvergne, particularly in connexion with the Origin of Trap Rocks and the Elevation Theory. Edinb. New Phil. Journ. XXI., 1836, pp. 1-21.

32. Researches on Heat. Edinb. Roy. Soc. Trans. XIII., 1836, pp. 446-471.

33. Observations on Terrestrial Magnetism, made in different parts of Europe. Edinb. Roy. Soc. Proc. I., 1836, p. 154.

34. On the Undulatory Theory of Heat, and on the circular Polarization of Heat by total Reflexion. Phil. Mag. VIII., 1836, pp. 246-249; Poggend. Annal. XXXVII., 1836, pp. 501-505.

35. On the Mathematical Form of the Gothic Pendent. Phil. Mag. VIII., 1836, pp. 449-455.

36. Note relative to the supposed Origin of the deficient Rays in the Solar Spectrum; being an Account of an Experiment made at Edinburgh during the Annular Eclipse of 1836. Phil. Trans. 1836, pp. 453-456.

37. On the Temperatures and geological Relations of certain Hot Springs, particularly those of the Pyrenees; and on the Verification of Thermometers. Phil. Trans. 1836, pp. 571-616.

38. On the muscular Effort required to ascend Planes of different Inclinations. Phil. Mag., April 1837, p. 261.

39. Researches on Heat. Phil. Mag. XII., 1838, pp. 545-559; XIII., 1838, pp. 97-113, 180-192; Poggend. Annal. XLV., 1838, pp. 64-85, 442-460.

40. Observations faites pendant une Année avec des Thermomètres enfouis à diverses Profondeurs dans différentes Localités du Voisinage d'Edimbourg. Bibl. Univ. XIX., 1839, pp. 196-200.

41. Notice respecting the Use of Mica in polarizing Light. Brit. Assoc. Rep., 1839 (*pt.* 2), pp. 6, 7.

42. Memorandum on the Intensity of reflected Light and Heat. Proc. Roy. Soc., Edinb. I., 1839, p. 254.

43. Account of an intermitting Brine Spring discharging Carbonic Acid Gas, near Kissingen, in Bavaria. Edinb. New Phil. Journ. XXVI., 1839, pp. 306-326; Brit. Assoc. Rep. 1838 (*pt.* 2), pp. 28, 29.

44. On the Colour of Steam under certain circumstances. Phil. Mag. XIV., 1839, pp. 121-126; Edinb. Roy. Soc. Trans. XIV., 1840, pp. 371-374; Froriep, Notizen, XI., 1839, col. 295-299; Poggend. Annal. XLVII., 1839, pp. 593-599.

45. On the Colours of the Atmosphere. Phil. Mag. XIV., 1839, pp. 419-426; XV., 1839, pp. 25-37; Edinb. Roy. Soc. Trans. XIV., 1840, pp. 375-392; Poggend. Annal. LI. (*Ergänz.*), 1842, pp. 49-78; Liebig, Annal. XXXVI., 1840, pp. 132-136.

46. Supplementary Report on Meteorology. Brit. Assoc. Rep. 1840, pp. 37-156.

47. On the Effect of the Mechanical Texture of Screens on the immediate Transmission of Radiant Heat. Proc. Roy. Soc., Edinb. I., 1840, p. 281.

48. On an Optical Illusion giving the idea of an Inversion of Perspective in viewing Objects through a Telescope. Proc. Roy. Soc., Edinb. I., 1840, p. 296.

49. Results of additional Experiments on Terrestrial Magnetism. Proc. Roy. Soc., Edinb. 1840, I., p. 299.

50. On the Temperature and Conducting Power of different Strata. Brit. Assoc. Rep. 1840, pp. 434, 435.

51. On excessive Falls of Rain. Brit. Assoc. Rep. 1840 (*pt.* 2), pp. 43, 44.

52. Notice respecting the Fall of an Aërolite near Juvenas, Ardêche, in France, 15th June, 1821. Edinb. New. Phil. Journ. XXVIII., 1840, pp. 385-387.

53. Account of some Experiments made in different parts of Europe, on Terrestrial Magnetic Intensity, particularly with reference to the Effect of Height. [1836.] Edinb. Roy. Soc. Trans. XIV., 1840, pp. 1-29; Brit. Assoc. Rep. 1836 (*pt.* 2), pp. 30, 31; Froriep, Notizen, XVII., 1841, col. 97-106; Phil. Mag. XI., 1837, pp. 58-66, 166-174, 254-260, 363-375.

54. Researches on Heat. Third Series. 1. On the unequally polarizable Nature of different kinds of Heat. 2. On the Depolarization of Heat. 3. On the Refrangibility of Heat. [1838.] Edinb. Roy. Soc. Trans. XIV., 1840, pp. 176-207; Phil. Mag. XIX., 1841, pp. 69-81, 109-125; Poggend. Annal LI., 1840, pp. 88-109, 387-405.

55. On the Diminution of Temperature with Height in the Atmosphere at different Seasons of the Year. [1839.] Edinb. Roy. Soc. Trans. XIV., 1840, pp. 489-496; Edinb. New Phil. Journ. XXIX., 1840, pp. 205-214.

56. Sur la Transmissibilité des divers Genres de Chaleur à travers la Surface des Corps. Paris, Comptes Rendus, X., 1840, pp. 19-21.

57. Contributions to Optical Meteorology. Edinb. Roy. Soc. Proc. I., 1841, p. 324.

58. On a peculiar Structure in the Ice of Glaciers. Edinb. Roy. Soc. Proc. I., 1841, 332*.

59. On the Results of the most recent Experiments on the Conducting Power for Heat of different Soils. Edinb. Roy. Soc. Proc. I., 1841, p. 343*.

60. On an apparent Inversion of Perspective in viewing Objects with a Telescope. Phil. Mag. XVI. 1840, pp. 506-509.

61. Théorie des Glaciers. [*Transl. fr. Edinb. Review.*] Annal. de Chimie VI., 1842, pp. 220-255, 257-301.

62. On a remarkable Structure observed in the Ice of Glaciers. [1841.] Edinb. New Phil. Journ. XXXII., 1842, pp. 84-91.

63. Remarks on a Paper by Dr. Scoresby, 'On the Colours of the Dew-Drop.' Edinb. New Phil. Journ. XXXII., 1842, pp. 391-393.

64. Recent Observations on Glaciers. Edinb. New Phil. Journ. XXXIII., 1842, pp. 338-352; XXXIV. pp. 1-10.

65. Résultats des Observations faites aux environs d'Edimbourg, sur la Propagation des Variations extérieurs de Température dans l'Intérieur du Sol, pendant les quatre Années 1837, '38, '39, et '40. Paris, Comptes Rendus, XIV., 1842, pp. 410, 411; Poggend. Annal. LVI., 1842, pp. 616, 617.

66. On the Transparency of the Atmosphere, and the Law of Distinction of the Solar Rays in passing through it. (Bakerian Lecture.) Phil. Trans. 1842, pp. 225-274.

67. Geological Notes on the Alps of Dauphiné. Proc. Roy. Soc., Edinb., I., 1842, p. 365.

68. Sur l'Eclipse totale de Soleil du 8 Juillet 1842. Bibl. Univ. XLVIII., 1843, pp. 361-368.

69. Historical Remarks on the first Discovery of the real Structure of Glacier Ice. Edinb. New Phil. Journ. XXXIV., 1843, pp. 133-152. (See Appendix B, above.)

70. An Attempt to explain the leading Phenomena of Glaciers. Edinburgh New Phil. Journ. XXXV., 1843, pp. 221-252.

71. Travels through the Alps of Savoy, and other parts of

the Pennine Chain; with Observations on the Phenomena of Glaciers. Geogr. Soc. Journ. XIII., 1843, pp. 133-147; Froriep, Notizen, XXVI., 1843, col. 230-232.

72. Ueber Gletscher. Bern, Mittheil, 1844, pp. 118-122; Bibl. Univ. XLIX., 1844, pp. 170-174.

73. Account of an Attempt to establish the Plastic Nature of Glacier Ice by Experiment. Brit. Assoc. Rep. 1844 (*pt.* 2), pp. 24, 25.

74. On Solar Radiation. Edinb. New Phil. Journ. XXXVI., 1844, pp. 113-118.

75. Fifth Letter on Glaciers. Edinb. New Phil. Journ. XXXVI., 1844, pp. 217-223.

76. On a possible Explanation of the Adaptation of the Eye to distinct Vision, 1844. Edinb. Roy. Soc. Proc.; Trans. Roy. Soc., Edinb., XVI., 1849, p. 1.

77. Sixth Letter on Glaciers. Edinb. New Phil. Journ. XXXVII., 1844, pp. 231-244.

78. Seventh Letter on Glaciers:—On the Veined Structure of the Ice. Edinb. New Phil. Journ. XXXVII., 1844, pp. 244-249; Bibl. Univ. LI., 1844, pp. 354-358.

79. Eighth Letter on Glaciers:—On the Plasticity of Glacier Ice. Edinb. New Phil. Journ. XXXVII., 1844, pp. 375-381; Bibl. Univ. LIII., 1844, pp. 140-146; LV., 1845, pp. 339-347.

80. Researches on Heat. Fourth Series:—On the Effect of the mechanical Texture of Screens on the immediate Transmission of Radiant Heat. [1839.] Edinb. Roy. Soc. Trans. XV., 1844, pp. 1-26.

81. Account of some additional Experiments on Terrestrial Magnetism made in different parts of Europe in 1837. Edinb. Roy. Soc. Trans. XV., 1844, pp. 27-36.

82. On the Theory and Construction of a Seismometer, or instrument for measuring Earthquake Shocks and other Concussions. [1841.] Edinb. Roy. Soc. Trans. XV., 1844, pp. 219-228.

83. On the Determination of Heights by the boiling-point of Water. [1843.] Edinb. Roy. Soc. Trans. XV., 1844, pp. 409-416; Bibl. Univ. XLVIII., 1843, pp. 383, 384; Edinb. New Phil. Journ. XXXVIII., 1845, pp. 286-294.

84. Remarques sur les nouvelles Observations sur le Glacier du Faulhorn de M. Charles Martins. Bibl. Univ. LVIII., 1845, pp. 142-146.

85. Ninth Letter on Glaciers:—Remarks on the recent Observations made on the Glaciers of the Aar, by M. Agassiz. Edinb. New Phil. Journ. XXXVIII., 1845, pp. 332-341; Bibl. Univ. LVI., 1845, pp. 134-144.

86. Notice of Experiments on the Diminution of Intensity sustained by the Sun's Rays in passing through the Atmosphere. Edinb. Roy. Soc. Proc. I., 1845, pp. 55, 56.

87. Notice respecting an intermitting Brine Spring discharging Carbonic Acid Gas, near Kissingen, in Bavaria. Edinb. Roy. Soc. Proc. I., 1845, pp. 233-287.

88. Memorandum on the Intensity of reflected Light and Heat. Edinb. Roy. Soc. Proc. I., 1845, pp. 254-257.

89. Contributions to Optical Meteorology. Edinb. Roy. Soc. Proc. I., 1845, pp. 324-326.

90. On the Results of the most recent Experiments on the Conducting Power for Heat of different Soils. Edinb. Roy. Soc. Proc. I., 1845, pp. 343-346.

91. Papers on Glaciers. Edinb. Roy. Soc. Proc. I., 1845, pp. 406, 407, 409-411, 414, 415.

92. Reply to Mr. Hopkins on the Motion of Glaciers. Phil. Mag. XXVI., 1845, pp. 404-418.

93. Notes on the Topography and Geology of the Cuchullin Hills in Skye, and on the Traces of ancient Glaciers which they present. [1845.] Edinburgh New Phil. Journ. XL., 1846, pp. 76-79.

94. Tenth Letter on Glaciers:—M. Agassiz's Adoption of the Plastic Theory. [1845.] Edinb. New Phil. Journ. XL., 1846, pp. 154-160.

95. Eleventh Letter on Glaciers:—Observations on their Depression and the relative Velocity of the Surface and Bottom of a Glacier. Edinb. New Phil. Journ. XLI., 1846, pp. 414-421; Bibl. Univ. Archives, III., 1846, pp. 107-114.

96. Illustrations of the Viscous Theory of Glacier-motion:—Part I., containing Experiments on the Flow of Plastic Bodies, and Observations on the Phenomena of Lava Streams. [1845.] Phil. Trans. 1846, pp. 143-156.

97. Illustrations of the Viscous Theory of Glacier-motion:—Part II., An Attempt to establish by observation the Plasticity of Glacier-ice. Phil. Trans. 1846, pp. 157-176.

98. Illustrations of the Viscous Theory of Glacier-motion:—Part III., On the Motion of Glaciers of the Second Order. On the Annual Motion of Glaciers, and on the Influence of Seasons. Summary of the Evidence adduced in favour of the Theory. Phil. Trans. 1846, pp. 177-210.

99. Biographical Notice of Sir John Robison. Edinb. Roy. Soc. Proc. II., 1846, p. 68.

100. New Observations on the Glaciers of Savoy, 1846. Edinb. Roy. Soc. Proc. II., p. 103.

101. Observations on the Temperature of the Earth at Trevan-

drum, from the Observations of John Caldecott, Esq. Brit. Assoc. Rep. 1847 (*pt.* 2), p. 40.

102. Twelfth Letter on Glaciers:—On the extraordinary Increase of the Glacier of La Brenva (1842-46); its Motion. Experimentum crucis respecting the Origin of the Veined Structure. Edinb. New Phil. Journ. XLII., 1847, pp. 94-104; Bibl. Univ. Archives, IV., 1847, pp. 163-166, 366-374; V., 1847, pp. 14-18.

103. Thirteenth Letter on Glaciers:—Acceleration of Surface Motion; new Determination of Velocity of different Glaciers; Conversion of Névé into Ice; Ejection of Stones from Glaciers. Edinb. New Phil. Journ. XLII., 1847, pp. 136-154.

104. Fourteenth Letter on Glaciers:—On the Variation of the Motion at different Seasons. Edinb. New Phil. Journ. XLII., 1847, pp. 327-343.

105. Fifteenth Letter on Glaciers; containing Observations on the Analogies derived from Mud Slides on a large Scale, and from some Processes in the Arts, in favour of the Viscous Theory of Glaciers. Edinb. New Phil. Journ. XLVI., 1840, pp. 139-149.

106. On a possible Explanation of the Adaptation of the Eye to distinct Vision at different distances. [1844-45.] Edinb. Roy. Soc. Trans. XVI., 1849, pp. 1-6.

107. Account of some Experiments on the Temperature of the Earth at different Depths and in different Soils, near Edinburgh. [1846.] Edinb. Roy. Soc. Trans. XVI., 1849, pp. 189-236.

108. Hints towards a Classification of Colours. Phil. Mag. XXXIV., 1849, pp. 161-178.

109. Notices. (1.) On an Instrument for measuring the Extensibility of Elastic Solids. (2.) Note respecting the refractive and dispersive power of Chloroform. (3.) Note regarding an Experiment suggested by Prof. Robison. Phil. Mag. XXXV., 1849, pp. 92-96.

110. On the Danger of superficial Knowledge, a Lecture., London, 1849.

111. On the alleged Evidence for a physical Connexion between Stars forming binary or multiple Groups, deduced from the Doctrine of Chances. Phil. Mag. XXXVII., 1850, pp. 401-427.

112. Note respecting the Dimensions and refracting power of the Eye. Goodsir, Ann. Anat. Phys. 1850-53, pp. 44-48; Edinb. Roy. Soc. Proc. II., 1851, pp. 251-253.

113. Sixteenth Letter on Glaciers:—On the Movement of the Mer de Glace; Observations by Balmat; on the gradual Passing of the Ice into the fluid State; an undescribed Pass of the Alps. Edinb. New Phil. Journ., L., 1851, pp. 167-174.

114. Further Remarks on the intermitting Brine Springs

of Kissingen. Edinb. New Phil. Journ., LI., 1851, pp. 139-142; Edinb. Roy. Soc. Proc. III., 1857, pp. 66-69.

115. On an Instrument for measuring the Extensibility of Elastic Solids. Edinb. Roy. Soc. Proc. II., 1851, pp. 172-175.

116. Note respecting the refractive and dispersive power of Chloroform. Edinb. Roy. Soc. Proc. II., 1851, pp. 187, 188.

117. On the Classification of Colours. Edinb. Roy. Soc. Proc. II., 1851, pp. 214-216.

118. On the Intensity of Heat reflected from Glass. Edinb. Roy. Soc. Proc. II., 1851, pp. 256, 257.

119. On the Volcanic Formations of the Alban Hills, Rome, Edinb. Roy. Soc. Proc. II., 1850, p. 259.

120. Account of a remarkable Meteor seen 19th December, 1849. Edinb. Roy. Soc. Proc. II., 1851, pp. 309-316.

121. Norway and its Glaciers. 8vo., Edinburgh, 1853.

122. On the Volcanic Geology of the Vivarais (Ardêche). [1848.] Edinb. Roy. Soc. Trans. XX., 1853, pp. 1-38.

123. Notes on the Geology of the Eildon Hills in Roxburghshire. [1851.] Edinb. Roy. Soc. Trans. XX., 1853, pp. 211-218.

124. Dissertation on the Progress of mathematical and physical Science. Encyclop. Britann., 8th edition, 1853.

125. Remarks on the Rev. H. Moseley's Theory of the Descent of Glaciers. Roy. Soc. Proc. VII., 1854-55, pp. 412-417.

126. Review of the Life and Writings of Thomas Young. Edinb. New Phil. Journ., 1855.

127. Tour of Mont Blanc and Monte Rosa. 12mo., Edinburgh, 1855.

128. On the geological Relations of the secondary and primary Rocks of the Chain of Mont Blanc. Edinb. New Phil. Journ. III., 1856, pp. 189-203; Bibl. Univ. Archives, XXXI., 1856, pp. 281-298.

129. On the Movement of Glaciers of Chamouni in Winter. Edinb. Roy. Soc. Proc., III., 1855, p. 285.

130. Notice respecting Father Secchi's Statical Barometer, and on the Origin of the Cathetometer. Edinb. New. Phil. Journ. V., 1857, pp. 316-319.

131. Further Observations on Glaciers. (1.) Observations on the Movement of the Mer de Glace down to 1850. (2.) Observations by Balmat, in continuation of those detailed in the Fourteenth Letter. (3.) On the gradual Passage of Ice into the fluid State. Edinb. Roy. Soc. Proc. III., 1857, p. 14.

132. Further Experiments and Remarks on the Measurement of Heights by the boiling-point of Water. [1854.] Edinb. Roy.

Soc. Trans. XXI., 1857, pp. 235-244; Bibl. Univ. Archives, XXX., 1855, pp. 290-304.

133. Sur quelques Propriétés que présente la Glace près de son Point de Fusion. Paris, Comptes Rendus, XLVII., 1858, pp. 367, 368; Edinb. Roy. Soc. Proc. IV., 1862, pp. 103-106; Pharmaceut. Journ. I., 1860, pp. 38-40.

134. Notes on certain Vibrations produced by Electricity. Edinb. New Phil. Journ. IX., 1859, pp. 266-269; Edinb. Roy. Soc. Proc. IV., 1862, pp. 151-153; Phil. Mag. XVII., 1859, pp. 358-360; Poggend. Annal. CVII., 1859, pp. 458-461.

135. On a Paper 'On Ice and Glaciers.' Phil. Mag. XVII., 1859, pp. 197-201.

136. Occasional Papers on the Theory of Glaciers. 8vo. Edinb., 1859.

137. Reply to Professor Tyndall on Rendu's 'Théorie des Glaciers.' Pamphlet, Edinburgh, 1860. (See Appendix A, above.)

138. An Account of two artificial Hemispheres representing graphically the Distribution of Temperature and Magnetism from the Earth's Equator to the North Pole. Edinb. New Phil. Journ. XIII., 1861, pp. 118-121.

139. Inquiries about terrestrial Temperature; to which is added an Index to Mr. Dove's Five Memoirs on the Temperature of the Globe. Edinb. Roy. Soc. Trans. XXII., 1861, pp. 75-100.

140. On the Climate of Edinburgh for Fifty-six Years, from 1795 to 1850, deduced principally from Mr. Adie's Observations, with an Account of other and earlier Registers. Edinb. Roy. Soc. Trans. XXII., 1861, pp. 327-356.

141. Account of a Thermometrical Register kept at Dunfermline by the Rev. Henry Fergus, from 1799 till 1837, with the principal Results. Edinb. Roy. Soc. Trans. XXII., 1861, pp. 357-360.

142. Experimental Inquiries into the Laws of the Conduction of Heat in Bars, and into the conducting power of Wrought Iron. Edinb. Roy. Soc. Trans. XXIII., 1861, pp. 133-146; Edinb. Roy. Soc. Proc. IV., 1862, pp. 607-610.

143. Note respecting Ampère's Experiment on the Repulsion of a rectilinear Electrical Current on itself. Phil. Mag. XXI., 1861, pp. 81-86; Edinb. Roy. Soc. Proc. IV., 1862, pp. 391, 392.

144. On the Climate of Palestine in modern compared to ancient Times. Edinb. New Phil. Journ. XV., 1862, pp. 169-179.

145. Opening Address to Royal Society, Edinburgh, 1862. Edinb. Roy. Soc. Proc., V. 2.

146. Biographical Notice of Professor Necker, 1863. Edinb. Roy. Soc. Proc., V. p. 53.

147. Experimental Inquiry into the Laws of Conduction of

Heat in Bars: Part II. 1865. Edinb. Roy. Soc. Proc., V. p. 369; Edinb. Roy. Soc. Trans., XXIV. p. 73.

148. Notice respecting Mr. Reilly's Survey of Mont Blanc, 1865. Edinb. Roy. Soc. Proc., V. 335.

149. The British Association considered with reference to its History, Plan, and Results, and to the approaching Meeting at Dundee. An Address to the United College of St. Andrews Nov. 1, 1866. 8vo.

[Principal Forbes contributed to the Quarterly, Edinburgh and North British Reviews, the Encyclopædia Britannica, Fraser's Magazine, &c., articles on various subjects connected with different branches of science, as on *Climate, the Antiquity of Man, Alpine Travelling, the Greenwich Observatory, Hooker's Himalayan Journals,* &c.]

THE END.

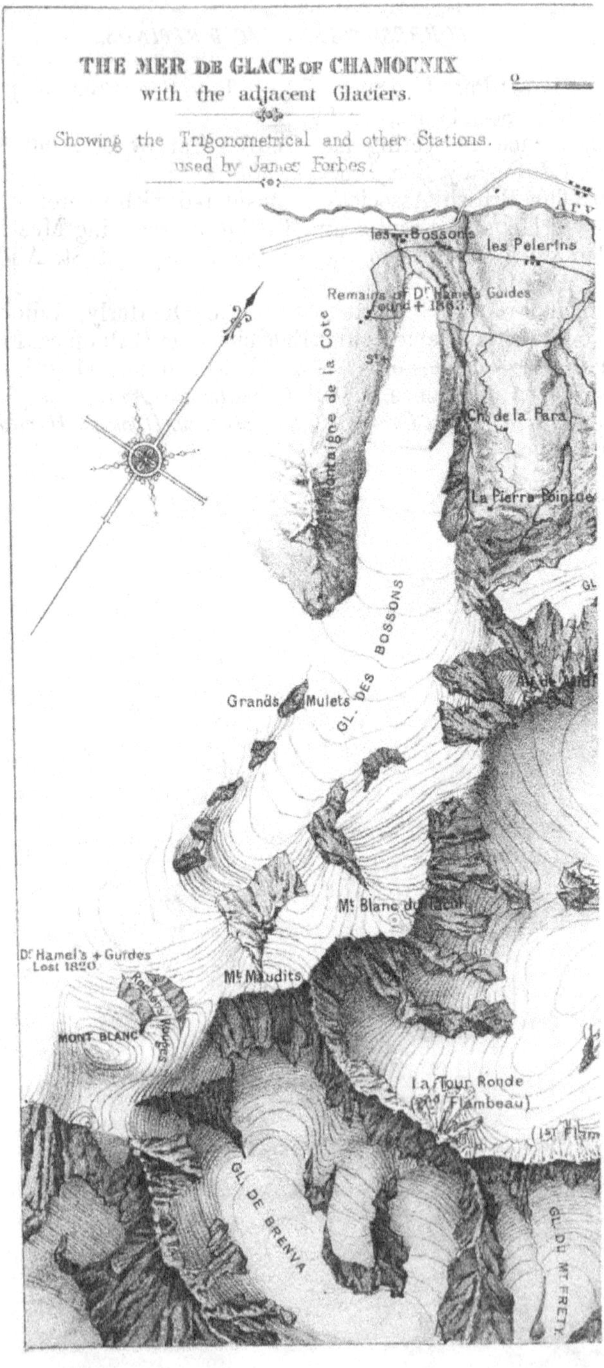

THE MER DE GLACE OF CHAMOUNIX
with the adjacent Glaciers.

Showing the Trigonometrical and other Stations.
used by James Forbes.

les Bossons
les Pelerins
Remains of Dr. Hamel's Guides found + 1863.
Montaigne de la Cote
Ch. de la Para
La Pierra Pointue
GL. DES BOSSONS
Grands Mulets
Mt. Blanc du Tacul
Dr. Hamel's + Guides Lost 1820.
Mt. Maudits
Rochers Rouges
MONT BLANC
La Tour Ronde
(2nd Flambeau)
GL. DE BRENVA
GL. DU MT. FRETY

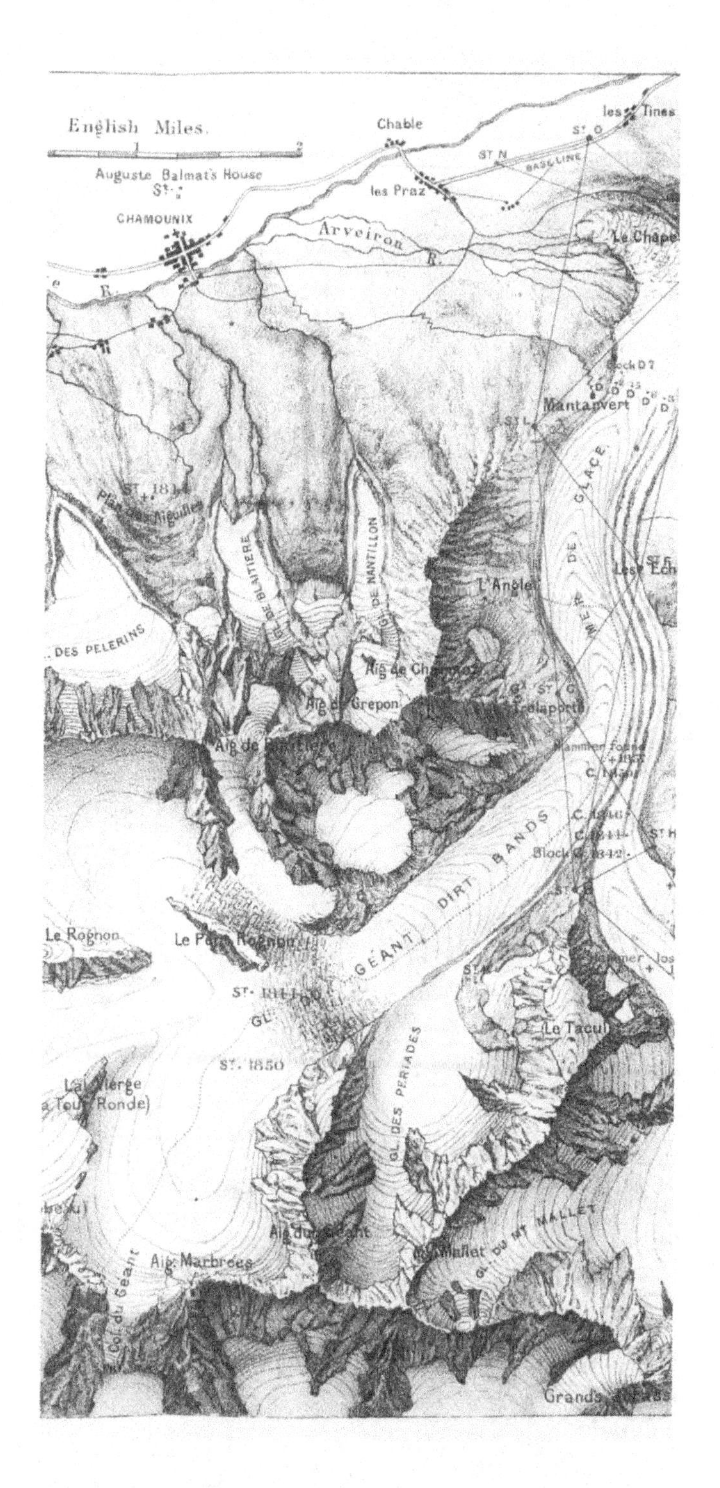
English Miles.
Auguste Balmat's House
CHAMOUNIX
Chable
les Tines
les Praz
Arveiron R.
BASE LINE
Le Chape
Mantanvert
MER DE GLACE
L'Angle
GL. DE BLAITIERE
GL. DE NANTILLON
DES PELERINS
Aig de Grepon
Aig de Blaitière
Le Rognon
GÉANT DIRT BANDS
Le Tacul
ST. 1850
GL. DES PERIADES
Aig. Marbrées
Col du Géant
GL. DU MT MALLET
Grands

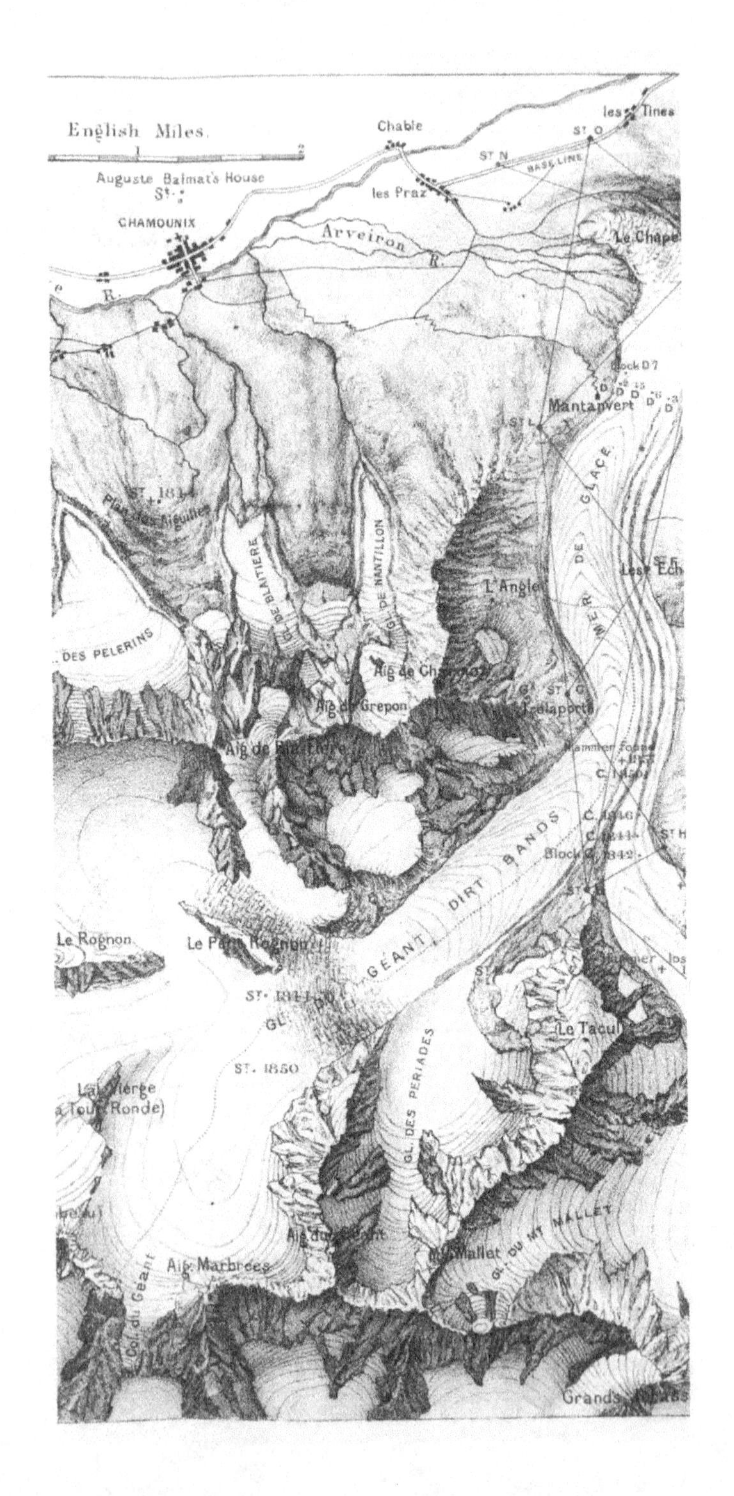

English Miles.
1
2
Auguste Balmat's House
CHAMOUNIX
Chable
les Praz
BASE LINE
Arveiron
R.
Mantanvert
MER DE GLACE
L'Angle
GL. DE BLAITIERE
GL. DE NANTILLON
DES PELERINS
Aig de Grepon
Le Rognon
GEANT DIRT BANDS
ST. 1850
Le Tacul
GL. DES PERIADES
GL. DU MT MALLET
Aig. Marbrees
Col du Geant

Ch. de la Pendant
Ch. de Lognan
Aig. Béchard
GL. DE LA PENDANT
GL. D'ARGENTIERE
GL. DU NANT BLANC
GL. DE CHARPUA
Aig. Verte
Les Droites
Aig. du Moine
Knapsack
+ lost 1836
Le Jardin
Les Courtes
GL. DE TALÈFRE
GL. DE LECHAUD
Aig. de Triolet
Aig. de l'Eboulement
Gl. de Pierre Joseph
Aig. de Léchaud
GL. DE TRIOLET
Petites Jorasses
ADAM REILLY, DELT.

Printed in the United States
By Bookmasters